Applied Mechanics

Applied Mechanics

A modern approach

F.P. Sayer
Formerly Senior Lecturer, Department of Engineering Mathematics, University of Bristol

J.A. Bones
Senior Lecturer, Department of Mechanical Engineering, University of Bristol

CHAPMAN AND HALL
University and Professional Division
LONDON • NEW YORK • TOKYO • MELBOURNE • MADRAS

UK	Chapman and Hall, 11 New Fetter Lane, London EC4P 4EE
USA	Van Nostrand Reinhold, 115 5th Avenue, New York NY10003
JAPAN	Chapman and Hall Japan, Thomson Publishing Japan, Hirakawacho Nemoto Building, 7F, 1-7-11 Hirakawa-cho, Chiyoda-ku, Tokyo 102
AUSTRALIA	Chapman and Hall Australia, Thomas Nelson Australia, 480 La Trobe Street, PO Box 4725, Melbourne 3000
INDIA	Chapman and Hall India, R. Sheshadri, 32 Second Main Road, CIT East, Madras 600 035

First edition 1990

Typeset in 10 pt Times by
Best-set Typesetter Ltd, Hong Kong
Printed in Great Britain by T. J. Press Ltd,
Padstow, Cornwall

ISBN 0 412 34140 9
0 442 31182 6 (USA)

British Library Cataloguing in Publication Data

Sayer, F.P.
Applied mechanics
1. Applied mechanics
I. Title II. Bones, J.A.
620.1

ISBN 0-412-34140-9

Library of Congress Cataloging in Publication Data
available

Contents

Preface

Vectors and matrices have been consistently used throughout the book to develop the theory of mechanics and to solve some of its many problems. Graphical methods, which at best have limited use, are not considered.

The requisite knowledge of vectors and matrices required for this book is set out in Chapter 1. Numerous worked examples give instruction in the use and manipulation of vectors and matrices and also provide results that are essential for the work of subsequent chapters.

It is not necessary for the reader to study the whole of Chapter 1 before proceeding to subsequent chapters of the book. A knowledge of vector algebra suffices for the work of Chapter 3. Both vector algebra and vector differentiation are required in the study of Chapter 4 while Chapter 5 demands that the reader be competent in the use of both vectors and matrices. The reader can proceed directly from Chapter 4 to Chapter 6 and study those parts of this chapter that relate to dynamical principles (Sections 6.1–6.7 and Section 6.10) and the Worked Examples 6.1–6.5 and 6.13. It would however be to the reader's advantage to work through problem 5.37 before starting on Chapter 6. This problem shows how the important formula (5.27) namely $\mathbf{v}_j = \mathbf{v}_i + \mathbf{w} \times (\mathbf{r}_j - \mathbf{r}_i)$ may be derived without consideration of rotation theory.

We have sought to encourage the fullest possible use of the computer by the provision of examples of computer applications in mechanics. These relate to mechanisms, cams and robots. Other areas which are recommended (here) for the reader's consideration are rotation theory, the determination of forces in mechanisms and the balancing of rotors.

The reader should note that the following devices have been adopted in respect of figures. If a figure contains a drawing of the vector pair **i**, **j** it is to be interpreted as a two–dimensional figure while if it contains a drawing of the triad **i**, **j**, **k** the figure is to be interpreted as a three–dimensional figure. The reader should also note that proofs are preceded by the symbol □ and solutions to worked examples by ▷.

In conclusion we record our appreciation of the skill exercised by Miss Christine Absolon in the preparation of the typescript for the publisher.

J.A. Bones
F.P. Sayer

1 Vectors, determinants and matrices

This chapter provides an introductory treatment of vectors, determinants and matrices designed to meet the needs of the rest of the book. It assumes no previous knowledge of these topics and, given its purpose, makes no claim to be comprehensive.

1.1 INTRODUCTORY IDEAS

All physical quantities can be divided into two classes. One class contains those quantities that require only a magnitude for their specification, a magnitude being a number (or measure) multiplied by a unit. Such quantities are referred to as scalar quantities or more simply as scalars. Examples are length, mass and time, their respective units of measurement in the SI system (see Chapter 2) being the metre, kilogram and second. The second class contains quantities that require both magnitude and direction for their specification. Common experience shows that force belongs to this class but so does, for example, rotation. To define the rotation of a body it is necessary to specify both the angle and the axis of rotation and these provide us with a magnitude and a direction. Thus it is possible to represent rotation and force in an identical manner, that is by a directed line (a straight line in a particular direction) of an appropriate length. The length of the line will for the former be proportional to the angle of rotation and for the latter be proportional to its magnitude.

There is, however, a fundamental difference between the two quantities. From repeated experiments it is known that two forces acting at a point have the same effect as a single force, known as the resultant, given by the parallelogram of forces. This is illustrated by Fig. 1.1 where forces of magnitudes P and Q, acting at the point O and represented by the directed

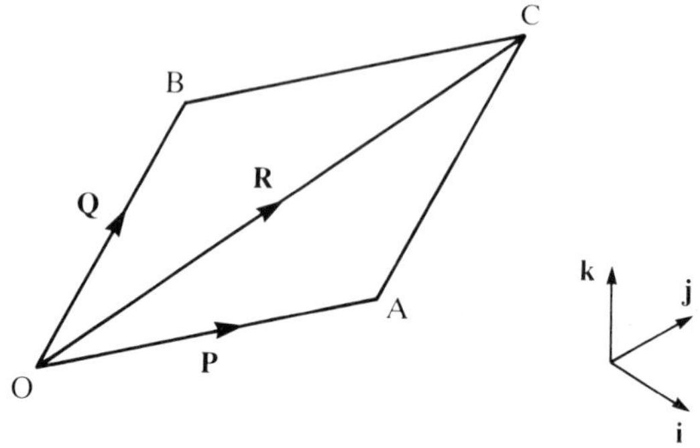

Fig. 1.1

lines OA and OB, have a resultant of magnitude R, represented by the directed line OC, where OC is the diagonal of the parallelogram with OA and OB as adjacent sides. Mathematically we say that the two forces have been added together by means of the parallelogram law. To find the resultant of several forces acting at a point, repeated use is made of the law. Firstly, any two forces are selected and their resultant found, this resultant then being combined with any one of the remaining forces to give a further resultant. The process is continued until all the forces have been combined into a single resultant force. This resultant is independent of the order in which the additions are performed, so that forces obey what is known as the commutative law of addition. The fact that they do so is a consequence of the parallelogram law, as will be shown in the course of Section 1.2.

However, the order in which several rotations about intersecting axes are combined is all-important, with different orders producing different outcomes, as will be demonstrated in Chapter 5. The parallelogram law cannot, therefore, provide the means whereby rotations are combined. It will be shown in Chapter 5 that a rotation can be represented by a matrix and the combined effect of two rotations by the product of two matrices. Our first objective in this chapter is not to discuss matrices but to develop an algebra of quantities that possess both magnitude and direction and are combined by means of the parallelogram law. This algebra is called vector algebra and is the subject of the next section.

1.2 VECTOR ALGEBRA

We begin with a formal definition of a vector. It is a physical quantity that possesses both magnitude and direction and is added to a (like) physical quantity by means of the parallelogram law.

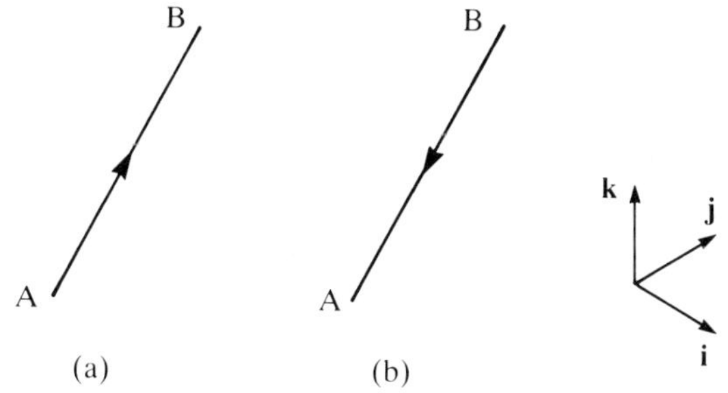

Fig. 1.2

Any vector can be represented by a directed line of the kind shown in Fig. 1.2(a). The magnitude of the vector is given (on some defined scale) by the length of the line, whilst its direction corresponds to that of the line. The arrow in Fig. 1.2(a) is important and indicates that the sense of the vector is from A to B. In Fig. 1.2(b) the sense of the vector is from B to A, that is in the opposite sense to that of the vector in Fig. 1.2(a).

Vectors subdivide into two groups, namely *free* vectors and *localized* vectors. We define the terms 'free' and 'localized' by means of two examples. From experiment it is known that a force of given magnitude and direction when applied at an appropriate point of a body can make all points of it move in the same direction, but if applied at a different point will cause the body to turn. Thus a complete description of force requires the specification of not only its magnitude and direction, but also its point of application. Force is an example of a localized vector, that is a vector linked with or tied to a point. In contrast the vector defining, for example, the velocity of the wind does not have to pass through a particular point: the vector can be represented by any one of an unlimited number of lines all parallel to the direction of the wind and having the same length and sense. When a vector can be represented in this way it is called a free vector.

The parallelogram law is applicable to localized vectors only if they are tied to the same point. No such restriction applies to free vectors because they may be moved parallel to themselves and we can, therefore, for mathematical convenience, take them to pass through a common point and add them by the parallelogram law. The vector algebra that is now developed is, accordingly, restricted to free vectors and to localized vectors tied to the same point. This restriction means that temporarily a number of questions will go unanswered. For example, how do we combine a pair of parallel localized vectors or localized vectors in skew lines, that is lines that are not parallel and do not intersect? These and other questions will be considered in Chapter 3.

Notation

We shall denote localized and free vectors in two ways. One way is by a single letter (lower case or upper case) in bold type, for example **A**, **B**, **X**, **a** or **b**: this is a universal notation. The second way is by underlined pairs of upper case letters (not bold) so that, for example, $\underline{\mathrm{OA}}$ and $\underline{\mathrm{OB}}$ are the vectors represented by the directed lines OA and OB. This notation is particularly useful when we are considering localized vectors, as the first letter (O in our example) can be used to indicate the point to which the vectors are tied.

Magnitude

The magnitude of a vector is indicated by modulus lines. Thus the magnitudes of **A** and $\underline{\mathrm{AB}}$ are denoted by $|\mathbf{A}|$ and $|\underline{\mathrm{AB}}|$ respectively.

Equality

Two free vectors are equal if they have the same magnitude, direction and sense. The same definition applies to localized vectors but in addition the vectors must be tied to the same point, and are thus coincident. Equality is expressed by the statement $\mathbf{A} = \mathbf{B}$ or $\underline{\mathrm{CD}} = \underline{\mathrm{EF}}$: it follows that $|\mathbf{A}| = |\mathbf{B}|$ and $|\underline{\mathrm{CD}}| = |\underline{\mathrm{EF}}|$.

Addition

The operation of adding the vectors **A** and **B** is denoted by $\mathbf{A} + \mathbf{B}$ or $\mathbf{B} + \mathbf{A}$. The outcome of the operation is a third vector **C** that is determined by the parallelogram law illustrated in Fig. 1.3. Vectors **A** and **B** are represented by the directed lines OA and OB while **C** is given by the (directed) diagonal OC of the parallelogram OACB. The point C is the point of intersection of lines drawn through A and B respectively parallel to OB and OA. Clearly the order in which the lines are drawn to find the point C is immaterial. It is, therefore, not possible to claim precedence for one vector over the other in the operation of addition and, accordingly, we have the option of writing $\mathbf{C} = \mathbf{A} + \mathbf{B}$ or $\mathbf{C} = \mathbf{B} + \mathbf{A}$: thus vectors obey what is known as the commutative law of addition. The vector **C** is called the resultant of the two vectors **A** and **B**.

Alternative means of finding $\mathbf{A} + \mathbf{B}$ are illustrated in Fig. 1.4(a) and (b). The line representing **B** (see Fig. 1.3) is moved parallel to itself so that the

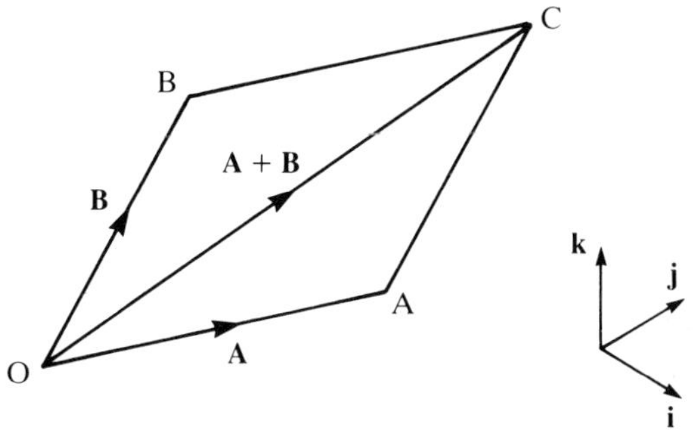

Fig. 1.3

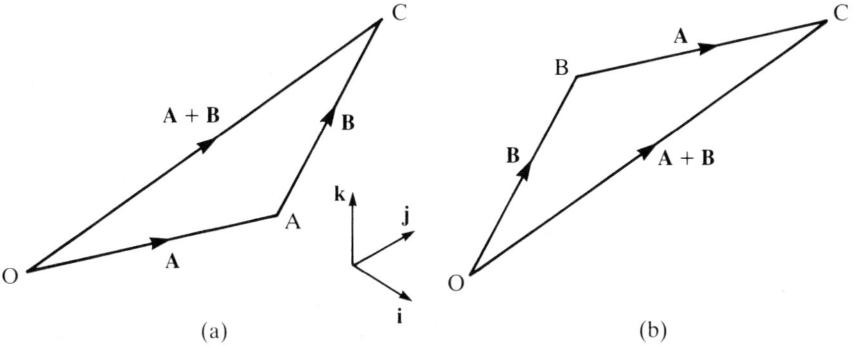

(a) (b)

Fig. 1.4

extremity of it, initially at O, is made to coincide with A (see Fig. 1.4(a)). The point C is situated at the other end of the displaced line and thus $\underline{OC}$ is determined. Similarly the line representing **A** may be moved parallel to itself to find the point C. This procedure of placing vectors end to end, in the manner described, to find their resultant is a very useful one. It leads to the following alternative form of the addition law, but it must be emphasized that this form treats all vectors as if they are *free* vectors. Using this assumption we have $\mathbf{A} = \underline{OA} = \underline{BC}$ and $\mathbf{B} = \underline{OB} = \underline{AC}$, and hence

$$\begin{aligned} \underline{OC} = \mathbf{C} = \mathbf{A} + \mathbf{B} = \underline{OA} + \underline{AC} \\ \underline{OC} = \mathbf{C} = \mathbf{B} + \mathbf{A} = \underline{OB} + \underline{BC} \end{aligned} \tag{1.1}$$

The addition of several vectors is achieved by the repeated application of the parallelogram law. The procedure is illustrated for the three vectors **A**, **B** and **C**. The outcomes of the successive additions $(\mathbf{A} + \mathbf{B}) + \mathbf{C}$ and

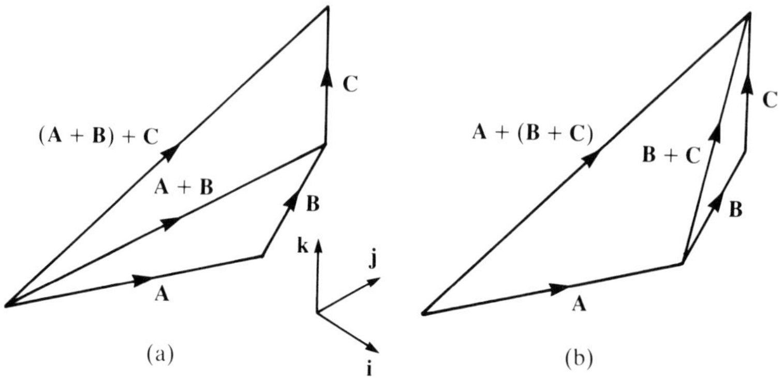

Fig. 1.5

$\mathbf{A} + (\mathbf{B} + \mathbf{C})$ are shown in Fig. 1.5(a) and (b) respectively. The brackets are essential in the first instance, because they indicate the order in which the operations are performed. It is evident that the vectors $(\mathbf{A} + \mathbf{B}) + \mathbf{C}$ and $\mathbf{A} + (\mathbf{B} + \mathbf{C})$ are equal. The third vector $(\mathbf{A} + \mathbf{C}) + \mathbf{B}$ can also be shown to be equal to them and hence the brackets are unnecessary and it is sufficient to write $\mathbf{A} + \mathbf{B} + \mathbf{C}$ for the addition of **A**, **B** and **C**, irrespective of the way in which it is performed. In a similar way the sum of several vectors **A**, **B**, **C**, **D**, **E**. . . is denoted by $\mathbf{A} + \mathbf{B} + \mathbf{C} + \mathbf{D} + \mathbf{E} \ldots$. The relation $(\mathbf{A} + \mathbf{B}) + \mathbf{C} = \mathbf{A} + (\mathbf{B} + \mathbf{C})$ establishes that vectors obey what is known as the associative law of addition.

Subtraction

As a preliminary to defining subtraction we introduce the vector $-\mathbf{A}$. For free vectors the vector $-\mathbf{A}$ is a vector equal in magnitude to **A** and parallel to it but in the opposite sense. If **A** is a localized vector the magnitude and sense of $-\mathbf{A}$ are defined in the same way as for a free vector, but $-\mathbf{A}$ is located in the same line and tied to the same point O as **A**: see Fig. 1.6(a) and (b). Figure 1.2(a) shows the vector $\underline{AB}$ while Fig. 1.2(b) shows the vector $\underline{BA}$ and it thus follows from the definition of $-\mathbf{A}$ that $\underline{BA} = -\underline{AB}$.

Using the procedure of placing vectors end to end to obtain their sum we see that $\mathbf{A} + (-\mathbf{A})$ is a vector of zero magnitude. This vector is called the zero or null vector and is denoted by the symbol $\mathbf{0}$. The zero vector has no particular direction associated with it and clearly $\mathbf{A} + \mathbf{0} = \mathbf{A}$.

The operation of subtracting **B** from **A** can now be defined. It produces a third vector **C**, denoted by $\mathbf{A} - \mathbf{B}$, and defined by $\mathbf{C} = \mathbf{A} + (-\mathbf{B})$. Subtraction is illustrated in Fig. 1.7.

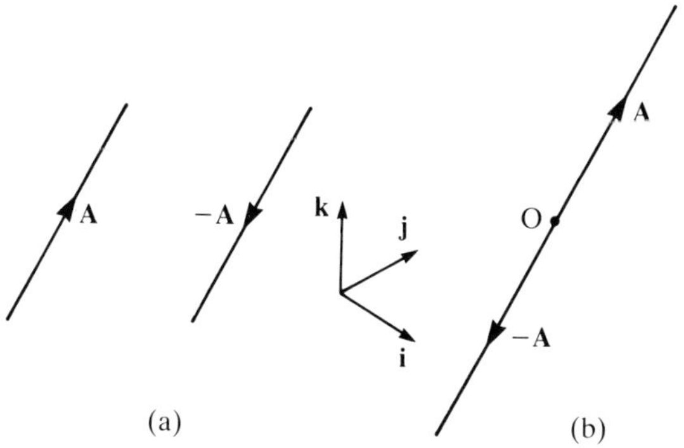

Fig. 1.6

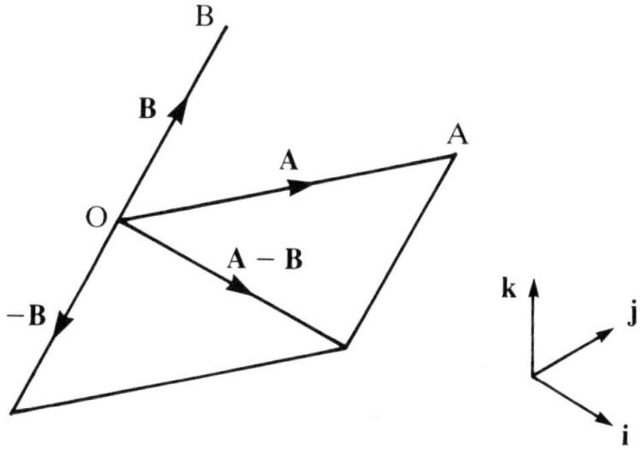

Fig. 1.7

Multiplication of a vector by a scalar

The operation of multiplying the vector **A** by the scalar quantity x is denoted by $x\mathbf{A}$. It is customary to write the scalar before the vector and we shall follow that practice. If $x > 0$ the vector $x\mathbf{A}$ is parallel to or in the same line as **A** and has the same sense as **A**, and its magnitude is equal to $x|\mathbf{A}|$. If $x < 0$ the vector $x\mathbf{A}$ is again parallel to or in the same line as **A** but its sense is opposite to that of **A** and its magnitude is equal to $-x|\mathbf{A}|$. If **A** is a localized vector then $x\mathbf{A}$ is tied to the same point as **A**. It is a consequence of the definition of $x\mathbf{A}$ that

$$|x\mathbf{A}| = |x|\,|\mathbf{A}| \tag{1.2}$$

We note in passing that the units of measurement of $x\mathbf{A}$ will be a combination of the units that measure x and those that measure **A**. This point is discussed in some detail in Chapter 2.

When manipulating vector quantities of the type $x\mathbf{A}$ we will find the rules (1.3) to (1.5) given below useful. They may be deduced from the definition of $x\mathbf{A}$ and the parallelogram law of addition. Proofs are not given here. The rules are:

$$(x + y)\mathbf{A} = x\mathbf{A} + y\mathbf{A} \tag{1.3}$$

$$x(\mathbf{A} + \mathbf{B}) = x\mathbf{A} + x\mathbf{B} \tag{1.4}$$

$$x(y\mathbf{A}) = y(x\mathbf{A}) = (xy)\mathbf{A} \tag{1.5}$$

In rule (1.5) the expressions are simply written as $xy\mathbf{A}$. Rules (1.3) to (1.5) can be used to expand expressions such as $(x + y)(\mathbf{A} + \mathbf{B})$ and $(x + y + z)\mathbf{A}$. In order to expand $(x + y)(\mathbf{A} + \mathbf{B})$ we first put $\mathbf{A} + \mathbf{B} = \mathbf{C}$. Using rules (1.3) and (1.4) yields

$$\begin{aligned}(x + y)(\mathbf{A} + \mathbf{B}) = (x + y)\mathbf{C} &= x\mathbf{C} + y\mathbf{C} \\ &= x(\mathbf{A} + \mathbf{B}) + y(\mathbf{A} + \mathbf{B}) \\ &= x\mathbf{A} + x\mathbf{B} + y\mathbf{A} + y\mathbf{B}\end{aligned}$$

Likewise

$$\begin{aligned}(x + y + z)\mathbf{A} = ((x + y) + z)\mathbf{A} &= (x + y)\mathbf{A} + z\mathbf{A} \\ &= x\mathbf{A} + y\mathbf{A} + z\mathbf{A}\end{aligned}$$

Planar resolution

Let PQ and PR be two lines, intersecting at the point P, that are not necessarily perpendicular. If $\mathbf{C}$ is any vector lying on a plane *parallel* to the plane defined by the lines PQ and PR, then it is possible to find two vectors $\mathbf{A}$ and $\mathbf{B}$ respectively parallel to PQ and PR such that $\mathbf{C} = \mathbf{A} + \mathbf{B}$. We say we have resolved $\mathbf{C}$ into two component vectors, namely $\mathbf{A}$ and $\mathbf{B}$, parallel to two given directions. The component vectors are unique as we shall shortly establish by a proof that illustrates the power of the ideas developed so far in our vector algebra. Firstly, however, we show how $\mathbf{A}$ and $\mathbf{B}$ are determined. Let $\mathbf{C} = \underline{\mathrm{OC}}$, where O and C are two points lying in the plane that contains $\mathbf{C}$ and is parallel to the plane of PQ and PR. Through O and C draw lines respectively parallel to PR and PQ, meeting in the point B. Likewise draw a second pair of lines through O and C parallel respectively to PQ and PR, meeting in the point A. The required component vectors are given by $\underline{\mathrm{OA}}$ and $\underline{\mathrm{OB}}$. These vectors are labelled $\mathbf{A}$ and $\mathbf{B}$ respectively, and clearly by the parallelogram law $\mathbf{C} = \mathbf{A} + \mathbf{B}$.

To prove that $\mathbf{A}$ and $\mathbf{B}$ are unique we assume that there exist two further vectors $\mathbf{A}'$ and $\mathbf{B}'$ parallel to PQ and PR respectively such that $\mathbf{C} = \mathbf{A}' + \mathbf{B}'$:

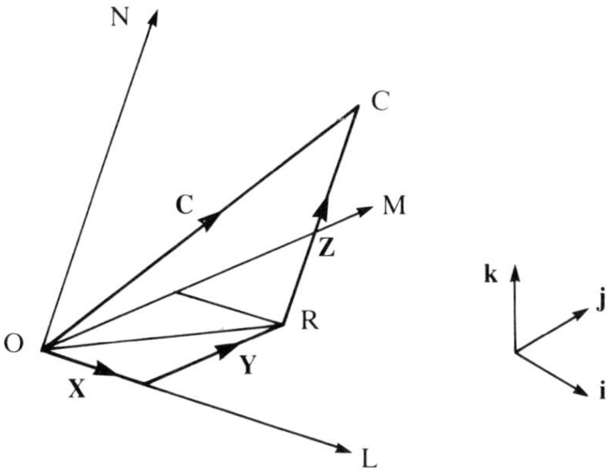

Fig. 1.8

this implies that $\mathbf{A}' + \mathbf{B}' = \mathbf{A} + \mathbf{B}$. To each side of this relation successively add the vectors $-\mathbf{A}$ and $-\mathbf{B}'$. This yields $\mathbf{A}' - \mathbf{A} + \mathbf{B}' - \mathbf{B}' = \mathbf{A} - \mathbf{A} + \mathbf{B} - \mathbf{B}'$ or $\mathbf{A}' - \mathbf{A} = \mathbf{B} - \mathbf{B}'$. The vector $\mathbf{A}' - \mathbf{A}$ is either parallel to PQ or equal to the null vector: similarly vector $\mathbf{B} - \mathbf{B}'$ is either parallel to PR or equal to the null vector. Since equal (non-zero) vectors have, by the definition of equality, the same direction, the only possibility is that $\mathbf{A}' - \mathbf{A} = \mathbf{B} - \mathbf{B}' = \mathbf{0}$ and therefore $\mathbf{A}$ and $\mathbf{B}$ are unique vectors.

Three-dimensional resolution

Let OL, OM, ON be three non-coplanar lines that are not necessarily mutually perpendicular (Fig. 1.8). For mathematical convenience one end of the vector $\mathbf{C}$ is tied to O so that $\mathbf{C} = \underline{OC}$. We now, through the point C, draw a line parallel to ON meeting the plane defined by the lines OL, OM in the point R. Equation (1.1) enables us to write $\underline{OC} = \underline{OR} + \underline{RC}$. The vector $\underline{OR}$ can be resolved into component vectors, one parallel to OL, the other parallel to OM. If these are designated $\mathbf{X}$ and $\mathbf{Y}$ respectively, and $\underline{RC}$ is put equal to $\mathbf{Z}$, then it follows that $\mathbf{C} = \mathbf{X} + \mathbf{Y} + \mathbf{Z}$. Hence $\mathbf{C}$ has been resolved into three component vectors. The resolution can be shown to be unique and hence we can now formally state the following important theorem:

Any vector can be resolved into three component vectors respectively parallel to three non-coplanar lines OL, OM, ON, the resolution being unique.

In practice we shall use this theorem in one of two ways. Either the lines OL, OM, ON will be mutually perpendicular (in which circumstances they

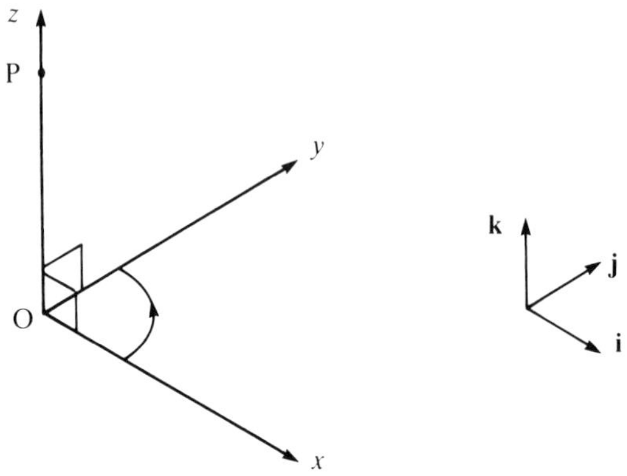

Fig. 1.9

will be relabelled as for example Ox, Oy, Oz); or ON will be perpendicular to the plane defined by the lines OL, OM with OL and OM inclined to each other at any (non-zero) angle.

Right-handed axes

Let Ox, Oy, Oz be a set of mutually perpendicular directed lines intersecting at the point O. The senses of the three lines are indicated by arrows, as shown in Fig. 1.9, where if Oz is in the plane of the page then Ox points out of the page and Oy into it. When they are oriented as shown in Fig. 1.9 we say that Ox, Oy, Oz form a right-handed set of directed lines. Once the directed lines Ox, Oy have been specified, the sense of the line Oz is determined by noting that to an observer at a point P on Oz the turning of Ox (in the plane Ox, Oy) through an angle of $\pi/2$ radians to coincide with Oy will be in the counter-clockwise sense. It follows at once from our definition that Oy, Oz, Ox and Oz, Ox, Oy are both right-handed sets of directed lines. If the sense of Oz (see Fig. 1.9) were reversed, then the lines Ox, Oy, Oz would be described as a left-handed set of directed lines.

Clearly Ox, Oy, Oz may be regarded as a set of coordinate axes. It the senses of x, y, z increasing correspond respectively to the senses of the directed lines Ox, Oy, Oz then we describe Ox, Oy, Oz as a mutually perpendicular set of right-handed axes with origin at the point O, or, more simply, a set of right-handed axes at O: in the rest of this book Ox, Oy, Oz will always mean such a set of axes. Right-handed axes have no particular advantage over left-handed axes; it is simply the current convention to prefer right-handed axes.

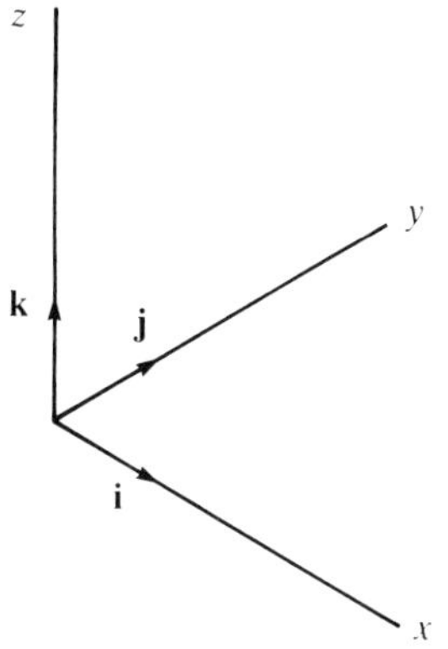

Fig. 1.10

Unit vectors

A unit vector is a vector whose magnitude is one unit of the physical quantity that it represents. Thus if the vector represents a displacement the unit will be one of length, whereas if it represents a velocity the unit will be that of length/time (speed). At this stage we do not introduce any particular system of units for the magnitudes of physical quantities: this is done is Chapter 2.

Unit vectors are of particular importance as it is possible to represent other vectors in terms of them, as we will show shortly. Throughout this book we will use the symbols **i**, **j**, **k** to denote unit vectors in the directions of the right-handed set of axes Ox, Oy, Oz respectively. These unit vectors are described, because of their relationship to the axes, as a right-handed set of unit vectors. The relationship is shown in Fig. 1.10. In the discussion of the resolution of vectors it was shown that any vector **C** can be expressed in the form $\mathbf{C} = \mathbf{X} + \mathbf{Y} + \mathbf{Z}$, where **X**, **Y**, **Z** are parallel to three non-coplanar directions. This representation is not the most useful but it does lead to a more satisfactory alternative as follows. Firstly we take the three non-coplanar directions to coincide with Ox, Oy, Oz so that, for example, **X** will be parallel to **i**. Thus we can write $\mathbf{X} = C_x\mathbf{i}$, where C_x is equal to $|\mathbf{X}|$ if **X** and **i** have the same sense but is equal to $-|\mathbf{X}|$ if the senses of **X** and **i** oppose each other. Similarly we can write $\mathbf{Y} = C_y\mathbf{j}$ and $\mathbf{Z} = C_z\mathbf{k}$. The quantities C_x, C_y, C_z are referred to as the *components* of **C**, and since they may be either positive or negative they are signed algebraic quantities. It is standard to write $[C_x, C_y, C_z]$ for $C_x\mathbf{i} + C_y\mathbf{j} + C_z\mathbf{k}$, and we shall use this notation as it has particular advantages when we wish to manipulate vectors. Thus if $\mathbf{A} = [A_x, A_y, A_z]$ and $\mathbf{B} = [B_x, B_y, B_z]$ then, using rule (1.3),

$$\begin{aligned}\mathbf{A}+\mathbf{B} &= A_x\mathbf{i}+A_y\mathbf{j}+A_z\mathbf{k}+B_x\mathbf{i}+B_y\mathbf{j}+B_z\mathbf{k}\\ &= (A_x+B_x)\mathbf{i}+(A_y+B_y)\mathbf{j}+(A_z+B_z)\mathbf{k}\\ &= [A_x+B_x, A_y+B_y, A_z+B_z] \qquad (1.6)\end{aligned}$$

Likewise the use of rules (1.4) and (1.5) or extensions of them yields the results

$$\mathbf{A}-\mathbf{B} = [A_x-B_x, A_y-B_y, A_z-B_z]$$
$$x\mathbf{A} = [xA_x, xA_y, xA_z]$$

The reader should note that in our use of the unit vectors **i**, **j**, **k** we have not had to distinguish between free and localized vectors. This is essentially because vector statements in terms of **i**, **j**, **k** are statements about magnitudes and/or directions and not points of application. In the remainder of this chapter we shall not make a distinction between free and localized vectors unless it is necessary for the work in hand.

Position vector of a point

If the point O in space is taken as origin and P is any other point in space, then the vector $\underline{\mathrm{OP}}$ is called the position vector of P with respect to the origin O. It is an example of a localized vector. Writing $\mathbf{r} = \underline{\mathrm{OP}} = x\mathbf{i} + y\mathbf{j} + z\mathbf{k}$ it follows that x, y, z are the coordinates of the point P with respect to the right-handed axes Ox, Oy, Oz.

At this stage we consider the first two of our worked examples on vectors. They illustrate the power of the vector algebra developed so far.

Example 1.1

Show that the four points A, B, C, D with coordinates [1, −1, 1], [2, 1, 4], [−1, 1, 1], [−2, 5, 4] referred to the axes Ox, Oy, Oz are coplanar.

▷ *Solution* If the vector $\underline{\mathrm{AD}}$ can be resolved into two component vectors, one parallel to $\underline{\mathrm{AB}}$ and the other parallel to $\underline{\mathrm{AC}}$, then the point D will lie in the plane defined by the lines AB and AC. Alternatively we have to show that $\underline{\mathrm{AD}} = \alpha\underline{\mathrm{AB}} + \beta\underline{\mathrm{AC}}$, where α and β are scalars to be found. Since the coordinates of the four points A, B, C, D are the components of the vectors $\underline{\mathrm{OA}}$, $\underline{\mathrm{OB}}$, $\underline{\mathrm{OC}}$, $\underline{\mathrm{OD}}$ referred to **i**, **j**, **k** it follows, as $\underline{\mathrm{OA}} + \underline{\mathrm{AB}} = \underline{\mathrm{OB}}$, that

$$\underline{AB} = \underline{OB} - \underline{OA} = [2, 1, 4] - [1, -1, 1] = [1, 2, 3]$$

Likewise

$$\underline{AC} = [-1, 1, 1] - [1, -1, 1] = [-2, 2, 0]$$

$$\underline{AD} = [-2, 5, 4] - [1, -1, 1] = [-3, 6, 3]$$

Thus α and β must be such that

$$[-3, 6, 3] = \alpha[1, 2, 3] + \beta[-2, 2, 0] = [\alpha - 2\beta, 2\alpha + 2\beta, 3\alpha]$$

The uniqueness theorem on the resolution of vectors enables us to assert that corresponding components are equal. Equating components yields the following three equations for the two unknowns α and β: $\alpha - 2\beta = -3$, $\alpha + \beta = 3$ and $\alpha = 1$. From the second and third equations we deduce $\alpha = 1$ and $\beta = 2$. Substitution of these values for α and β into the left-hand side of the first equation reveals that this equation is also satisfied. Hence the vector $\underline{AD}$ can be resolved into two components, one parallel to $\underline{AB}$ and the other to $\underline{AC}$, and we can therefore assert that the points A, B, C, D are coplanar.

Consistency of equations

An important point to note in Example 1.1 is that the number of equations exceeds the number of unknowns. A discussion of the systematic techniques used for solving such equations is beyond the scope of this book and any reader who wishes to pursue the matter further is recommended to consult A. Jeffrey, *Mathematics for Engineers and Scientists* (Van Nostrand Reinhold (International), 1990). Our technique, adequate for later problems, is based on the procedure just used in Example 1.1. In this a number of equations equal to the number of unknowns is selected and solved. The values obtained for the unknowns are then substituted in the remaining equations. If these are satisfied then a solution of the complete set of equations exists, and the equations are said to be *consistent*. If the remaining equations are not satisfied then there is no solution and the equations are described as *inconsistent*.

Example 1.2

The vectors **A** and **B** are defined (see Fig. 1.11(a)) by $\mathbf{A} = \underline{OA}$ and $\mathbf{B} = \underline{OB}$. Show, if α and β are positive scalars, that the vector $\alpha\mathbf{A} + \beta\mathbf{B}$ is equal to $(\alpha + \beta)\underline{OQ}$, where Q is an interior point of the line AB (that is Q lies between A and B) such that $\alpha|\underline{AQ}| = \beta|\underline{QB}|$. Explain how the result would have to be modified if $\alpha > 0$ and $\beta < 0$.

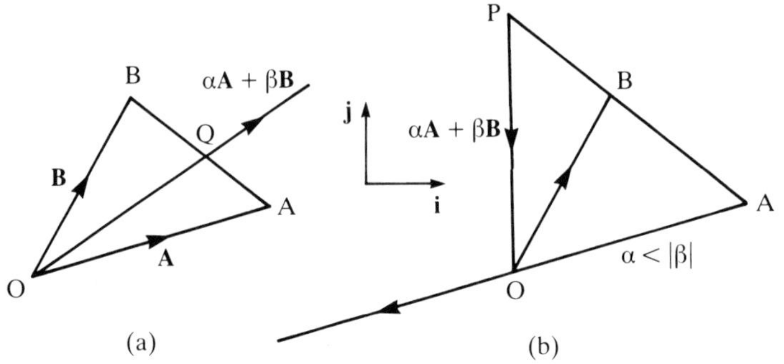

Fig. 1.11

▷ *Solution* Let P be any interior point of the line AB. Then

$$\alpha\mathbf{A} + \beta\mathbf{B} = \alpha\underline{OA} + \beta\underline{OB} = \alpha(\underline{OP} + \underline{PA}) + \beta(\underline{OP} + \underline{PB})$$
$$= (\alpha + \beta)\underline{OP} + \alpha\underline{PA} + \beta\underline{PB}$$

If $\mathbf{n}$ is a unit vector in the direction of $\underline{AB}$ it follows that $\underline{PA} = -|\underline{PA}|\mathbf{n}$ and $\underline{PB} = |\underline{PB}|\mathbf{n}$. Hence

$$\alpha\mathbf{A} + \beta\mathbf{B} = (\alpha + \beta)\underline{OP} + (-\alpha|\underline{PA}| + \beta|\underline{PB}|)\mathbf{n}$$

Now choose P to be the particular point of AB such that $\alpha|\underline{AP}| = \beta|\underline{PB}|$. Thus P coincides with Q so that $\alpha\mathbf{A} + \beta\mathbf{B} = (\alpha + \beta)\underline{OQ}$ as required.

When $\alpha > 0$ and $\beta < 0$ we take the point P to be an exterior (that is a non-interior) point of AB, as for example in Fig. 1.11(b). This leads to

$$\alpha\mathbf{A} + \beta\mathbf{B} = (\alpha + \beta)\underline{OP} \pm (\alpha|\underline{PA}| + \beta|\underline{PB}|)\mathbf{n}$$

the plus sign applying when the point P lies on BA produced and the minus when it lies on AB produced. Choosing P so that $\alpha|\underline{AP}| = -\beta|\underline{PB}|$ gives $\alpha\mathbf{A} + \beta\mathbf{B} = (\alpha + \beta)\underline{OP}$. When $\alpha > |\beta|$ the point P lies on BA produced, while if $\alpha < |\beta|$ the point P lies on AB produced. The analysis breaks down when $\alpha = -\beta$ since there is no exterior point of AB such that $|\underline{PA}| = |\underline{PB}|$. In this case the resultant of $\alpha\mathbf{A}$ and $\beta\mathbf{B}$ is given by the localized vector $\alpha(\mathbf{A} - \mathbf{B})$ tied to the point O. ◂

The angle between two vectors

Figure 1.12(a) shows two vectors **A** and **B** that are represented by the directed lines OA and OB. Also shown in the figure are two angles θ and ϕ, either of which could be described as the angle between the two vectors **A** and **B**. In order to avoid ambiguity we adopt the convention of choosing

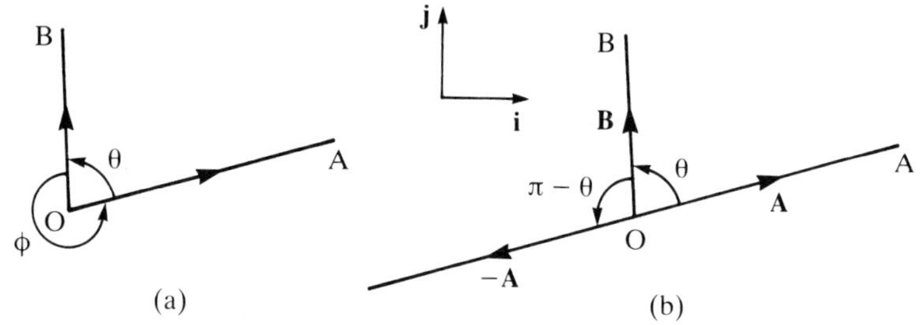

Fig. 1.12

the angle whose value lies in the inclusive range 0 to π. Thus in Fig. 1.12(a) the angle between **A** and **B** is θ. The angle between $-\mathbf{A}$ and **B** is $\pi - \theta$, as in Fig. 1.12(b).

The direction cosines of a vector

If θ_1, θ_2, θ_3 are the angles that the vector $\underline{\mathrm{OC}}$ makes with **i**, **j**, **k** then the direction cosines of $\underline{\mathrm{OC}}$ with respect to the axes Ox, Oy, Oz are defined to be the three quantities l, m, n, where

$$l = \cos\theta_1, \qquad m = \cos\theta_2, \qquad n = \cos\theta_3$$

Determination of the components of a vector

In Fig. 1.13(a) and (b) the vector $\underline{\mathrm{OR}}$ is the projection of $\mathbf{C} = \underline{\mathrm{OC}}$ onto the plane Ox, Oy, while $\underline{\mathrm{OP}}$ and $\underline{\mathrm{OQ}}$ are respectively the projections of $\underline{\mathrm{OR}}$ onto Ox and Oy. Thus $\underline{\mathrm{RC}} = C_z\mathbf{k}$ while $\underline{\mathrm{OP}} = C_x\mathbf{i}$ and $\underline{\mathrm{OQ}} = C_y\mathbf{j}$. In Fig. 1.13(a) the component C_z is equal to $|\underline{\mathrm{RC}}|$ but in Fig. 1.13(b) it is equal to $-|\underline{\mathrm{RC}}|$. Now $|\underline{\mathrm{RC}}|$ is $|\underline{\mathrm{OC}}|\cos\theta_3$ in the former figure but is $-|\underline{\mathrm{OC}}|\cos\theta_3$ in the latter. In either case therefore $C_z = |\mathbf{C}|\cos\theta_3$. Similar expressions can be derived for C_x and C_y. Hence $C_x = l|\mathbf{C}|$, $C_y = m|\mathbf{C}|$, $C_z = n|\mathbf{C}|$, where l, m, n are the direction cosines of **C** with respect to Ox, Oy, Oz.

The magnitude of **C** is readily determined when its components referred to **i**, **j**, **k** have been found. We have

$$\begin{aligned}|\mathbf{C}|^2 = |\underline{\mathrm{OC}}|^2 = |\underline{\mathrm{OR}}|^2 + |\underline{\mathrm{RC}}|^2 &= |\underline{\mathrm{OP}}|^2 + |\underline{\mathrm{OQ}}|^2 + |\underline{\mathrm{RC}}|^2 \\ &= C_x^2 + C_y^2 + C_z^2\end{aligned}$$

Alternatively $|\mathbf{C}|^2 = |\mathbf{C}|^2(l^2 + m^2 + n^2)$, which yields the result that $l^2 + m^2 + n^2 = 1$. If **C** is a unit vector then $\mathbf{C} = [l, m, n]$.

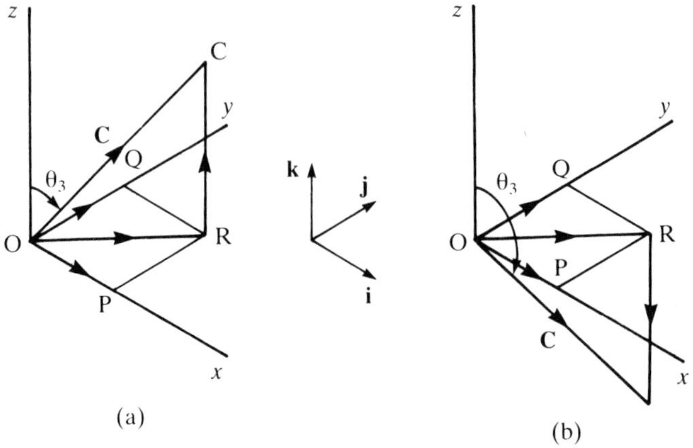

Fig. 1.13

The scalar product of two vectors

The scalar, or dot, product of two vectors **A** and **B** is denoted by **A . B** and defined by $\mathbf{A} \cdot \mathbf{B} = |\mathbf{A}|\,|\mathbf{B}| \cos\theta$, where θ is the angle between the vectors **A** and **B**. The scalar product arises in a variety of physical contexts as readers who are familiar with such concepts as flux and work will be aware. For our present purpose it is sufficient to regard the scalar product as a mathematical quantity whose usefulness will be demonstrated in subsequent work. It follows from the definition of the scalar product that $\mathbf{B} \cdot \mathbf{A} = \mathbf{A} \cdot \mathbf{B}$, and that the scalar product of two vectors is zero if they are perpendicular. Thus we have in particular $\mathbf{i} \cdot \mathbf{j} = \mathbf{j} \cdot \mathbf{k} = \mathbf{k} \cdot \mathbf{i} = 0$. Further, $\mathbf{A} \cdot \mathbf{A} = |\mathbf{A}|^2$ since the angle between **A** and itself is zero. It is customary to write $\mathbf{A}^2$ for **A . A** and we will adhere to this practice. Hence $\mathbf{i}^2 = \mathbf{j}^2 = \mathbf{k}^2 = 1$.

The following rules (and extensions of them) enable us to manipulate expressions containing the scalar product of two vectors:

$$(\mathbf{A} + \mathbf{B}) \cdot \mathbf{C} = \mathbf{A} \cdot \mathbf{C} + \mathbf{B} \cdot \mathbf{C} \tag{1.7}$$

$$(\mathbf{A} + \mathbf{B}) \cdot (\mathbf{C} + \mathbf{D}) = \mathbf{A} \cdot \mathbf{C} + \mathbf{A} \cdot \mathbf{D} + \mathbf{B} \cdot \mathbf{C} + \mathbf{B} \cdot \mathbf{D} \tag{1.8}$$

$$(\mathbf{A} + \mathbf{B}) \cdot (\mathbf{A} - \mathbf{B}) = \mathbf{A}^2 - \mathbf{B}^2 = |\mathbf{A}|^2 - |\mathbf{B}|^2 \tag{1.9}$$

$$(x\mathbf{A}) \cdot \mathbf{B} = \mathbf{A} \cdot (x\mathbf{B}) = x(\mathbf{A} \cdot \mathbf{B}) \tag{1.10}$$

$$\mathbf{A} \cdot \mathbf{B} = A_x B_x + A_y B_y + A_z B_z \tag{1.11}$$

We now provide proofs of the rules (1.7) and (1.11).

Proofs It follows from the definition $\mathbf{A} \cdot \mathbf{B} = |\mathbf{A}|\,|\mathbf{B}| \cos\theta$ that if $\mathbf{B}$ is resolved into components in three mutually perpendicular directions, one of these being that of the vector $\mathbf{A}$, then

$$\mathbf{A} \cdot \mathbf{B} = |\mathbf{A}| \text{ (component of } \mathbf{B} \text{ in the direction of } \mathbf{A})$$

Hence

$$\begin{aligned}(\mathbf{A} + \mathbf{B}) \cdot \mathbf{C} &= |\mathbf{C}| \text{ (component of } (\mathbf{A} + \mathbf{B}) \text{ in the direction of } \mathbf{C}) \\ &= |\mathbf{C}| \text{ (component of } \mathbf{A} \text{ in the direction of } \mathbf{C} + \text{component} \\ &\qquad \text{of } \mathbf{B} \text{ in the direction of } \mathbf{C})\end{aligned}$$

by virtue of the result (1.6). The right-hand side of the above relation is seen on expansion to be equal to $\mathbf{A} \cdot \mathbf{C} + \mathbf{B} \cdot \mathbf{C}$, so that rule (1.7) is proved.

Now using an extension of rule (1.8) we have

$$\begin{aligned}\mathbf{A} \cdot \mathbf{B} &= (A_x\mathbf{i} + A_y\mathbf{j} + A_z\mathbf{k}) \cdot (B_x\mathbf{i} + B_y\mathbf{j} + B_z\mathbf{k}) \\ &= (A_x\mathbf{i}) \cdot (B_x\mathbf{i}) + (A_x\mathbf{i}) \cdot (B_y\mathbf{j}) + (A_x\mathbf{i}) \cdot (B_z\mathbf{k}) + \text{six further terms.}\end{aligned}$$

But $\mathbf{i}^2 = \mathbf{j}^2 = \mathbf{k}^2 = 1$ while $\mathbf{i} \cdot \mathbf{j} = \mathbf{j} \cdot \mathbf{k} = \mathbf{k} \cdot \mathbf{i} = 0$, and hence we deduce, with the help of rule (1.10), that

$$\mathbf{A} \cdot \mathbf{B} = A_x B_x + A_y B_y + A_z B_z$$

Thus rule (1.11) is proved. By putting $\mathbf{B} = \mathbf{A}$ in this result we obtain a result already established, namely $\mathbf{A}^2 = |\mathbf{A}|^2 = A_x^2 + A_y^2 + A_z^2$.

Example 1.3

The vectors $\mathbf{A}$ and $\mathbf{B}$ are defined by $\mathbf{A} = -4\mathbf{i} - 3\mathbf{j} + 2\mathbf{k}$ and $\mathbf{B} = 8\mathbf{i} - 3\mathbf{j} - \mathbf{k}$. Find

(a) The angle between the two vectors
(b) A unit vector perpendicular to both $\mathbf{A}$ and $\mathbf{B}$.

Solution
(a) Using rule (1.11) we find that $\mathbf{A} \cdot \mathbf{B} = -25$. Now $|\mathbf{A}| = \sqrt{29}$ and $|\mathbf{B}| = \sqrt{74}$, and hence $\cos\theta = (\mathbf{A} \cdot \mathbf{B})/|\mathbf{A}|\,|\mathbf{B}| = -0.5397$. Thus the angle between the two vectors is 2.14 radians.
(b) Let $\mathbf{C} = [C_x, C_y, C_z]$ be a unit vector perpendicular to $\mathbf{A}$ and $\mathbf{B}$. Since $\mathbf{A} \cdot \mathbf{C} = \mathbf{B} \cdot \mathbf{C} = 0$ it follows that $-4C_x - 3C_y + 2C_z = 0$ and $8C_x - 3C_y - C_z = 0$. These equations yield the relations $C_z = 4C_x = C_y$, and hence using the further relation $C_x^2 + C_y^2 + C_z^2 = 1$ we find that $C_x = \pm 3/13$. There are therefore, as we would anticipate, two possible expressions for the vector $\mathbf{C}$, namely $[3, 4, 12]/13$ and $-[3, 4, 12]/13$. These two vectors have opposite senses.

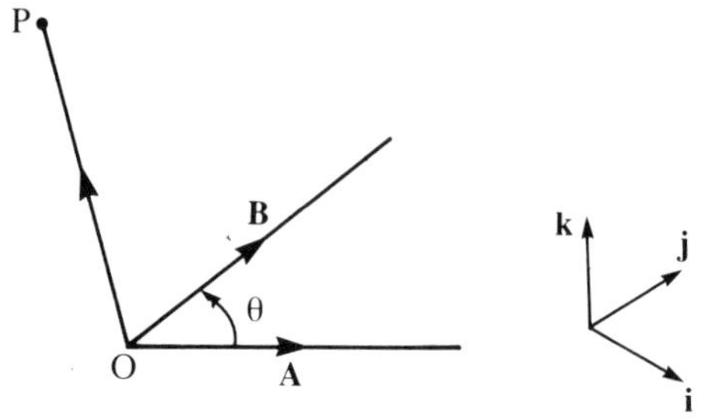

Fig. 1.14

The vector product of two vectors

We now associate with two vectors a second quantity whose usefulness, like that of the scalar product, will be demonstrated in future work. This quantity is called the vector product. The vector product of the vector **A** with the vector **B** is a third vector denoted by $\mathbf{A} \times \mathbf{B}$. Its magnitude is defined by $|\mathbf{A} \times \mathbf{B}| = |\mathbf{A}|\,|\mathbf{B}|\sin\theta$ where θ is the angle between **A** and **B**. The direction of $\mathbf{A} \times \mathbf{B}$ is perpendicular to both **A** and **B**, its sense being determined by the following rule. If the turning of **A** to **B** (in the plane of **A** and **B**) through the angle θ appears to be counter-clockwise to an observer at the point P in the line perpendicular to **A** and **B** (see Fig. 1.14) then the sense of $\mathbf{A} \times \mathbf{B}$ is that of $\underline{OP}$; otherwise it is opposite to it. This rule is known as the right-hand thread rule since the turning of **A** to **B** through θ would advance a right-hand screw thread in the sense of $\mathbf{A} \times \mathbf{B}$.

It is a consequence of the definition of $\mathbf{A} \times \mathbf{B}$ that the senses of $\mathbf{A} \times \mathbf{B}$ and $\mathbf{B} \times \mathbf{A}$ are opposite to one another. Since the two vectors are, by definition, perpendicular to **A** and **B** and have the same magnitude it follows that $\mathbf{B} \times \mathbf{A} = -\mathbf{A} \times \mathbf{B}$. Thus we note for future work that the order in which the vectors **A** and **B** are written in the expression $\mathbf{A} \times \mathbf{B}$ (or $\mathbf{B} \times \mathbf{A}$) is most important. We also note that if **A** and **B** are parallel, that is $\theta = 0$ or π, the vectors $\mathbf{A} \times \mathbf{B}$ and $\mathbf{B} \times \mathbf{A}$ are both equal to the zero vector.

A number of important relations between the unit vectors **i**, **j**, **k** can be written down directly using the vector product definition. They are:

$$\mathbf{i} \times \mathbf{i} = \mathbf{j} \times \mathbf{j} = \mathbf{k} \times \mathbf{k} = \mathbf{0}$$

$$\mathbf{j} \times \mathbf{k} = \mathbf{i}, \qquad \mathbf{k} \times \mathbf{i} = \mathbf{j}, \qquad \mathbf{i} \times \mathbf{j} = \mathbf{k}$$

These relations, coupled with the following rules (or extensions of them), enable us to manipulate expressions involving the vector product of two vectors:

$$(x\mathbf{A}) \times (y\mathbf{B}) = xy(\mathbf{A} \times \mathbf{B}) \tag{1.12}$$

$$(\mathbf{A} + \mathbf{B}) \times \mathbf{C} = \mathbf{A} \times \mathbf{C} + \mathbf{B} \times \mathbf{C} \tag{1.13}$$

The reader should note that since **A** precedes **C** in the left-hand side of (1.13) it must do so in the right-hand side. Likewise for the vector **B**. We give a proof of rule (1.13)

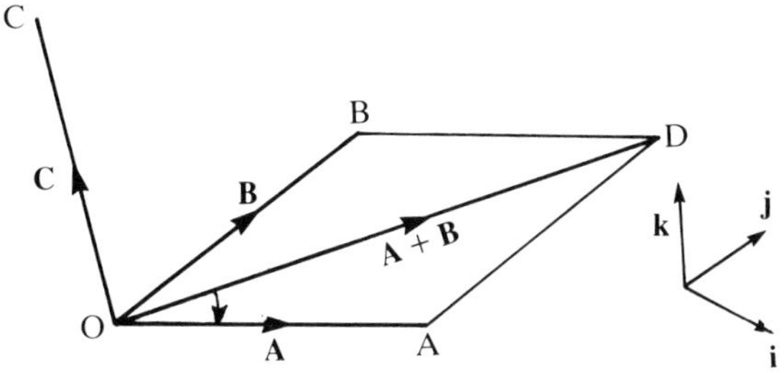

Fig. 1.15

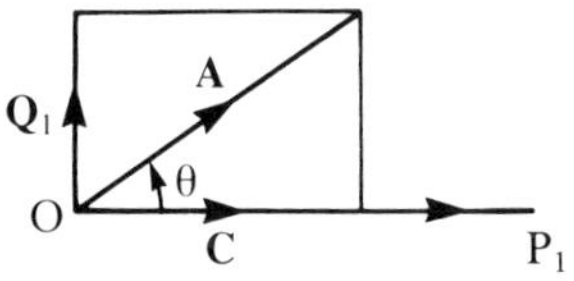

Fig. 1.16

□ *Proof* We start by establishing the rule for the particular case when **A** and **B** lie in a plane that has the vector **C** as a normal (see Fig. 1.15). It is evident that the vectors $\mathbf{A} \times \mathbf{C}$, $\mathbf{B} \times \mathbf{C}$, $(\mathbf{A} + \mathbf{B}) \times \mathbf{C}$ will be represented by directed lines obtained by turning the lines OA, OB, OD through an angle $\pi/2$ in the plane of **A** and **B** and in the sense shown and magnifying them by a factor equal to $|\mathbf{C}|$. Thus by the parallelogram law, $(\mathbf{A} + \mathbf{B}) \times \mathbf{C} = \mathbf{A} \times \mathbf{C} + \mathbf{B} \times \mathbf{C}$. In the general case we proceed as follows. We resolve **A** into two component vectors $\mathbf{P}_1$, $\mathbf{Q}_1$, the former being parallel to **C** and the latter being perpendicular to **C** and in the plane of **A** and **C** (see Fig. 1.16). By virtue of the results for determining components, $|\mathbf{Q}_1| = |\mathbf{A}| \sin\theta$, and therefore we can infer from the definition of the vector product that $\mathbf{A} \times \mathbf{C} = \mathbf{Q}_1 \times \mathbf{C}$. Likewise we put $\mathbf{B} = \mathbf{P}_2 + \mathbf{Q}_2$, where $\mathbf{P}_2$ is parallel to **C** and $\mathbf{Q}_2$ is perpendicular to **C** and lies in the plane of **B** and **C**, so that $\mathbf{B} \times \mathbf{C} = \mathbf{Q}_2 \times \mathbf{C}$. Since $\mathbf{Q}_1$ and $\mathbf{Q}_2$ lie in a plane with **C** as a normal, we have

$$(\mathbf{Q}_1 + \mathbf{Q}_2) \times \mathbf{C} = \mathbf{Q}_1 \times \mathbf{C} + \mathbf{Q}_2 \times \mathbf{C}$$
$$= \mathbf{A} \times \mathbf{C} + \mathbf{B} \times \mathbf{C}$$

by virtue of the results just obtained. Now

$$\mathbf{A} + \mathbf{B} = (\mathbf{P}_1 + \mathbf{P}_2) + (\mathbf{Q}_1 + \mathbf{Q}_2)$$

where $\mathbf{P}_1 + \mathbf{P}_2$ is a vector parallel to $\mathbf{C}$ and $\mathbf{Q}_1 + \mathbf{Q}_2$ is a vector lying in the plane of $\mathbf{Q}_1$ and $\mathbf{Q}_2$ and is therefore perpendicular to $\mathbf{C}$. Hence

$$(\mathbf{Q}_1 + \mathbf{Q}_2) \times \mathbf{C} = (\mathbf{A} + \mathbf{B}) \times \mathbf{C}$$

and the proof is therefore complete.

It has already been seen that the scalar product of two vectors can be expressed in terms of their components. The same is possible for the vector product of two vectors:

$$\mathbf{A} \times \mathbf{B} = (A_x\mathbf{i} + A_y\mathbf{j} + A_z\mathbf{k}) \times (B_x\mathbf{i} + B_y\mathbf{j} + B_z\mathbf{k})$$
$$= (A_x\mathbf{i}) \times (B_x\mathbf{i}) + (A_x\mathbf{i}) \times (B_y\mathbf{j}) + (A_x\mathbf{i}) \times (B_z\mathbf{k})$$
$$+ \text{ six further terms}$$

Using the relations that exist between the vectors $\mathbf{i}$, $\mathbf{j}$, $\mathbf{k}$ and the rule (1.12) we obtain, after some simplification,

$$\mathbf{A} \times \mathbf{B} = (A_y B_z - A_z B_y)\mathbf{i} + (A_z B_x - A_x B_z)\mathbf{j} + (A_x B_y - A_y B_x)\mathbf{k} \quad (1.14)$$

The result contained in (1.14) can be put in a concise form. In order to do this it is first necessary to define a determinant.

Determinants

We begin by considering second-order determinants. The expression

$$\begin{vmatrix} a & b \\ c & d \end{vmatrix}$$

where a, b, c, d are signed numbers is called a second-order determinant. It represents a number defined by

$$\begin{vmatrix} a & b \\ c & d \end{vmatrix} = ad - bc$$

The numbers a, b, c, d constitute the elements of the determinant. The elements a and b form the first row of the determinant while c and d form the second. Similarly a and c constitute the first column and b and d the second column of the determinant. Recalling the relation (1.14) we see that

$$\mathbf{A}\times\mathbf{B} = \begin{vmatrix} A_y & A_z \\ B_y & B_z \end{vmatrix}\mathbf{i} + \begin{vmatrix} A_z & A_x \\ B_z & B_x \end{vmatrix}\mathbf{j} + \begin{vmatrix} A_x & A_y \\ B_x & B_y \end{vmatrix}\mathbf{k}$$

which is abbreviated to

$$\mathbf{A}\times\mathbf{B} = \begin{vmatrix} \mathbf{i} & \mathbf{j} & \mathbf{k} \\ A_x & A_y & A_z \\ \mathrm{B_x} & B_y & B_z \end{vmatrix} \qquad (1.15)$$

The expression on the right-hand side of (1.15) is an example of a third-order determinant. A third-order determinant may, as here, be equal to a vector or have a scalar value. When it is equal to a scalar the determinant is an expression of the form

$$\begin{vmatrix} a_1 & a_2 & a_3 \\ b_1 & b_2 & b_3 \\ c_1 & c_2 & c_3 \end{vmatrix}$$

whose value is defined by

$$\begin{vmatrix} a_1 & a_2 & a_3 \\ b_1 & b_2 & b_3 \\ c_1 & c_2 & c_3 \end{vmatrix} = a_1\begin{vmatrix} b_2 & b_3 \\ c_2 & c_3 \end{vmatrix} + a_2\begin{vmatrix} b_3 & b_1 \\ c_3 & c_1 \end{vmatrix} + a_3\begin{vmatrix} b_1 & b_2 \\ c_1 & c_2 \end{vmatrix} \qquad (1.16)$$

The properties of third-order determinants are numerous. Some of the simpler ones are given in Problem 1.9 with suggestions for their proof.

We now consider worked examples involving the scalar and vector product of two vectors.

Example 1.4

Show for any two vectors **a** and **b** that

$$\mathbf{a}^2\mathbf{b}^2 = (\mathbf{a}\,.\,\mathbf{b})^2 + (\mathbf{a}\times\mathbf{b})^2$$

> *Solution* If θ is the angle between **a** and **b** then

$$(\mathbf{a}\,.\,\mathbf{b})^2 + (\mathbf{a}\times\mathbf{b})^2 = (|\mathbf{a}|\,|\mathbf{b}|\cos\theta)^2 + (|\mathbf{a}|\,|\mathbf{b}|\sin\theta)^2 = \mathbf{a}^2\mathbf{b}^2$$

Example 1.5

The vectors **A**, **B**, **C** are defined in terms of **i**, **j**, **k** by $\mathbf{A} = [5, 4, 8]$, $\mathbf{B} = [2, 3, 1]$ and $\mathbf{C} = [1, 4, 2]$. Express **A** in terms of **B**, **C**, $\mathbf{B}\times\mathbf{C}$.

▷ *Solution* The directions of the three vectors **B**, **C**, **B** × **C** are non-coplanar and therefore **A** can be resolved into component vectors parallel to the three directions. Moreover the resolution is unique. We put

$$\mathbf{A} = \alpha\mathbf{B} + \beta\mathbf{C} + \mu\mathbf{B} \times \mathbf{C} \tag{1.17}$$

where α, β and μ are scalars to be determined. Now **B** . (**B** × **C**) and **C** . (**B** × **C**) are each zero since **B** and **C** are perpendicular to **B** × **C**, and therefore the value of μ can be found directly by taking the scalar product of both sides of (1.17) with **B** × **C**. Thus

$$\mathbf{A} \cdot (\mathbf{B} \times \mathbf{C}) = \mu(\mathbf{B} \times \mathbf{C})^2 \tag{1.18}$$

Now, using (1.15),

$$\mathbf{B} \times \mathbf{C} = \begin{vmatrix} \mathbf{i} & \mathbf{j} & \mathbf{k} \\ 2 & 3 & 1 \\ 1 & 4 & 2 \end{vmatrix} = [2, -3, 5]$$

Hence $\mathbf{A} \cdot (\mathbf{B} \times \mathbf{C}) = 10 - 12 + 40 = 38$ and $|\mathbf{B} \times \mathbf{C}|^2 = 38$ so that $\mu = 1$.

Expressing the vector equation (1.17) in component form yields

$$[5, 4, 8] = \alpha[2, 3, 1] + \beta[1, 4, 2] + [2, -3, 5]$$

from which we obtain the three equations

$$2\alpha + \beta = 3, \qquad 3\alpha + 4\beta = 7, \qquad \alpha + 2\beta = 3$$

These equations are consistent and have the solution $\alpha = \beta = 1$. Hence $\mathbf{A} = \mathbf{B} + \mathbf{C} + \mathbf{B} \times \mathbf{C}$.

Example 1.6

The vectors **a**, **b**, **c** are given in terms of **i**, **j**, **k** by $3\mathbf{a} = [2, -1, 2]$, $3\mathbf{b} = [-1, 2, 2]$ and $3\mathbf{c} = [-2, -2, 1]$. Show that **a**, **b**, **c** form a right-handed set of mutually perpendicular unit vectors.

▷ *Solution* Clearly $|\mathbf{a}| = |\mathbf{b}| = |\mathbf{c}| = 1$ so that **a**, **b**, **c** are all unit vectors. Further $9\mathbf{a} \cdot \mathbf{b} = -2 - 2 + 4 = 0$ and therefore **a** and **b** are perpendicular. To complete the solution we have to show that $\mathbf{c} = \mathbf{a} \times \mathbf{b}$. Now

$$9\mathbf{a} \times \mathbf{b} = \begin{vmatrix} \mathbf{i} & \mathbf{j} & \mathbf{k} \\ 2 & -1 & 2 \\ -1 & 2 & 2 \end{vmatrix} = [-6, -6, 3] = 9\mathbf{c}$$

and hence $\mathbf{c} = \mathbf{a} \times \mathbf{b}$ as required.

Example 1.7

The vectors **a** and **b** are projected (perpendicularly) onto a plane whose normal is in the direction of the unit vector **n**. Obtain an expression for the angle between the two projections.

Solution The vector **a** can be resolved into two component vectors, one parallel to **n** and the other perpendicular to **n** and in the plane of **a** and **n**. The former is given by $(\mathbf{a}\,.\,\mathbf{n})\mathbf{n}$ and hence the latter is equal to $\mathbf{a} - (\mathbf{a}\,.\,\mathbf{n})\mathbf{n}$. Thus the projection of the vector **a** onto the plane is the vector $\mathbf{a} - (\mathbf{a}\,.\,\mathbf{n})\mathbf{n}$. Since **n** is a unit vector we have

$$(\mathbf{a} - (\mathbf{a}\,.\,\mathbf{n})\mathbf{n})^2 = (\mathbf{a} - (\mathbf{a}\,.\,\mathbf{n})\mathbf{n})\,.\,(\mathbf{a} - (\mathbf{a}\,.\,\mathbf{n})\mathbf{n}) = \mathbf{a}^2 - (\mathbf{a}\,.\,\mathbf{n})^2 = |\mathbf{a} \times \mathbf{n}|^2$$

noting the identity contained in Example 1.4. If α is the angle between the two projections then

$$|\mathbf{a} \times \mathbf{n}|\,|\mathbf{b} \times \mathbf{n}| \cos\alpha = (\mathbf{a} - (\mathbf{a}\,.\,\mathbf{n})\mathbf{n})\,.\,(\mathbf{b} - (\mathbf{b}\,.\,\mathbf{n})\mathbf{n}) = \mathbf{a}\,.\,\mathbf{b} - (\mathbf{a}\,.\,\mathbf{n})(\mathbf{b}\,.\,\mathbf{n})$$

so that α can now be found.

Triple scalar and triple vector products

The expression $\mathbf{A}\,.\,(\mathbf{B} \times \mathbf{C})$ which occurs on the left-hand side of equation (1.18) is an example of a triple-scalar product. The brackets are frequently omitted, the triple product being written as $\mathbf{A}\,.\,\mathbf{B} \times \mathbf{C}$. For a meaning to be attached to the expression it is first necessary to form the vector product $\mathbf{B} \times \mathbf{C}$ and then take the scalar product of this vector with **A**. Omission of the brackets therefore does not lead to any ambiguity. In contrast the quantity $\mathbf{A} \times \mathbf{B} \times \mathbf{C}$ is capable of two interpretations, namely $(\mathbf{A} \times \mathbf{B}) \times \mathbf{C}$ or $\mathbf{A} \times (\mathbf{B} \times \mathbf{C})$. Since these two vectors are not, in general, equal the brackets are essential. The vectors $(\mathbf{A} \times \mathbf{B}) \times \mathbf{C}$ and $\mathbf{A} \times (\mathbf{B} \times \mathbf{C})$ are examples of triple-vector products.

We now state and prove three important identities concerning triple products:

$$\mathbf{A}\,.\,\mathbf{B} \times \mathbf{C} = \begin{vmatrix} A_x & A_y & A_z \\ B_x & B_y & B_z \\ C_x & C_y & C_z \end{vmatrix} \tag{1.19}$$

$$\mathbf{A}\,.\,\mathbf{B} \times \mathbf{C} = \mathbf{B}\,.\,\mathbf{C} \times \mathbf{A} = \mathbf{C}\,.\,\mathbf{A} \times \mathbf{B} \tag{1.20}$$

$$\mathbf{A} \times (\mathbf{B} \times \mathbf{C}) = (\mathbf{A}\,.\,\mathbf{C})\mathbf{B} - (\mathbf{A}\,.\,\mathbf{B})\mathbf{C} \tag{1.21}$$

Proofs Using the second-order determinant definition of the components of the vector product $\mathbf{A} \times \mathbf{B}$ (see (1.15)) yields

$$\mathbf{A}\,.\,\mathbf{B}\times\mathbf{C} = A_x\begin{vmatrix} B_y & B_z \\ C_y & C_z \end{vmatrix} + A_y\begin{vmatrix} B_z & B_x \\ C_z & C_x \end{vmatrix} + A_z\begin{vmatrix} B_x & B_y \\ C_x & C_y \end{vmatrix}$$
$$= \begin{vmatrix} A_x & A_y & A_z \\ B_x & B_y & B_z \\ C_x & C_y & C_z \end{vmatrix}$$

by virtue of the definition given in (1.16). This proves (1.19).

Next,

$$\mathbf{B}\,.\,\mathbf{C}\times\mathbf{A} = \begin{vmatrix} B_x & B_y & B_z \\ C_x & C_y & C_z \\ A_x & A_y & A_z \end{vmatrix}$$
$$= B_x(C_yA_z - C_zA_y) + B_y(C_zA_x - C_xA_z) + B_z(C_xA_y - C_yA_x)$$
$$= A_x(B_yC_z - B_zC_y) + A_y(B_zC_x - B_xC_z) + A_z(B_xC_y - B_yC_x)$$
$$= \mathbf{A}\,.\,\mathbf{B}\times\mathbf{C}$$

Likewise $\mathbf{C}\,.\,\mathbf{A}\times\mathbf{B} = \mathbf{A}\,.\,\mathbf{B}\times\mathbf{C}$, so that the identity (1.20) is established.

The identity (1.21) may be proved in a number of ways. The direct way is to expand $\mathbf{A}\times(\mathbf{B}\times\mathbf{C})$ in component form by repeated use of the relation (1.14) and then verify that it is equal to the right-hand side of (1.21). An alternative way is suggested in Problem 1.12(b).

We now illustrate, through examples, some of the manipulative techniques that will be used in subsequent work.

Example 1.8

Show that

$$(\mathbf{A}\times\mathbf{B})\,.\,(\mathbf{C}\times\mathbf{D}) = (\mathbf{A}\,.\,\mathbf{C})(\mathbf{B}\,.\,\mathbf{D}) - (\mathbf{A}\,.\,\mathbf{D})(\mathbf{B}\,.\,\mathbf{C})$$

▷ *Solution* Let $\mathbf{E} = \mathbf{A}\times\mathbf{B}$. Hence

$$(\mathbf{A}\times\mathbf{B})\,.\,(\mathbf{C}\times\mathbf{D}) = \mathbf{E}\,.\,(\mathbf{C}\times\mathbf{D}) = \mathbf{D}\,.\,(\mathbf{E}\times\mathbf{C})$$

using the identity (1.20) for triple-scalar products. But by identity (1.21)

$$\mathbf{E}\times\mathbf{C} = (\mathbf{A}\times\mathbf{B})\times\mathbf{C} = (\mathbf{A}\,.\,\mathbf{C})\mathbf{B} - (\mathbf{A}\,.\,\mathbf{B})\mathbf{C}$$

and thus

$$(\mathbf{A}\times\mathbf{B})\,.\,(\mathbf{C}\times\mathbf{D}) = \mathbf{D}\,.\,(\mathbf{E}\times\mathbf{C}) = \mathbf{D}\,.\,((\mathbf{A}\,.\,\mathbf{C})\mathbf{B} - (\mathbf{A}\,.\,\mathbf{B})\mathbf{C})$$
$$= (\mathbf{A}\,.\,\mathbf{C})(\mathbf{B}\,.\,\mathbf{D}) - (\mathbf{A}\,.\,\mathbf{D})(\mathbf{B}\,.\,\mathbf{C})$$

as required.

Example 1.9

Expand $(\mathbf{A} \times \mathbf{B}) \times (\mathbf{C} \times \mathbf{D})$ in two ways, and deduce that any four vectors are related by an identity of the form

$$a\mathbf{A} + b\mathbf{B} + c\mathbf{C} + d\mathbf{D} = \mathbf{0}$$

where a, b, c, d are scalars. Write down possible expressions for these scalars.

Solution Let $\mathbf{E} = \mathbf{A} \times \mathbf{B}$. Hence

$$(\mathbf{A} \times \mathbf{B}) \times (\mathbf{C} \times \mathbf{D}) = \mathbf{E} \times (\mathbf{C} \times \mathbf{D}) = (\mathbf{E} \,.\, \mathbf{D})\mathbf{C} - (\mathbf{E} \,.\, \mathbf{C})\mathbf{D}$$

Thus

$$(\mathbf{A} \times \mathbf{B}) \times (\mathbf{C} \times \mathbf{D}) = (\mathbf{D} \,.\, \mathbf{A} \times \mathbf{B})\mathbf{C} - (\mathbf{C} \,.\, \mathbf{A} \times \mathbf{B})\mathbf{D}$$

Likewise if we put $\mathbf{F} = \mathbf{C} \times \mathbf{D}$ we can deduce that

$$(\mathbf{A} \times \mathbf{B}) \times (\mathbf{C} \times \mathbf{D}) = (\mathbf{A} \,.\, \mathbf{C} \times \mathbf{D})\mathbf{B} - (\mathbf{B} \,.\, \mathbf{C} \times \mathbf{D})\mathbf{A}$$

Equating the two expressions for $(\mathbf{A} \times \mathbf{B}) \times (\mathbf{C} \times \mathbf{D})$ leads, after some rearrangement of the triple-scalar products, to

$$(\mathbf{B} \,.\, \mathbf{C} \times \mathbf{D})\mathbf{A} + (\mathbf{C} \,.\, \mathbf{A} \times \mathbf{D})\mathbf{B} + (\mathbf{D} \,.\, \mathbf{A} \times \mathbf{B})\mathbf{C} + (\mathbf{A} \,.\, \mathbf{C} \times \mathbf{B})\mathbf{D} = \mathbf{0}$$

The expressions for a, b, c, d are therefore given apart from a scalar multiplier by

$$a = \mathbf{B} \,.\, \mathbf{C} \times \mathbf{D}, \qquad b = \mathbf{C} \,.\, \mathbf{A} \times \mathbf{D}, \qquad c = \mathbf{D} \,.\, \mathbf{A} \times \mathbf{B}, \qquad d = \mathbf{A} \,.\, \mathbf{C} \times \mathbf{B}$$

Example 1.10

The scalars p, q, r satisfy the vector equation $p\mathbf{a} + q\mathbf{b} + r\mathbf{a} \times \mathbf{b} = \mathbf{c}$ where $\mathbf{a}$, $\mathbf{b}$, $\mathbf{c}$ are known vectors such that $\mathbf{a} \times \mathbf{b} \neq \mathbf{0}$. Determine the values of p, q, r in terms of the vectors $\mathbf{a}$, $\mathbf{b}$, $\mathbf{c}$.

Solution We begin by taking the vector product of both sides of the equation with $\mathbf{b}$. This gives

$$p\mathbf{a} \times \mathbf{b} + r(\mathbf{a} \times \mathbf{b}) \times \mathbf{b} = \mathbf{c} \times \mathbf{b}$$

Taking the scalar product of both sides of this equation with $\mathbf{a} \times \mathbf{b}$ yields

$$p|\mathbf{a} \times \mathbf{b}|^2 = (\mathbf{c} \times \mathbf{b}) \,.\, (\mathbf{a} \times \mathbf{b})$$

the scalar product of $\mathbf{a} \times \mathbf{b}$ with $(\mathbf{a} \times \mathbf{b}) \times \mathbf{b}$ being zero as these two vectors are perpendicular.

Using the identity contained in Example 1.8 we deduce that

$$p|\mathbf{a} \times \mathbf{b}|^2 = (\mathbf{a} \,.\, \mathbf{c})\mathbf{b}^2 - (\mathbf{c} \,.\, \mathbf{b})(\mathbf{a} \,.\, \mathbf{b})$$

Similarly q is given by

$$q|\mathbf{a} \times \mathbf{b}|^2 = (\mathbf{b} \,.\, \mathbf{c})\mathbf{a}^2 - (\mathbf{c} \,.\, \mathbf{a})(\mathbf{a} \,.\, \mathbf{b})$$

The value of r can be written down directly (see Example 1.5). We have

$$r|\mathbf{a} \times \mathbf{b}|^2 = \mathbf{c} \,.\, \mathbf{a} \times \mathbf{b}$$

Vector algebra is a powerful tool in the study of three-dimensional geometry. For the purposes of this book our study is confined to obtaining the vector equations of a straight line and plane and making elementary use of them.

Vector equation of a straight line

The word 'line' is usually taken to imply something that is of limited extent, and so far we have used the word in this context. However, when we speak of the equation of a line the convention is to regard the line as being infinite in extent. If we are interested in a particular segment or part of it then this segment will be defined by restricting the parameter that occurs in the equation of the line to an appropriate range of values.

A straight line may be specified in more than one way. We can, for example, give the positions of two points of it, or specify the position of one point and the direction of a unit vector parallel to it. Either specification enables us to determine the vector equation of the line. Let A and B be two given points of the line and P any other point of it. Further, let $\mathbf{r} = \underline{\mathrm{OP}}$ be the position vector of P with respect to the origin O, so that $\mathbf{r} = \underline{\mathrm{OP}} = \underline{\mathrm{OA}} + \underline{\mathrm{AP}}$. Now the points A, B and P are collinear and therefore $\underline{\mathrm{AP}} = \mu\underline{\mathrm{AB}}$, where μ is a scalar that may be positive or negative. Hence

$$\mathbf{r} = \underline{\mathrm{OA}} + \mu\underline{\mathrm{AB}} = \underline{\mathrm{OA}} + \mu|\underline{\mathrm{AB}}|\frac{\underline{\mathrm{AB}}}{|\underline{\mathrm{AB}}|}$$

On writing α for $\mu|\underline{\mathrm{AB}}|$ and $\mathbf{n}$ for the unit vector in the direction of $\underline{\mathrm{AB}}$ the relation becomes

$$\mathbf{r} = \underline{\mathrm{OA}} + \alpha\mathbf{n} \tag{1.22}$$

The relation (1.22) is called the vector equation of the line through the points A and B. The position vectors of all points in the line, when it is taken to be infinite in extent, are obtained by allowing the parameter α to range from $-\infty$ to ∞.

Vector equation of a plane

We assume, when determining the vector equation of a plane, that the position of one point of it together with the direction of a vector perpendicular to it are known. Let A be a given point of the plane and $\mathbf{n}$ any vector perpendicular to it. If P is any point in the plane it follows that $\underline{\mathrm{AP}}$ and $\mathbf{n}$ are perpendicular and therefore $\underline{\mathrm{AP}} \cdot \mathbf{n} = 0$. Now $\underline{\mathrm{AP}} = \underline{\mathrm{OP}} - \underline{\mathrm{OA}} = \mathbf{r} - \underline{\mathrm{OA}}$ and hence

$$(\mathbf{r} - \underline{\mathrm{OA}}) \cdot \mathbf{n} = 0 \tag{1.23}$$

which is the vector equation of the plane. If $\mathbf{r} = [x, y, z]$ and $\mathbf{n} = [a, b, c]$ referred to the unit vectors $\mathbf{i}, \mathbf{j}, \mathbf{k}$ then equation (1.23) has the alternative form $ax + by + cz = p$ where $p = \underline{\mathrm{OA}} \cdot \mathbf{n}$. This form is frequently referred to as the cartesian equation of the plane.

If the positions of three non-collinear points A, B, C of the plane are given then equation (1.23) can be determined from the fact that $\mathbf{n}$ is parallel to $\underline{\mathrm{AB}} \times \underline{\mathrm{AC}}$. There is a vector alternative to equation (1.23) that is also referred to as the vector equation of the plane. As before, $\mathbf{r} = \underline{\mathrm{OA}} + \underline{\mathrm{AP}}$. Now $\underline{\mathrm{AP}}$ can be resolved into two component vectors parallel to the directions of $\underline{\mathrm{AB}}$ and $\underline{\mathrm{AC}}$ since the lines AB, AC, AP are coplanar. Writing $\underline{\mathrm{AP}} = \alpha\underline{\mathrm{AB}} + \beta\underline{\mathrm{AC}}$ we have

$$\mathbf{r} = \underline{\mathrm{OA}} + \alpha\underline{\mathrm{AB}} + \beta\underline{\mathrm{AC}}$$

The points of the plane, assumed to be infinite in extent, are generated by allowing the quantities α, β (known as parameters) to range (independently of each other) over all values from $-\infty$ to ∞.

Example 1.11

The four points A, B, C, D have coordinates [3, 1, 0], [0, 1, 2], [1, 2, 1] and [2, 0, 1] referred to the axes Ox, Oy, Oz. Find the cartesian equation of the plane defined by the points A, B, C, D and hence deduce that the four points A, B, C, D are coplanar. Two further points E and F have coordinates [1, 1, 1] and [−5, 3, 5]. Find where the line EF meets the plane ABC.

▷ *Solution* Firstly, $\underline{\mathrm{AB}} = [0, 1, 2] - [3, 1, 0] = [-3, 0, 2]$ and $\underline{\mathrm{AC}} = [-2, 1, 1]$ referred to the unit vectors $\mathbf{i}, \mathbf{j}, \mathbf{k}$. Hence

$$\underline{\mathrm{AB}} \times \underline{\mathrm{AC}} = \begin{vmatrix} \mathbf{i} & \mathbf{j} & \mathbf{k} \\ -3 & 0 & 2 \\ -2 & 1 & 1 \end{vmatrix} = [-2, -1, -3]$$

Therefore the normal to the plane ABC is parallel to the vector [2, 1, 3]. The cartesian equation of the plane ABC is thus given by $2x + y + 3z = 6 + 1 + 0 = 7$. Clearly the point D with coordinates [2, 0, 1] lies in the plane.

Next, $\underline{EF} = [-5, 3, 5] - [1, 1, 1] = [-6, 2, 4] = 2[-3, 1, 2]$. The vector equation of the line EF is therefore $\mathbf{r} = [1, 1, 1] + \alpha[-3, 1, 2]$ from which it follows that the coordinates of any point on EF are of the form $[1 - 3\alpha, 1 + \alpha, 1 + 2\alpha]$. Hence EF meets the plane ABC where α has the value given by $2(1 - 3\alpha) + 1 + \alpha + 3(1 + 2\alpha) = 7$, that is $\alpha = 1$. The point of intersection of EF and the plane ABC thus has coordinates [−2, 2, 3].

Example 1.12

Show vectorially how the position vector of the foot of the perpendicular from a given point to a given line can be found.

▷ *Solution* Let $\mathbf{r} = \mathbf{a} + \alpha\mathbf{n}$, where $\mathbf{n}$ is a unit vector, be the vector equation of the given line; and let $\mathbf{b}$ be the position vector, with respect to the origin O, of the given point. If $\mathbf{N}$ is a unit vector in the direction of the perpendicular from the given point to the given line, then the vector equation of the perpendicular is $\mathbf{r} = \mathbf{b} + \beta\mathbf{N}$ where β is a parameter. The perpendicular and the given line intersect where $\mathbf{a} + \alpha\mathbf{n} = \mathbf{b} + \beta\mathbf{N}$. Taking the scalar product of both sides of this equation with $\mathbf{n}$ yields, since $\mathbf{n} \,.\, \mathbf{N} = 0$, an equation for α, namely $\mathbf{n} \,.\, \mathbf{a} + \alpha\mathbf{n}^2 = \mathbf{n} \,.\, \mathbf{b}$. Since $\mathbf{n}^2 = 1$ the value of α is $(\mathbf{b} - \mathbf{a}) \,.\, \mathbf{n}$, and thus the position vector of the foot of the perpendicular is found to be $\mathbf{a} + ((\mathbf{b} - \mathbf{a}) \,.\, \mathbf{n})\mathbf{n}$.

The remainder of this section on vector algebra relates to the change of sets of unit vectors and the solution of certain vector equations. Since these topics are not required until Chapter 5, readers may defer the study of them and pass to Section 1.3 on the differentiation of vectors. Section 1.3 is essential for the work of Chapter 4 but not that of Chapter 3.

Change of unit vectors

The process of replacing one right-handed set of mutually perpendicular unit vectors by another is an example of a vector transformation. The usefulness of such transformations will be shown in later chapters. For the present we confine ourselves to a discussion of their algebra properties.

Let $\mathbf{i}, \mathbf{j}, \mathbf{k}$ and $\mathbf{I}, \mathbf{J}, \mathbf{K}$ be two sets of right-handed mutually perpendicular unit vectors with associated axes Ox, Oy, Oz and OX, OY, OZ. The vectors $\mathbf{I}, \mathbf{J}, \mathbf{K}$ can each be resolved into components in the directions of $\mathbf{i}$,

j, **k**. We write

$$\begin{aligned}\mathbf{I} &= l_{11}\mathbf{i} + l_{12}\mathbf{j} + l_{13}\mathbf{k}\\ \mathbf{J} &= l_{21}\mathbf{i} + l_{22}\mathbf{j} + l_{23}\mathbf{k}\\ \mathbf{K} &= l_{31}\mathbf{i} + l_{32}\mathbf{j} + l_{33}\mathbf{k}\end{aligned} \tag{1.24}$$

Since **I** is a unit vector the quantities l_{11}, l_{12}, l_{13} are the direction cosines of **I** with respect to the axes Ox, Oy, Oz. Taking the scalar product of **I** successively with **i**, **j**, **k** yields

$$l_{11} = \mathbf{I}\,.\,\mathbf{i}, \qquad l_{12} = \mathbf{I}\,.\,\mathbf{j}, \qquad l_{13} = \mathbf{I}\,.\,\mathbf{k} \tag{1.25}$$

Similar expressions can be written down for the remaining quantities l_{21}, l_{22}, ..., l_{33}. The coefficients of **i**, **j**, **k** in the transformation defined by (1.24) are collectively denoted by l_{pq} $(p, q = 1, 2, 3)$. They are connected by a number of relations, six of which can be derived from the fact that

$$\mathbf{I}^2 = \mathbf{J}^2 = \mathbf{K}^2 = 1, \qquad \mathbf{I}\,.\,\mathbf{J} = \mathbf{J}\,.\,\mathbf{K} = \mathbf{K}\,.\,\mathbf{I} = 0$$

The relations are

$$l_{11}^2 + l_{12}^2 + l_{13}^2 = l_{21}^2 + l_{22}^2 + l_{23}^2 = l_{31}^2 + l_{32}^2 + l_{33}^2 = 1$$

and

$$l_{11}l_{21} + l_{12}l_{22} + l_{13}l_{23} = l_{11}l_{31} + l_{12}l_{32} + l_{13}l_{33} = l_{21}l_{31} + l_{22}l_{32} + l_{23}l_{33} = 0$$

These six relations can be put into a single form by the introduction of the nine quantities δ_{pq} $(p, q = 1, 2, 3)$ collectively known as the Kronecker delta. The value of δ_{pq} is defined to be zero if p and q are different and one if p and q are equal. Hence we have

$$\sum_{r=1}^{3} l_{pr}l_{qr} = \delta_{pq} \qquad (p, q = 1, 2, 3) \tag{1.26}$$

It follows, since **I**, **J**, **K** form a mutually perpendicular right-handed set of unit vectors, that **K** is completely defined when **I** and **J** have been specified. Hence the components of **K** with respect to **i**, **j**, **k** must be functions of the components of **I** and **J**. To find expressions for these functions we could use the relation $\mathbf{K} = \mathbf{I} \times \mathbf{J}$ but prefer to proceed as follows.

The relations

$$l_{11}l_{31} + l_{12}l_{32} + l_{13}l_{33} = l_{21}l_{31} + l_{22}l_{32} + l_{23}l_{33} = 0$$

can be regarded as two equations for the three unknowns l_{31}, l_{32}, l_{33} when the values of l_{pq} $(p = 1, 2;\ q = 1, 2, 3)$ are known. The equations can be solved for the ratio $l_{31} : l_{32}$ by the elimination of l_{33}. Further elimination yields the ratios $l_{31} : l_{32} : l_{33}$ and hence that

$$l_{31} = \mu(l_{12}l_{23} - l_{13}l_{22}), \qquad l_{32} = \mu(l_{13}l_{21} - l_{11}l_{23}),$$
$$l_{33} = \mu(l_{11}l_{22} - l_{12}l_{21}) \tag{1.27}$$

where μ is a scalar whose value we now determine. Substitution of the expressions for l_{31}, l_{32}, l_{33} into the relation $l_{31}^2 + l_{32}^2 + l_{33}^2 = 1$ leads, after some algebraic rearrangement, to the equation for μ, namely

$$\mu^2\{(l_{11}^2 + l_{12}^2 + l_{13}^2)(l_{21}^2 + l_{22}^2 + l_{23}^2) - (l_{11}l_{21} + l_{12}l_{22} + l_{13}l_{23})^2\} = 1$$

Using the relations (1.26) we deduce that $\mu^2 = 1$ or $\mu = \pm 1$. In order to determine the appropriate value for μ we observe that in the particular case when the vectors **I** and **J** respectively coincide with **i** and **j**, the vector **K** must coincide with **k**. When $\mathbf{I} = \mathbf{i}$, $\mathbf{J} = \mathbf{j}$ it follows from (1.24) that $l_{11} = l_{22} = 1$ and $l_{12} = l_{13} = l_{21} = l_{23} = 0$ and thus from the relations (1.27) we obtain $l_{31} = l_{32} = 0$ and $l_{33} = \mu$. The value of μ is therefore one. The relations (1.27) can be put in the form

$$l_{31} = \begin{vmatrix} l_{12} & l_{13} \\ l_{22} & l_{23} \end{vmatrix}, \qquad l_{32} = -\begin{vmatrix} l_{11} & l_{13} \\ l_{21} & l_{23} \end{vmatrix}, \qquad l_{33} = \begin{vmatrix} l_{11} & l_{12} \\ l_{21} & l_{22} \end{vmatrix} \tag{1.28}$$

These expressions can be written down directly by means of the following rule: strike out from the third-order determinant

$$\begin{vmatrix} l_{11} & l_{12} & l_{13} \\ l_{21} & l_{22} & l_{23} \\ l_{31} & l_{32} & l_{33} \end{vmatrix}$$

the row and column that contain l_{pq} and then multiply the resulting second-order determinant by $(-1)^{p+q}$.

Expressions can also be found for the components of the vector **I** in terms of the components of **J** and **K** and for the components of **J** in terms of those of **K** and **I**. They can also be written down by the rule just described, and therefore

$$l_{11} = l_{22}l_{33} - l_{23}l_{32}, \qquad l_{12} = l_{23}l_{31} - l_{21}l_{33}, \qquad l_{13} = l_{21}l_{32} - l_{22}l_{31} \tag{1.29}$$

and

$$l_{21} = l_{13}l_{32} - l_{12}l_{33}, \qquad l_{22} = l_{11}l_{33} - l_{13}l_{31}, \qquad l_{23} = l_{12}l_{31} - l_{11}l_{32} \tag{1.30}$$

The roles of **i**, **j**, **k** and **I**, **J**, **K** in the relations (1.24) can be interchanged. We write

$$\mathbf{i} = m_{11}\mathbf{I} + m_{12}\mathbf{J} + m_{13}\mathbf{K}$$
$$\mathbf{j} = m_{21}\mathbf{I} + m_{22}\mathbf{J} + m_{23}\mathbf{K} \tag{1.31}$$
$$\mathbf{k} = m_{31}\mathbf{I} + m_{32}\mathbf{J} + m_{33}\mathbf{K}$$

which is known as the inverse transformation of (1.24). It follows directly from (1.26) that

$$\sum_{r=1}^{3} m_{pr} m_{qr} = \delta_{pq} \qquad (p, q = 1, 2, 3) \tag{1.32}$$

We can now determine the transformation law for the components of a vector **A**. Let **A** be given in terms of the two systems of unit vectors by

$$A_x\mathbf{i} + A_y\mathbf{j} + A_z\mathbf{k} = \mathbf{A} = A_X\mathbf{I} + A_Y\mathbf{J} + A_Z\mathbf{K}$$

Taking the scalar product throughout with **I** yields

$$A_X = (\mathbf{I}\,.\,\mathbf{i})A_x + (\mathbf{I}\,.\,\mathbf{j})A_y + (\mathbf{I}\,.\,\mathbf{k})A_z = l_{11}A_x + l_{12}A_y + l_{13}A_z$$

from which we infer that

$$\begin{aligned} A_X &= l_{11}A_x + l_{12}A_y + l_{13}A_z \\ A_Y &= l_{21}A_x + l_{22}A_y + l_{23}A_z \\ A_Z &= l_{31}A_x + l_{32}A_y + l_{33}A_z \end{aligned} \tag{1.33}$$

The inverse of (1.33) is readily found to be

$$\begin{aligned} A_x &= m_{11}A_X + m_{12}A_Y + m_{13}A_Z \\ A_y &= m_{21}A_X + m_{22}A_Y + m_{23}A_Z \\ A_z &= m_{31}A_X + m_{32}A_Y + m_{33}A_Z \end{aligned} \tag{1.34}$$

Vector equations

In anticipation of work on rotation theory we consider two examples of vector equations. The method of solution of these equations makes use of the result that any vector **X** can be expressed in the form

$$\mathbf{X} = \alpha\mathbf{A} + \beta\mathbf{B} + \mu\mathbf{A} \times \mathbf{B}$$

provided the vectors **A** and **B** are not parallel.

Example 1.13

The vector **X** satisfies the vector equation $\mathbf{X} \times \mathbf{A} = \mathbf{B}$, where **A** and **B** are known non-zero vectors such that $\mathbf{A}\,.\,\mathbf{B} = 0$. Determine **X**.

Solution The three vectors **A**, **B** and $\mathbf{A} \times \mathbf{B}$ are non-coplanar and accordingly the unknown vector **X** can be expressed in the form

$$\mathbf{X} = \alpha\mathbf{A} + \beta\mathbf{B} + \mu\mathbf{A} \times \mathbf{B}$$

where α, β and μ are scalars to be found. Substituting for $\mathbf{X}$ in the equation $\mathbf{X} \times \mathbf{A} = \mathbf{B}$ yields

$$(\alpha\mathbf{A} + \beta\mathbf{B} + \mu\mathbf{A} \times \mathbf{B}) \times \mathbf{A} = \mathbf{B}$$

or

$$\beta\mathbf{B} \times \mathbf{A} + \mu(\mathbf{A}^2\mathbf{B} - (\mathbf{A} \,.\, \mathbf{B})\mathbf{A}) = \mathbf{B}$$

which reduces, since $\mathbf{A} \,.\, \mathbf{B} = 0$, to

$$\beta\mathbf{B} \times \mathbf{A} + \mu\mathbf{A}^2\mathbf{B} = \mathbf{B}$$

Using the uniqueness theorem on the resolution of vectors we can equate the coefficients of $\mathbf{B}$ and $\mathbf{B} \times \mathbf{A}$ on the left-hand side of the equation to the corresponding coefficients on the right-hand side. Hence $\beta = 0$ and $\mathbf{A}^2\mu = 1$, while α can take any value. The solution for $\mathbf{X}$ is therefore given by

$$\mathbf{X} = \alpha\mathbf{A} + (\mathbf{A} \times \mathbf{B})/\mathbf{A}^2$$

where α is arbitrary.

Example 1.14

The unknown vector $\mathbf{X}$ satisfies the two vector equations

$$\mathbf{X} \times \mathbf{A} = \mathbf{B}, \qquad \mathbf{X} \times \mathbf{C} = \mathbf{D}$$

where $\mathbf{A}$, $\mathbf{B}$, $\mathbf{C}$, $\mathbf{D}$ are known vectors such that $\mathbf{A} \,.\, \mathbf{B} = \mathbf{C} \,.\, \mathbf{D} = 0$. Show that:

(a) If $\mathbf{A}$ and $\mathbf{C}$ are parallel, that is $\mathbf{A} = \alpha\mathbf{C}$ where α is some scalar, then the vector equations are either inconsistent or $\mathbf{B} = \alpha\mathbf{D}$.
(b) If $\mathbf{A}$ and $\mathbf{C}$ are not parallel then a solution to the equations only exists if $\mathbf{A} \,.\, \mathbf{D} + \mathbf{B} \,.\, \mathbf{C} = 0$.

Determine the solution of the equations when the relation in (b) is satisfied, showing that it is unique. Write down the particular form of the solution when $\mathbf{A} \,.\, \mathbf{D} = \mathbf{B} \,.\, \mathbf{C} = 0$.

▷ *Solution*
(a) We have $\mathbf{B} - \alpha\mathbf{D} = \mathbf{X} \times (\mathbf{A} - \alpha\mathbf{C}) = \mathbf{0}$. Hence $\mathbf{B} = \alpha\mathbf{D}$ or the equations are inconsistent. When $\mathbf{A} = \alpha\mathbf{C}$ and $\mathbf{B} = \alpha\mathbf{D}$ the equation $\mathbf{X} \times \mathbf{A} = \mathbf{B}$ is equivalent to the equation $\mathbf{X} \times \mathbf{C} = \mathbf{D}$ and the method of Example 1.13 can be used to obtain the vector $\mathbf{X}$.
(b) Since $\mathbf{A}$ and $\mathbf{C}$ are not parallel the three vectors $\mathbf{B}$, $\mathbf{D}$, $\mathbf{X}$ can be expressed in terms of $\mathbf{A}$, $\mathbf{C}$, $\mathbf{A} \times \mathbf{C}$. Using results contained in Example 1.10 with $\mathbf{a}$ replaced by $\mathbf{A}$, $\mathbf{b}$ replaced by $\mathbf{C}$, and $\mathbf{c}$ put successively equal to $\mathbf{B}$ and $\mathbf{D}$, leads, as $\mathbf{A} \,.\, \mathbf{B} = \mathbf{C} \,.\, \mathbf{D} = 0$, to

$$|\mathbf{A}\times\mathbf{C}|^2\mathbf{B} = (\mathbf{B}\,.\,\mathbf{C})(A^2\mathbf{C} - (\mathbf{A}\,.\,\mathbf{C})\mathbf{A}) + (\mathbf{B}\,.\,\mathbf{A}\times\mathbf{C})\mathbf{A}\times\mathbf{C}$$
$$|\mathbf{A}\times\mathbf{C}|^2\mathbf{D} = (\mathbf{D}\,.\,\mathbf{A})(C^2\mathbf{A} - (\mathbf{A}\,.\,\mathbf{C})\mathbf{C}) + (\mathbf{D}\,.\,\mathbf{A}\times\mathbf{C})\mathbf{A}\times\mathbf{C} \qquad (1.35)$$

Putting $\mathbf{X} = \alpha\mathbf{A} + \beta\mathbf{C} + \mu\mathbf{A}\times\mathbf{C}$ in the equations $\mathbf{X}\times\mathbf{A} = \mathbf{B}$ and $\mathbf{X}\times\mathbf{C} = \mathbf{D}$ and expanding the triple-vector products that arise yields the following equations for α, β, μ:

$$\beta\mathbf{C}\times\mathbf{A} + \mu(A^2\mathbf{C} - (\mathbf{A}\,.\,\mathbf{C})\mathbf{A}) = \mathbf{B}$$
$$\alpha\mathbf{A}\times\mathbf{C} + \mu((\mathbf{A}\,.\,\mathbf{C})\mathbf{C} - C^2\mathbf{A}) = \mathbf{D}$$

These two vector equations are equivalent to six scalar equations for the three unknowns α, β, μ. Comparison of the expression for $\mathbf{B}$ in (1.35) with the first of these equations shows that β and μ are given by

$$|\mathbf{A}\times\mathbf{C}|^2\beta = \mathbf{B}\,.\,\mathbf{C}\times\mathbf{A}, \qquad |\mathbf{A}\times\mathbf{C}|^2\mu = \mathbf{B}\,.\,\mathbf{C}$$

Likewise comparing the expression for $\mathbf{D}$ with the second equation leads to

$$|\mathbf{A}\times\mathbf{C}|^2\alpha = \mathbf{D}\,.\,\mathbf{A}\times\mathbf{C}, \qquad |\mathbf{A}\times\mathbf{C}|^2\mu = -\mathbf{A}\,.\,\mathbf{D}$$

Thus for consistency we require that $-\mathbf{A}\,.\,\mathbf{D} = \mathbf{B}\,.\,\mathbf{C}$ or $\mathbf{A}\,.\,\mathbf{D} + \mathbf{B}\,.\,\mathbf{C} = 0$. When this condition holds the solution for $\mathbf{X}$ is given by

$$|\mathbf{A}\times\mathbf{C}|^2\mathbf{X} = (\mathbf{D}\,.\,\mathbf{A}\times\mathbf{C})\mathbf{A} + (\mathbf{B}\,.\,\mathbf{C}\times\mathbf{A})\mathbf{C} + (\mathbf{B}\,.\,\mathbf{C})(\mathbf{A}\times\mathbf{C})$$

To show that the solution for $\mathbf{X}$ is unique we assume two solutions $\mathbf{X}_1$, $\mathbf{X}_2$ exist. Hence

$$\mathbf{X}_1\times\mathbf{A} = \mathbf{X}_2\times\mathbf{A} = \mathbf{B}, \qquad \mathbf{X}_1\times\mathbf{C} = \mathbf{X}_2\times\mathbf{C} = \mathbf{D}$$

so that

$$(\mathbf{X}_1 - \mathbf{X}_2)\times\mathbf{A} = \mathbf{0}, \qquad (\mathbf{X}_1 - \mathbf{X}_2)\times\mathbf{C} = \mathbf{0}$$

Either $\mathbf{X}_1 - \mathbf{X}_2$ is parallel to $\mathbf{A}$ and $\mathbf{C}$, or $\mathbf{X}_1 - \mathbf{X}_2$ is the null vector. The former is not possible as $\mathbf{A}$ and $\mathbf{C}$ are not parallel, and therefore $\mathbf{X}_1 - \mathbf{X}_2 = \mathbf{0}$ which implies the solution is unique.

When $\mathbf{A}\,.\,\mathbf{D} = -\mathbf{B}\,.\,\mathbf{C} = 0$ the solution for $\mathbf{X}$ reduces to

$$|\mathbf{A}\times\mathbf{C}|^2\mathbf{X} = (\mathbf{D}\,.\,\mathbf{A}\times\mathbf{C})\mathbf{A} + (\mathbf{B}\,.\,\mathbf{C}\times\mathbf{A})\mathbf{C}$$

1.3 VECTOR DIFFERENTIATION

The vectors considered so far have, for the most part, been constant in both magnitude and direction. An important exception is the expression $\mathbf{a} + \alpha\mathbf{n}$ for the position vector $\mathbf{r}$ of a point in a given line, where α is a parameter. Two further examples of variable vectors are given by $\cos\theta\mathbf{i} + \sin\theta\mathbf{j}$

and $\cos\theta\mathbf{i} + \sin\theta\mathbf{j} + \theta\mathbf{k}$, where θ is also a parameter that takes all values in a given range. Both vectors have simple physical interpretations.

If we take O as origin and put $\underline{OP} = \cos\theta\mathbf{i} + \sin\theta\mathbf{j}$, where $0 \leq \theta \leq 2\pi$, then $\underline{OP}$ represents, for a given value of θ, the position vector of a point P on the circle centre O and unit radius. The angle of inclination of the line OP to Ox is θ, and as θ varies from 0 to 2π the point P will describe the circle once. The second vector $\cos\theta\mathbf{i} + \sin\theta\mathbf{j} + \theta\mathbf{k}$ is the position vector of a point P on a circular helix. This becomes evident when we consider the projection Q of the point P onto the plane Ox, Oy. The point Q lies on the circle defined by $\underline{OQ} = \cos\theta\mathbf{i} + \sin\theta\mathbf{j}$ and as Q describes the circle once the $\mathbf{k}$ component of $\underline{OP}$ changes by 2π.

A vector may depend on more than one parameter. One example is given by $\underline{OP} = \cos\phi\sin\theta\mathbf{i} + \sin\phi\sin\theta\mathbf{j} + \cos\theta\mathbf{k}$ where $0 \leq \phi < 2\pi$ and $0 \leq \theta \leq \pi$. Elementary manipulation shows that $|\underline{OP}| = 1$ and hence P lies on the surface of a sphere of unit radius whose centre is at the origin O. A second example is provided by the equation of a plane $\underline{OP} = \underline{OA} + \alpha\underline{AB} + \beta\underline{AC}$, where A, B, C are points on the plane and α, β are parameters that vary independently from $-\infty$ to ∞. It is evident that a point P whose position vector depends on a single parameter lies on a curve and that a point with a position vector dependent on two parameters lies on a surface.

In order to indicate that a vector $\mathbf{A}$ depends on a single parameter t we write $\mathbf{A}(t)$. This notation is consistent with the practice, followed in the calculus and elsewhere, of denoting by $f(x)$ the value of the function f corresponding to the argument x. Again in analogy with the calculus we write $\mathbf{A}(t + \delta t) = \mathbf{A}(t) + \delta\mathbf{A}(t)$ where $\delta\mathbf{A}(t)$ is the change in $\mathbf{A}(t)$ consequent on the change δt in t. The quantity δt may be positive or negative. If we take $\mathbf{A}(t)$ to represent the position vector of a point on a curve, as illustrated in Fig. 1.17, then $\delta\mathbf{A}(t)$, subsequently abbreviated to $\delta\mathbf{A}$, is represented by the vector P_1P_2. As δt tends to zero the direction of $\delta\mathbf{A}$ tends to coincidence with that of the tangent at P_1. This assumes that the tangent has a well-

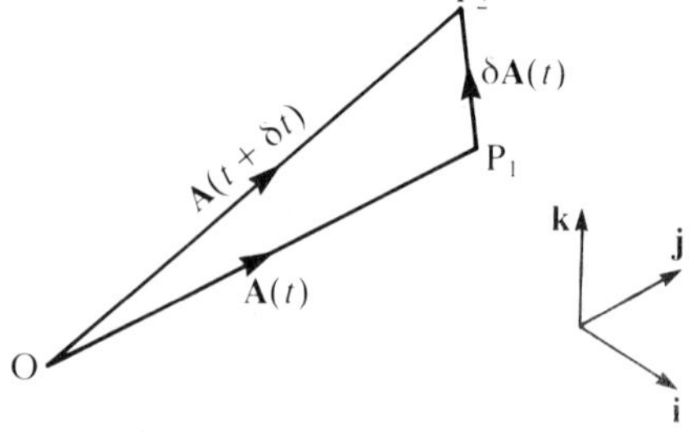

Fig. 1.17

defined direction, which is not always the case. Readers will discover this for themselves by sketching the curve defined by $\underline{OP} = t\mathbf{i} + |t|\mathbf{j}$ for $-1 \leq t \leq 1$, considering in particular the point corresponding to $t = 0$. Such exceptions possess a physical significance or interpretation. If we take the curve to represent the path of a particle then $t = 0$ may be interpreted as the point where the particle experiences a blow that instantaneously changes its direction of motion.

If the vector $(1/\delta t)\delta\mathbf{A}$, which is more usually written as $\delta\mathbf{A}/\delta t$, tends to a (vector) limit as δt tends to zero, then this limit is called the derivative of $\mathbf{A}(t)$ with respect to t. The limit is written as $\mathrm{d}\mathbf{A}(t)/\mathrm{d}t$ or more simply as $\mathrm{d}\mathbf{A}/\mathrm{d}t$. It is customary to omit the parameter t in the expression $\mathbf{A}(t)$ once the dependence of the vector on t has been noted.

The following rules, which are stated without proof, enable us to differentiate vectors:

$$\begin{aligned}
&\frac{\mathrm{d}\mathbf{A}}{\mathrm{d}t} = \mathbf{0} \qquad \text{if } \mathbf{A} \text{ fixed in magnitude and direction}\\
&\frac{\mathrm{d}}{\mathrm{d}t}(\mathbf{A} + \mathbf{B}) = \frac{\mathrm{d}\mathbf{A}}{\mathrm{d}t} + \frac{\mathrm{d}\mathbf{B}}{\mathrm{d}t}\\
&\frac{\mathrm{d}}{\mathrm{d}t}(x\mathbf{A}) = \frac{\mathrm{d}x}{\mathrm{d}t}\mathbf{A} + x\frac{\mathrm{d}\mathbf{A}}{\mathrm{d}t}\\
&\frac{\mathrm{d}}{\mathrm{d}t}(\mathbf{A} \cdot \mathbf{B}) = \frac{\mathrm{d}\mathbf{A}}{\mathrm{d}t} \cdot \mathbf{B} + \mathbf{A} \cdot \frac{\mathrm{d}\mathbf{B}}{\mathrm{d}t}\\
&\frac{\mathrm{d}}{\mathrm{d}t}(\mathbf{A} \times \mathbf{B}) = \frac{\mathrm{d}\mathbf{A}}{\mathrm{d}t} \times \mathbf{B} + \mathbf{A} \times \frac{\mathrm{d}\mathbf{B}}{\mathrm{d}t}\\
&\frac{\mathrm{d}\mathbf{A}}{\mathrm{d}t} = \frac{\mathrm{d}u}{\mathrm{d}t}\frac{\mathrm{d}\mathbf{A}}{\mathrm{d}u}
\end{aligned} \tag{1.36}$$

In the final rule, $\mathbf{A}$ is a function of u which in turn is a function of t.

Applying the rules (and extensions of them) we see that

$$\frac{\mathrm{d}\mathbf{A}}{\mathrm{d}t} = \frac{\mathrm{d}}{\mathrm{d}t}(A_x\mathbf{i} + A_y\mathbf{j} + A_z\mathbf{k}) = \frac{\mathrm{d}A_x}{\mathrm{d}t}\mathbf{i} + \frac{\mathrm{d}A_y}{\mathrm{d}t}\mathbf{j} + \frac{\mathrm{d}A_z}{\mathrm{d}t}\mathbf{k}$$

We write this result as $\dot{\mathbf{A}} = [\dot{A}_x, \dot{A}_y, \dot{A}_z]$, the dot being used, in accordance with standard convention, to denote differentiation with respect to the parameter t. Higher derivatives of $\mathbf{A}$ can be found by repeated use of this formula. Thus $(\mathrm{d}/\mathrm{d}t)(\mathrm{d}\mathbf{A}/\mathrm{d}t)$, written as $\mathrm{d}^2\mathbf{A}/\mathrm{d}t^2$ or $\ddot{\mathbf{A}}$, is given by $\ddot{\mathbf{A}} = [\ddot{A}_x, \ddot{A}_y, \ddot{A}_z]$.

The rules for differentiating a vector dependent on a single parameter can be used to find the partial derivatives of a vector that is dependent on two or more variables. If $\mathbf{A}$ is a function of the two variables θ and ϕ (we write $\mathbf{A}(\theta, \phi)$ to denote this dependence) then the expression $\partial\mathbf{A}/\partial\theta$ denotes the partial derivative of $\mathbf{A}$ with respect to θ, the variable ϕ for this operation being regarded as a constant. For our present purposes it is sufficient to note that

$$\frac{\partial\mathbf{A}}{\partial\theta} = \left[\frac{\partial A_x}{\partial\theta}, \frac{\partial A_y}{\partial\theta}, \frac{\partial A_z}{\partial\theta}\right]$$

referred to the unit vectors $\mathbf{i}$, $\mathbf{j}$, $\mathbf{k}$. A similar expression holds for $\partial\mathbf{A}/\partial\phi$.

Example 1.15

The vector $\mathbf{A}$ is given by

$$\mathbf{A} = \cos\omega t\mathbf{i} + \sin\omega t\mathbf{j}$$

where ω is a constant. Show that:

(a) $\mathbf{A}$ and $\dot{\mathbf{A}}$ are perpendicular
(b) $\mathbf{A}$ and $\ddot{\mathbf{A}}$ are parallel
(c) $\dot{\mathbf{A}} = \boldsymbol{\omega} \times \mathbf{A}$ where $\boldsymbol{\omega} = \omega\mathbf{k}$.

▷ *Solution*
(a) Differentiating $\mathbf{A}$ with respect to t yields

$$\dot{\mathbf{A}} = -\omega\sin\omega t\mathbf{i} + \omega\cos\omega t\mathbf{j}$$

so that $\mathbf{A} \,.\, \dot{\mathbf{A}} = 0$. Hence the vectors $\mathbf{A}$ and $\dot{\mathbf{A}}$ are perpendicular.
(b) A second differentiation of $\mathbf{A}$ gives

$$\ddot{\mathbf{A}} = -\omega^2(\cos\omega t\mathbf{i} + \sin\omega t\mathbf{j}) = -\omega^2\mathbf{A}$$

Thus $\mathbf{A}$ and $\ddot{\mathbf{A}}$ are parallel.
(c) We have

$$\begin{aligned}\omega \times \mathbf{A} &= \omega\mathbf{k} \times (\cos\omega t\mathbf{i} + \sin\omega t\mathbf{j}) \\ &= \omega\cos\omega t\mathbf{j} - \omega\sin\omega t\mathbf{i} = \dot{\mathbf{A}}\end{aligned}$$

as required.

Example 1.16

The vector $\mathbf{A}$ is a function of the parameter t. Show that $\mathrm{d}\mathbf{A}^2/\mathrm{d}t = 2\mathbf{A} \,.\, \dot{\mathbf{A}}$. Deduce if

(a) $\mathbf{A}$ and $\dot{\mathbf{A}}$ are always perpendicular that $|\mathbf{A}|$ is constant
(b) $|\mathbf{A}|$ is constant that $\mathbf{A}$ and $\dot{\mathbf{A}}$ are perpendicular provided $\dot{\mathbf{A}} \neq \mathbf{0}$.

▷ *Solution* Using the rule for differentiating the scalar products of two vectors we see that

$$\frac{dA^2}{dt} = \frac{d}{dt}(\mathbf{A} \cdot \mathbf{A}) = \frac{d\mathbf{A}}{dt} \cdot \mathbf{A} + \mathbf{A} \cdot \frac{d\mathbf{A}}{dt} = 2\mathbf{A} \cdot \dot{\mathbf{A}}$$

(a) If $\mathbf{A}$ and $\dot{\mathbf{A}}$ are perpendicular then $\mathbf{A} \cdot \dot{\mathbf{A}} = 0$ and hence $dA^2/dt = 0$ so that $|\mathbf{A}|$ is constant.
(b) If $|\mathbf{A}|$ is constant we have $\mathbf{A} \cdot \dot{\mathbf{A}} = 0$. It follows if $\dot{\mathbf{A}} \neq 0$ that the vectors $\mathbf{A}$ and $\dot{\mathbf{A}}$ must be perpendicular.

Example 1.17

The three unit vectors $\mathbf{a}$, $\mathbf{b}$, $\mathbf{N}$ lie in a plane that has the unit vector $\mathbf{k}$ as normal. The direction of $\mathbf{a}$ is inclined to the direction of $\mathbf{N}$ (which is fixed) at an angle θ, measured as shown in Fig. 1.18, and lying in the range 0 to 2π. The unit vector $\mathbf{b}$ is defined by $\mathbf{b} = \mathbf{k} \times \mathbf{a}$. Show that $d\mathbf{a}/d\theta = \mathbf{b}$ and $d\mathbf{b}/d\theta = -\mathbf{a}$.

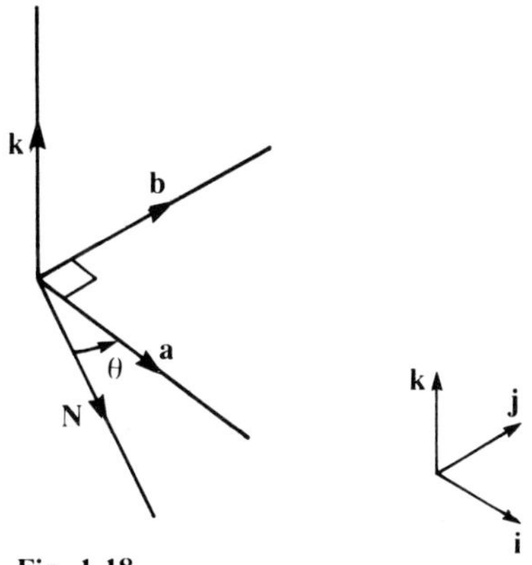

Fig. 1.18

▷ *Solution* It follows, if we define the vector $\mathbf{T}$ by $\mathbf{T} = \mathbf{k} \times \mathbf{N}$, that

$$\mathbf{a} = \cos\theta\mathbf{N} + \sin\theta\mathbf{T}, \qquad \mathbf{b} = -\sin\theta\mathbf{N} + \cos\theta\mathbf{T}$$

Differentiating with respect to θ we obtain, since $\mathbf{N}$ and $\mathbf{T}$ are constant vectors,

$$\frac{d\mathbf{a}}{d\theta} = -\sin\theta\mathbf{N} + \cos\theta\mathbf{T} = \mathbf{b}$$

$$\frac{d\mathbf{b}}{d\theta} = -\cos\theta\mathbf{N} - \sin\theta\mathbf{T} = -\mathbf{a}$$

as required.

Example 1.18

The position vector **r** of a point P of a curve lying in the plane Ox, Oy is a function of the parameter t. Show that d**r**/dt = (ds/dt)**t**, where s denotes arc length and **t** = d**r**/ds is a unit vector, located in the tangent, whose sense is that of s increasing. Show further that **n** = d**t**/dψ, where ψ is the angle between **t** and Ox, is a unit vector whose direction and sense are defined by **n** = **k** × **t**. Deduce that

$$\frac{\mathrm{d}^2\mathbf{r}}{\mathrm{d}t^2}=\frac{\mathrm{d}^2s}{\mathrm{d}t^2}\mathbf{t}+\left\{\left(\frac{\mathrm{d}s}{\mathrm{d}t}\right)^2\bigg/\left(\frac{\mathrm{d}s}{\mathrm{d}\psi}\right)\right\}\mathbf{n}$$

▷ *Solution* Since arc length s is a function of the parameter t we have, using the rule for differentiating a function of a function,

$$\frac{\mathrm{d}\mathbf{r}}{\mathrm{d}t}=\frac{\mathrm{d}s}{\mathrm{d}t}\frac{\mathrm{d}\mathbf{r}}{\mathrm{d}s}$$

If **r** = x**i** + y**i** then

$$\frac{\mathrm{d}\mathbf{r}}{\mathrm{d}s}=\frac{\mathrm{d}x}{\mathrm{d}s}\mathbf{i}+\frac{\mathrm{d}y}{\mathrm{d}s}\mathbf{j}$$

Hence

$$\left|\frac{\mathrm{d}\mathbf{r}}{\mathrm{d}s}\right|^2=\left(\frac{\mathrm{d}x}{\mathrm{d}s}\right)^2+\left(\frac{\mathrm{d}y}{\mathrm{d}s}\right)^2=1$$

using standard results relating to arc length. It follows from the definition of the derivative that d**r**/ds is located in the tangent to the curve and that its sense is that of s increasing.

Using the results of Example 1.17 with **N** = **i**, **a** = **t**, θ = ψ it is evident that d**t**/dψ = **n** where **n** = **k** × **t**. Thus **n** is a unit vector.

Differentiating both sides of the relation d**r**/dt = (ds/dt)**t** with respect to the parameter t yields

$$\begin{aligned}\frac{\mathrm{d}^2\mathbf{r}}{\mathrm{d}t^2}&=\frac{\mathrm{d}^2s}{\mathrm{d}t^2}\mathbf{t}+\frac{\mathrm{d}s}{\mathrm{d}t}\frac{\mathrm{d}\mathbf{t}}{\mathrm{d}t}\\&=\frac{\mathrm{d}^2s}{\mathrm{d}t^2}\mathbf{t}+\frac{\mathrm{d}s}{\mathrm{d}t}\frac{\mathrm{d}\psi}{\mathrm{d}t}\frac{\mathrm{d}\mathbf{t}}{\mathrm{d}\psi}=\frac{\mathrm{d}^2s}{\mathrm{d}t^2}\mathbf{t}+\left\{\left(\frac{\mathrm{d}s}{\mathrm{d}t}\right)^2\bigg/\left(\frac{\mathrm{d}s}{\mathrm{d}\psi}\right)\right\}\mathbf{n}\end{aligned}$$

as required.

Example 1.19

The vector **r**(t) is defined in terms of the parameter t by

$$\mathbf{r}(t) = \mathbf{r} = \cos\omega t\mathbf{a} + (1 - \cos\omega t)(\mathbf{a} \cdot \mathbf{n})\mathbf{n} + \sin\omega t\mathbf{n} \times \mathbf{a}$$

where **a** and **n** are fixed unit vectors and ω is a constant. Show that $\dot{\mathbf{r}} = \boldsymbol{\omega} \times \mathbf{r}$ where $\boldsymbol{\omega} = \omega\mathbf{n}$. Deduce that **r** is a unit vector.

Solution We have

$$\dot{\mathbf{r}} = -\omega\sin\omega t\mathbf{a} + \omega\sin\omega t(\mathbf{a} \cdot \mathbf{n})\mathbf{n} + \omega\cos\omega t\mathbf{n} \times \mathbf{a}$$

Now

$$\begin{aligned}\boldsymbol{\omega} \times \mathbf{r} &= \omega\mathbf{n} \times (\cos\omega t\mathbf{a} + (1 - \cos\omega t)(\mathbf{a} \cdot \mathbf{n})\mathbf{n} + \sin\omega t\mathbf{n} \times \mathbf{a}) \\ &= \omega\cos\omega t\mathbf{n} \times \mathbf{a} + \omega\sin\omega t((\mathbf{n} \cdot \mathbf{a})\mathbf{n} - \mathbf{n}^2\mathbf{a}) \\ &= \dot{\mathbf{r}}\end{aligned}$$

as $\mathbf{n}^2 = 1$. It follows that $\dot{\mathbf{r}}$ and **r** are always perpendicular, and therefore by virtue of Example 1.16(a) the magnitude of **r** is constant. When $t = 0$ we have $\mathbf{r} = \mathbf{a}$ so that

$$|\mathbf{r}| = |\mathbf{a}| = 1.$$

The two sections on matrix algebra and the differentiation of matrices that now follow are not prerequisites for the study of Chapters 2, 3 and 4. They may therefore be omitted at a first reading of this chapter.

1.4 MATRIX ALGEBRA

We begin by defining a matrix. A rectangular array of elements containing m rows and n columns, where m and n may have any (integer) values, is called a matrix. The elements, which for our purposes will be numbers, are enclosed in (square) brackets thus:

$$\begin{bmatrix} a_{11} & a_{12} & a_{13} & \dots & a_{1n} \\ a_{21} & a_{22} & a_{23} & \dots & a_{2n} \\ \cdot & \cdot & \cdot & & \cdot \\ \cdot & \cdot & \cdot & & \cdot \\ a_{m1} & a_{m2} & a_{m3} & \dots & a_{mn} \end{bmatrix}$$

In the course of our work on vectors we have implicitly dealt with such arrays, one example being provided by the relation (1.24). If we are given the array

$$\begin{bmatrix} l_{11} & l_{12} & l_{13} \\ l_{21} & l_{22} & l_{23} \\ l_{31} & l_{32} & l_{33} \end{bmatrix}$$

then the relation between the two sets of unit vectors **i**, **j**, **k** and **I**, **J**, **K** in (1.24) is completely defined.

A matrix is denoted in one of two ways: either by a single letter in bold type as for a vector, thus **A** or **a**; or by $[a_{ij}]$, where $i = 1, 2, \ldots, m$ and $j = 1, 2, \ldots, n$. The first suffix i indicates the row containing the element a_{ij} while the second suffix j refers to the column. The matrix $[a_{ij}]$ is frequently described as an $m \times n$ matrix and when $m = n$ it is said to be square.

A vector is, in the algebraic sense, a particular example of a matrix. In order to manipulate vectors we express them in component form such as $[A_x, A_y, A_z]$ which may be regarded as a 1×3 matrix. It is not surprising, therefore, as the reader will see, that certain aspects of matrix algebra are extensions of ideas contained in vector algebra.

Conformable matrices

Two matrices are said to be conformable if they possess the same number of rows and the same number of columns. Thus

$$\begin{bmatrix} 1 & 2 \\ 3 & 4 \\ 5 & 6 \end{bmatrix} \quad \text{and} \quad \begin{bmatrix} 1 & 0 \\ 0 & 1 \\ 1 & 0 \end{bmatrix}$$

are conformable matrices but

$$\begin{bmatrix} 1 & 2 \\ 3 & 4 \end{bmatrix} \quad \text{and} \quad \begin{bmatrix} 1 & 0 & 1 \\ 0 & 1 & 0 \end{bmatrix}$$

are not. The first two matrices are 3×2 matrices, the next is a 2×2, and the last is a 2×3 matrix.

Equality

Two conformable matrices **A** and **B** are said to be equal if all corresponding elements are equal. We write $\mathbf{A} = \mathbf{B}$.

Addition

If $\mathbf{A} = [a_{ij}]$ and $\mathbf{B} = [b_{ij}]$ are two conformable $m \times n$ matrices then the result of adding **B** to **A** is defined to be the $m \times n$ matrix $\mathbf{C} = [c_{ij}]$, where $c_{ij} = a_{ij} + b_{ij}$ for $i = 1, 2, \ldots, m, j = 1, 2, \ldots, n$. We write $\mathbf{C} = \mathbf{A} + \mathbf{B}$. It is evident that $\mathbf{A} + \mathbf{B} = \mathbf{B} + \mathbf{A}$. An immediate consequence of the definition of addition is that

$$\mathbf{A} + (\mathbf{B} + \mathbf{D}) = (\mathbf{A} + \mathbf{B}) + \mathbf{D} = (\mathbf{A} + \mathbf{D}) + \mathbf{B}$$

where $\mathbf{D}$ is a further $m \times n$ matrix. The brackets are therefore superfluous and each expression is written as $\mathbf{A} + \mathbf{B} + \mathbf{D}$. In thc same manner the addition of several conformable matrices is written as $\mathbf{A} + \mathbf{B} + \mathbf{C} + \mathbf{D} + \mathbf{E} \ldots$.

Subtraction

If $\mathbf{C} = [c_{ij}]$ denotes the result of subtracting $\mathbf{B}$ from $\mathbf{A}$ then $\mathbf{C}$ is defined by $c_{ij} = a_{ij} - b_{ij}$ for $i = 1, 2, \ldots, m$, $j = 1, 2, \ldots, n$. We write $\mathbf{C} = \mathbf{A} - \mathbf{B}$.

Multiplication of a matrix by a scalar

Let α be some scalar. The operation of multiplying the matrix $\mathbf{A}$ by the scalar α is denoted by $\alpha\mathbf{A}$. The outcome of the operation is the matrix $\mathbf{C}$ whose elements c_{ij} are given by $c_{ij} = \alpha a_{ij}$, $i = 1, 2, \ldots, m$, $j = 1, 2, \ldots, n$.

Unit matrix

Let $\mathbf{A}$ be a square $n \times n$ matrix. The elements $a_{11}, a_{22}, \ldots, a_{nn}$ of $\mathbf{A}$ are called its diagonal elements. A square matrix whose diagonal elements are all equal to one and whose non-diagonal elements are zero is called a unit matrix. Thus

$$\begin{bmatrix} 1 & 0 & 0 \\ 0 & 1 & 0 \\ 0 & 0 & 1 \end{bmatrix} \quad \text{and} \quad \begin{bmatrix} 1 & 0 \\ 0 & 1 \end{bmatrix}$$

are unit matrices. It is customary to denote a unit matrix by $\mathbf{I}$. If we wish to indicate the particular number of rows (or columns) in a unit matrix we add a subscript n to $\mathbf{I}$ and write $\mathbf{I}_n$.

We can extend the definition of the Kronecker delta δ_{ij}, given just prior to the relations (1.26), by putting

$$\delta_{ij} = \begin{cases} 0 & \text{if } i \neq j \\ 1 & \text{if } i = j \end{cases} \qquad i, j = 1, 2, \ldots, n$$

This enables us to write $\mathbf{I}_n = [\delta_{ij}]$.

Null matrix

Any matrix whose elements are all zero is called a null matrix. A null matrix is denoted by $\mathbf{0}$.

Column and row vectors

A matrix consisting of one row is called a row vector, while a matrix which has only one column is called a column vector. Thus

$$\mathbf{x} = [x_1, x_2, \ldots, x_p]$$

is a row vector with p elements, while

$$\mathbf{y} = \begin{bmatrix} y_1 \\ y_2 \\ \cdot \\ \cdot \\ \cdot \\ y_q \end{bmatrix}$$

is a column vector with q elements.

Although in our vector algebra we have used only row vectors, it transpires that for many purposes it is more convenient to work with column vectors. In fact there will be occasions when we refer to either simply as 'vectors', and the particular context or way in which they are used will reveal whether they are row or column vectors.

Scalar product

If

$$\mathbf{x} = [x_1, x_2, \ldots, x_n]$$

is a row vector with n columns, and

$$\mathbf{y} = \begin{bmatrix} y_1 \\ y_2 \\ \cdot \\ \cdot \\ \cdot \\ y_n \end{bmatrix}$$

is a column vector with n rows, then the scalar product of $\mathbf{x}$ with $\mathbf{y}$ is denoted by $\langle \mathbf{x}, \mathbf{y} \rangle$ and defined by

$$\langle \mathbf{x}, \mathbf{y} \rangle = x_1 y_1 + x_2 y_2 + \ldots + x_n y_n = \sum_{i=1}^{n} x_i y_i$$

Multiplication of matrices

The expressions on the right-hand sides of the relations (1.33) may be interpreted as the scalar products of each of the row vectors $[l_{11}, l_{12}, l_{13}]$, $[l_{21}, l_{22}, l_{23}]$, $[l_{31}, l_{32}, l_{33}]$ with the column vector $\mathbf{a}$ defined by

$$\mathbf{a} = \begin{bmatrix} A_x \\ A_y \\ A_z \end{bmatrix}$$

With this observation in mind we represent the relations contained in (1.33) in the following way:

$$\begin{bmatrix} A_X \\ A_Y \\ A_Z \end{bmatrix} = \begin{bmatrix} l_{11} & l_{12} & l_{13} \\ l_{21} & l_{22} & l_{23} \\ l_{31} & l_{32} & l_{33} \end{bmatrix} \begin{bmatrix} A_x \\ A_y \\ A_z \end{bmatrix} \tag{1.37}$$

Letting

$$\mathbf{a}_1 = \begin{bmatrix} A_X \\ A_Y \\ A_Z \end{bmatrix}, \qquad \mathbf{L} = [l_{ij}]$$

the relation (1.37) can be written more simply as

$$\mathbf{a}_1 = \mathbf{La}. \tag{1.38}$$

The expression $\mathbf{La}$ is a particular example of a matrix product. We say $\mathbf{a}$ is pre-multiplied by $\mathbf{L}$, or $\mathbf{L}$ is post-multiplied by $\mathbf{a}$.

An extension of the work here leads to a more general definition of matrix multiplication. Let $\mathbf{I}'$, $\mathbf{J}'$, $\mathbf{K}'$ be a set of right-handed mutually perpendicular unit vectors related to $\mathbf{I}$, $\mathbf{J}$, $\mathbf{K}$ by

$$\begin{aligned} \mathbf{I}' &= n_{11}\mathbf{I} + n_{12}\mathbf{J} + n_{13}\mathbf{K} \\ \mathbf{J}' &= n_{21}\mathbf{I} + n_{22}\mathbf{J} + n_{23}\mathbf{K} \\ \mathbf{K}' &= n_{31}\mathbf{I} + n_{32}\mathbf{J} + n_{33}\mathbf{K} \end{aligned}$$

If the components of $\mathbf{A}$ referred to $\mathbf{I}'$, $\mathbf{J}'$, $\mathbf{K}'$ are A'_x, A'_y, A'_z then the analysis that leads to (1.37) enables us to assert that

$$\begin{bmatrix} A'_X \\ A'_Y \\ A'_Z \end{bmatrix} = \begin{bmatrix} n_{11} & n_{12} & n_{13} \\ n_{21} & n_{22} & n_{23} \\ n_{31} & n_{32} & n_{33} \end{bmatrix} \begin{bmatrix} A_X \\ A_Y \\ A_Z \end{bmatrix} \tag{1.39}$$

or

$$\mathbf{a}_2 = \mathbf{Na}_1 \tag{1.40}$$

where

$$\mathbf{a}_2 = \begin{bmatrix} A'_X \\ A'_Y \\ A'_Z \end{bmatrix}, \qquad \mathbf{N} = [n_{ij}], \qquad i, j = 1, 2, 3$$

We now determine A'_X, A'_Y, A'_Z in terms of A_x, A_y, A_z. Using (1.39) and (1.37) we have

$$\begin{aligned} A'_X &= n_{11}A_X + n_{12}A_Y + n_{13}A_Z \\ &= n_{11}(l_{11}A_x + l_{12}A_y + l_{13}A_z) + n_{12}(l_{21}A_x + l_{22}A_y + l_{23}A_z) \\ &\quad + n_{13}(l_{31}A_x + l_{32}A_y + l_{33}A_z) \\ &= c_{11}A_x + c_{12}A_y + c_{13}A_z \end{aligned}$$

where

$$\begin{aligned} c_{11} &= n_{11}l_{11} + n_{12}l_{21} + n_{13}l_{31} \\ c_{12} &= n_{11}l_{12} + n_{12}l_{22} + n_{13}l_{32} \\ c_{13} &= n_{11}l_{13} + n_{12}l_{23} + n_{13}l_{33} \end{aligned}$$

It follows that c_{11}, c_{12}, c_{13} are respectively equal to the scalar product of the row vector $[n_{11}, n_{12}, n_{13}]$ with the column vectors

$$\begin{bmatrix} l_{11} \\ l_{21} \\ l_{31} \end{bmatrix}, \qquad \begin{bmatrix} l_{12} \\ l_{22} \\ l_{32} \end{bmatrix}, \qquad \begin{bmatrix} l_{13} \\ l_{23} \\ l_{33} \end{bmatrix}$$

Likewise

$$\begin{aligned} A'_Y &= c_{21}A_x + c_{22}A_y + c_{23}A_z \\ A'_Z &= c_{31}A_x + c_{32}A_y + c_{33}A_z \end{aligned}$$

where c_{ij} $(i = 2, 3,\ j = 1, 2, 3)$ is the scalar product of the vector $[n_{i1}, n_{i2}, n_{i3}]$ with

$$\begin{bmatrix} l_{1j} \\ l_{2j} \\ l_{3j} \end{bmatrix}$$

The relationship between the two sets of components $[A'_X, A'_Y, A'_Z]$ and $[A_x, A_y, A_z]$ is therefore given by

$$\mathbf{a}_2 = \mathbf{Ca} \tag{1.41}$$

where $\mathbf{C} = [c_{ij}]$ $(i, j = 1, 2, 3)$.

Formal substitution of the expression **La** for $\mathbf{a}_1$ (see (1.38)) in the right-hand side of (1.40) leads to

$$\mathbf{a}_2 = \mathbf{NLa} \tag{1.42}$$

Comparing the relations (1.42) and (1.41) we see that the matrix product **NL** must be equal to the matrix **C**, whose ijth element is, as we have just found, the scalar product of the ith row of $\mathbf{N}$ with the jth column of $\mathbf{L}$. This observation leads to the definition of matrix multiplication.

Let $\mathbf{A}$ be an $m \times n$ matrix and $\mathbf{B}$ be an $n \times p$ matrix (note that the number of columns in $\mathbf{A}$ must be the same as the number of rows in $\mathbf{B}$). If $\mathbf{C}$ denotes the result of multiplying $\mathbf{A}$ by $\mathbf{B}$, written as $\mathbf{C} = \mathbf{AB}$, then $\mathbf{C} = [c_{ij}]$, where

$$c_{ij} = \sum_{k=1}^{n} a_{ik} b_{kj}, \qquad i = 1, 2, \ldots, m, \qquad j = 1, 2, \ldots, p \tag{1.43}$$

That is, c_{ij} is the scalar product of the ith row of $\mathbf{A}$ with the jth column of $\mathbf{B}$. The matrix $\mathbf{C}$ is an $m \times p$ matrix.

The matrix product $\mathbf{AB}$ is undefined when the number of columns of $\mathbf{A}$ and the number of rows of $\mathbf{B}$ are not equal. It therefore follows that, even when $\mathbf{AB}$ exists, $\mathbf{BA}$ may not as the existence of $\mathbf{BA}$ requires that $p = m$. Further we shall see by example that even when both matrix products exist they are not necessarily equal and thus the order in which the two matrices are written is all important. If

$$\mathbf{A} = \begin{bmatrix} 1 & 2 & 3 & 4 \\ 5 & 6 & 7 & 8 \end{bmatrix}, \qquad \mathbf{B} = \begin{bmatrix} 1 & 0 \\ 0 & 1 \\ 1 & 0 \\ 0 & 1 \end{bmatrix}$$

then

$$\mathbf{AB} = \begin{bmatrix} 4 & 6 \\ 12 & 14 \end{bmatrix}, \qquad \mathbf{BA} = \begin{bmatrix} 1 & 2 & 3 & 4 \\ 5 & 6 & 7 & 8 \\ 1 & 2 & 3 & 4 \\ 5 & 6 & 7 & 8 \end{bmatrix}$$

The matrix $\mathbf{AB}$ is a 2×2 matrix while $\mathbf{BA}$ is a 4×4 matrix.

Provided the operations are possible the following rules can be readily deduced from the definitions of addition and multiplication of matrices. Their extension to more involved expressions is immediate.

$$\left.\begin{aligned} &(\mathbf{A} + \mathbf{B})\mathbf{C} = \mathbf{AC} + \mathbf{BC} \\ &(\mathbf{A} + \mathbf{B})(\mathbf{C} + \mathbf{D}) = \mathbf{AC} + \mathbf{AD} + \mathbf{BC} + \mathbf{BD} \\ &\mathbf{AI} = \mathbf{IA} = \mathbf{A} \\ &(\mathbf{AB})\mathbf{C} = \mathbf{A}(\mathbf{BC}) \qquad \text{written as } \mathbf{ABC} \end{aligned}\right. \tag{1.44}$$

Powers of a matrix

The product $\mathbf{AA}$ can only exist if $\mathbf{A}$ is a square matrix. It is denoted by $\mathbf{A}^2$. Likewise $\mathbf{A}^3 = \mathbf{AAA}$; while $\mathbf{A}^n$, where n is a positive integral power, is the product of n matrices all equal to $\mathbf{A}$. It is evident that the law of indices $\mathbf{A}^p\mathbf{A}^q = \mathbf{A}^{p+q}$ holds for any positive (integral) powers p, q.

Example 1.20

The matrices $\mathbf{A}$, $\mathbf{B}$, $\mathbf{C}$ are defined by

$$\mathbf{A} = \begin{bmatrix} 3 & -1 & 1 \\ 2 & 1 & 0 \\ 1 & 2 & -1 \end{bmatrix}, \qquad \mathbf{B} = \begin{bmatrix} 1 & 2 & 3 \\ 4 & 5 & 6 \\ 7 & 8 & 9 \end{bmatrix},$$

$$\mathbf{C} = \begin{bmatrix} 1 & -1 & 1 \\ -1 & 1 & -1 \\ 1 & -1 & 1 \end{bmatrix}$$

Verify that

(a) $(\mathbf{A}+\mathbf{B})\mathbf{C} = \mathbf{AC} + \mathbf{BC}$
(b) $(\mathbf{AB})\mathbf{C} = \mathbf{A}(\mathbf{BC})$.

▷ *Solution*
(a) The following matrices are readily obtained:

$$\mathbf{A}+\mathbf{B} = \begin{bmatrix} 4 & 1 & 4 \\ 6 & 6 & 6 \\ 8 & 10 & 8 \end{bmatrix}, \qquad (\mathbf{A}+\mathbf{B})\mathbf{C} = \begin{bmatrix} 7 & -7 & 7 \\ 6 & -6 & 6 \\ 6 & -6 & 6 \end{bmatrix}$$

$$\mathbf{AC} = \begin{bmatrix} 5 & -5 & 5 \\ 1 & -1 & 1 \\ -2 & 2 & -2 \end{bmatrix}, \qquad \mathbf{BC} = \begin{bmatrix} 2 & -2 & 2 \\ 5 & -5 & 5 \\ 8 & -8 & 8 \end{bmatrix},$$

$$\mathbf{AC}+\mathbf{BC} = \begin{bmatrix} 7 & -7 & 7 \\ 6 & -6 & 6 \\ 6 & -6 & 6 \end{bmatrix}$$

Hence the rule $(\mathbf{A}+\mathbf{B})\mathbf{C} = \mathbf{AC} + \mathbf{BC}$ is confirmed.
(b) First,

$$\mathbf{AB} = \begin{bmatrix} 6 & 9 & 12 \\ 6 & 9 & 12 \\ 2 & 4 & 6 \end{bmatrix}$$

so that

$$(\mathbf{AB})\mathbf{C} = \begin{bmatrix} 6 & 9 & 12 \\ 6 & 9 & 12 \\ 2 & 4 & 6 \end{bmatrix}\begin{bmatrix} 1 & -1 & 1 \\ -1 & 1 & -1 \\ 1 & -1 & 1 \end{bmatrix} = \begin{bmatrix} 9 & -9 & 9 \\ 9 & -9 & 9 \\ 4 & -4 & 4 \end{bmatrix}$$

Next,

$$\mathbf{A}(\mathbf{BC}) = \begin{bmatrix} 3 & -1 & 1 \\ 2 & 1 & 0 \\ 1 & 2 & -1 \end{bmatrix} \begin{bmatrix} 2 & -2 & 2 \\ 5 & -5 & 5 \\ 8 & -8 & 8 \end{bmatrix} = \begin{bmatrix} 9 & -9 & 9 \\ 9 & -9 & 9 \\ 4 & -4 & 4 \end{bmatrix}$$

Thus $(\mathbf{AB})\mathbf{C} = \mathbf{A}(\mathbf{BC})$.

Example 1.21

The two matrices **A** and **B** are defined by

$$\mathbf{A} = \begin{bmatrix} 1 & 1 & -1 \\ 2 & 1 & -3 \\ 4 & 5 & -2 \end{bmatrix}, \qquad \mathbf{B} = \begin{bmatrix} -13 & 3 & 2 \\ 8 & -2 & -1 \\ -6 & 1 & 1 \end{bmatrix}$$

Find **AB** and **BA**. Hence solve the two sets of equations

$$\begin{aligned} x_1 + x_2 - x_3 &= 2 \\ 2x_1 + x_2 - 3x_3 &= 2 \\ 4x_1 + 5x_2 - 2x_3 &= 7 \end{aligned} \qquad \text{and} \qquad \begin{aligned} -13y_1 + 3y_2 + 2y_3 &= 1 \\ 8y_1 - 2y_2 - y_3 &= -1 \\ -6y_1 + y_2 + y_3 &= -4 \end{aligned}$$

▷ *Solution* Applying the rule for the multiplication of two matrices yields $\mathbf{AB} = \mathbf{BA} = \mathbf{I}_3$. The two sets of equations can be written as $\mathbf{Ax} = \mathbf{b}$ and $\mathbf{By} = \mathbf{c}$, where

$$\mathbf{x} = \begin{bmatrix} x_1 \\ x_2 \\ x_3 \end{bmatrix}, \quad \mathbf{y} = \begin{bmatrix} y_1 \\ y_2 \\ y_3 \end{bmatrix}, \quad \mathbf{b} = \begin{bmatrix} 2 \\ 2 \\ 7 \end{bmatrix}, \quad \mathbf{c} = \begin{bmatrix} 1 \\ -1 \\ -4 \end{bmatrix}$$

Pre-multiplication of both sides of $\mathbf{Ax} = \mathbf{b}$ by **B** gives $\mathbf{BAx} = \mathbf{Bb}$, but $\mathbf{BA} = \mathbf{I}$ and hence

$$\begin{bmatrix} x_1 \\ x_2 \\ x_3 \end{bmatrix} = \begin{bmatrix} -13 & 3 & 2 \\ 8 & -2 & -1 \\ -6 & 1 & 1 \end{bmatrix} \begin{bmatrix} 2 \\ 2 \\ 7 \end{bmatrix} = \begin{bmatrix} -6 \\ 5 \\ -3 \end{bmatrix}$$

In like manner, pre-multiplication of both sides of $\mathbf{By} = \mathbf{C}$ by **A** yields the solution $y_1 = 4$, $y_2 = 13$, $y_3 = 7$.

This example leads us to the definition of the inverse of a square matrix.

Inverse of a square matrix

If **A** is an $n \times n$ matrix and **B** is a second matrix such that $\mathbf{AB} = \mathbf{BA} = \mathbf{I}_n$ then **B** is said to be the inverse of **A**. It follows directly from the definition of the inverse that **B** must be an $n \times n$ matrix whose inverse is **A**. The inverse of **A** is denoted by $\mathbf{A}^{-1}$ so that $\mathbf{B} = \mathbf{A}^{-1}$ and $\mathbf{A} = \mathbf{B}^{-1}$. In Example 1.21 the matrices **A** and **B** are the inverses of each other.

The reader should note that not all square matrices have inverses. Let $\mathbf{B} = [b_{ij}]$, $i, j = 1, 2, 3$ be the inverse of the matrix **A** given by

$$\mathbf{A} = \begin{bmatrix} 0 & -1 & 1 \\ 1 & 0 & -1 \\ -1 & 1 & 0 \end{bmatrix}$$

To find **B** we equate elements of **AB**, that is of

$$\begin{bmatrix} 0 & -1 & 1 \\ 1 & 0 & -1 \\ -1 & 1 & 0 \end{bmatrix} \begin{bmatrix} b_{11} & b_{12} & b_{13} \\ b_{21} & b_{22} & b_{23} \\ b_{31} & b_{32} & b_{33} \end{bmatrix}$$

to corresponding elements of the unit matrix $\mathbf{I}_3$. Equating the elements of the first column of **AB** with those of the first column of $\mathbf{I}_3$ yields the three equations

$$-b_{21} + b_{31} = 1, \qquad b_{11} - b_{31} = 0, \qquad -b_{11} + b_{21} = 0$$

Addition of corresponding sides of the second and third equations leads to an equation that is inconsistent with the first of the three equations. Clearly the inverse of **A** cannot exist. The existence or otherwise of the inverse of a square matrix depends on the value of the associated determinant of the matrix. This determinant is now defined.

Determinant of a square matrix

The determinants

$$\begin{vmatrix} a_{11} & a_{12} \\ a_{21} & a_{22} \end{vmatrix} \quad \text{and} \quad \begin{vmatrix} a_{11} & a_{12} & a_{13} \\ a_{21} & a_{22} & a_{23} \\ a_{31} & a_{32} & a_{33} \end{vmatrix}$$

are respectively the associated determinants of the matrices

$$\begin{bmatrix} a_{11} & a_{12} \\ a_{21} & a_{22} \end{bmatrix} \quad \text{and} \quad \begin{bmatrix} a_{11} & a_{12} & a_{13} \\ a_{21} & a_{22} & a_{23} \\ a_{31} & a_{32} & a_{33} \end{bmatrix}$$

These definitions are readily extended to include square matrices of any size once determinants of any order have been defined. The associated determinant of a square matrix **A** is denoted by det **A**. We now state two of its most important properties:

(a) If **A** and **B** are two $n \times n$ matrices then $\det(\mathbf{AB}) = \det\mathbf{A}\det\mathbf{B}$.
(b) If $\det\mathbf{A} \neq 0$ then **A** has an inverse.

Property (a) enables us to deduce if $\mathbf{A}$ has an inverse that $\det\mathbf{A} \neq 0$. Let $\mathbf{B}$ be the inverse of $\mathbf{A}$. Hence $1 = \det\mathbf{I} = \det(\mathbf{AB}) = \det\mathbf{A}\det\mathbf{B}$, so that $\det\mathbf{A}$ cannot be zero. It follows therefore that when $\det\mathbf{A} = 0$ the inverse of $\mathbf{A}$ does not exist. Thus the matrix $\mathbf{A}$ defined by

$$\mathbf{A} = \begin{bmatrix} 0 & -1 & 1 \\ 1 & 0 & -1 \\ -1 & 1 & 0 \end{bmatrix}$$

does not (as we have seen already) have an inverse, as $\det\mathbf{A} = 0$.

With the help of property (b) we can now assert that $\mathbf{A}$ has an inverse if and only if $\det\mathbf{A} \neq 0$. When $\det\mathbf{A} \neq 0$ the matrix $\mathbf{A}$ is said to be non-singular. The means whereby the inverses of non-singular 2×2 and 3×3 matrices can be found are given in the course of Problems 1.40 and 1.41. These involve the notion of the transpose of a matrix.

Transpose of a matrix

The transpose of the $m \times n$ matrix $\mathbf{A}$ is the $n \times m$ matrix whose rows are equal to the columns of $\mathbf{A}$. It is denoted by $\mathbf{A}^\mathrm{T}$. The definition of the transpose implies that the columns of $\mathbf{A}^\mathrm{T}$ are the rows of $\mathbf{A}$.

If we put $\mathbf{A}^\mathrm{T} = [a'_{ij}]$, $i = 1, 2, \ldots, n$, $j = 1, 2, \ldots, m$, then the elements of $\mathbf{A}^\mathrm{T}$ are given by $a'_{ij} = a_{ji}$.

The following rules relating to the transpose and inverse enable us to manipulate expressions in which they arise:

$$\begin{aligned}
&(\mathbf{A}^\mathrm{T})^\mathrm{T} = \mathbf{A} && \\
&(\mathbf{A}+\mathbf{B})^\mathrm{T} = \mathbf{A}^\mathrm{T} + \mathbf{B}^\mathrm{T} && \text{where } \mathbf{A} \text{ and } \mathbf{B} \text{ are conformable} \\
&(\mathbf{AB})^\mathrm{T} = \mathbf{B}^\mathrm{T}\mathbf{A}^\mathrm{T} && \text{when } \mathbf{AB} \text{ exists} \\
&\det\mathbf{A}^\mathrm{T} = \det\mathbf{A} && \text{where } \mathbf{A} \text{ is a square matrix} \\
&(\mathbf{A}^{-1})^{-1} = \mathbf{A} && \text{where } \mathbf{A} \text{ is non-singular} \\
&(\mathbf{A}^{-1})^\mathrm{T} = (\mathbf{A}^\mathrm{T})^{-1} && \\
&(\mathbf{AB})^{-1} = \mathbf{B}^{-1}\mathbf{A}^{-1} && \text{where } \mathbf{B} \text{ is also non-singular} \\
&(\mathbf{A}^n)^{-1} = (\mathbf{A}^{-1})^n && \text{written } \mathbf{A}^{-n}
\end{aligned} \qquad (1.45)$$

We give the proofs of two of these rules in the example that now follows.

Example 1.22

Show the following:

(a) $(\mathbf{AB})^{-1} = \mathbf{B}^{-1}\mathbf{A}^{-1}$
(b) $(\mathbf{A}^{-1})^\mathrm{T} = (\mathbf{A}^\mathrm{T})^{-1}$.

▷ *Solution*

(a) Since $\mathbf{A}$ and $\mathbf{B}$ possess inverses, det$\mathbf{A}$ and det$\mathbf{B}$ are both non-zero. Hence $\det(\mathbf{AB}) = \det\mathbf{A}\det\mathbf{B} \neq 0$, so that the matrix $\mathbf{AB}$ has an inverse which we denote by $\mathbf{C}$. Successive post-multiplication of each side of the relation $\mathbf{CAB} = \mathbf{I}$ by $\mathbf{B}^{-1}$ and $\mathbf{A}^{-1}$ yields $\mathbf{CABB}^{-1}\mathbf{A}^{-1} = \mathbf{IB}^{-1}\mathbf{A}^{-1}$ or $\mathbf{C} = \mathbf{B}^{-1}\mathbf{A}^{-1}$ as required.

(b) Taking the transpose of each side of the relation $\mathbf{A}^{-1}\mathbf{A} = \mathbf{I}$ gives, noting the third rule of (1.45),

$$\mathbf{I}^{\mathrm{T}} = \mathbf{I} = (\mathbf{A}^{-1}\mathbf{A})^{\mathrm{T}} = \mathbf{A}^{\mathrm{T}}(\mathbf{A}^{-1})^{\mathrm{T}}$$

Hence it follows from the definition of the inverse that $(\mathbf{A}^{-1})^{\mathrm{T}}$ is the inverse of $\mathbf{A}^{\mathrm{T}}$ or $(\mathbf{A}^{-1})^{\mathrm{T}} = (\mathbf{A}^{\mathrm{T}})^{-1}$.

Example 1.23

The elements of the matrix $\mathbf{B}$ are given by $b_{ij} = l_i l_j$ $(i, j = 1, 2, 3)$, where l_1, l_2, l_3 are the components of the unit vector $\mathbf{n}$ referred to be unit vectors $\mathbf{i}, \mathbf{j}, \mathbf{k}$. In addition, $\mathbf{C}$ is defined by

$$\mathbf{C} = \begin{bmatrix} 0 & -l_3 & l_2 \\ l_3 & 0 & -l_1 \\ -l_2 & l_1 & 0 \end{bmatrix}$$

Show that

(a) $\mathbf{B}^2 = \mathbf{B}$
(b) $\mathbf{CB} = \mathbf{BC} = \mathbf{0}$
(c) $\mathbf{C}^2 = \mathbf{B} - \mathbf{I}$.

▷ *Solution*

(a) Let $\mathbf{B}^2 = \mathbf{D} = [d_{ij}]$. Hence

$$d_{ij} = \sum_{k=1}^{3} b_{ik} b_{kj} = \sum_{k=1}^{3} l_i l_k l_k l_j = l_i l_j \sum_{k=1}^{3} l_k^2 = l_i l_j$$

since $|\mathbf{n}| = 1$ and therefore $l_1^2 + l_2^2 + l_3^2 = 1$. Thus $\mathbf{B}^2 = \mathbf{B}$.

(b) We have

$$\mathbf{CB} = \begin{bmatrix} 0 & -l_3 & l_2 \\ l_3 & 0 & -l_1 \\ -l_2 & l_1 & 0 \end{bmatrix} \begin{bmatrix} l_1^2 & l_1 l_2 & l_1 l_3 \\ l_2 l_1 & l_2^2 & l_2 l_3 \\ l_3 l_1 & l_3 l_2 & l_3^2 \end{bmatrix}$$

$$= \begin{bmatrix} -l_3 l_2 l_1 + l_2 l_3 l_1 & -l_3 l_2^2 + l_3 l_2^2 & 0 \\ 0 & 0 & 0 \\ 0 & 0 & 0 \end{bmatrix} = \mathbf{0}$$

Likewise $\mathbf{BC} = \mathbf{0}$.

(c) We have

$$\mathbf{C}^2 = \begin{bmatrix} -l_2^2 - l_3^2 & l_1 l_2 & l_1 l_3 \\ l_1 l_2 & -l_1^2 - l_3^2 & l_2 l_3 \\ l_1 l_3 & l_2 l_3 & -l_1^2 - l_2^2 \end{bmatrix}$$

$$= \begin{bmatrix} l_1^2 - 1 & l_1 l_2 & l_1 l_3 \\ l_1 l_2 & l_2^2 - 1 & l_2 l_3 \\ l_1 l_3 & l_2 l_3 & l_3^2 - 1 \end{bmatrix} = \mathbf{B} - \mathbf{I}$$

Symmetric and skew-symmetric matrices

In Example 1.23 the matrices **B** and **C** are respectively examples of symmetric and skew-symmetric matrices.

A matrix **A** is said to be *symmetric* if $\mathbf{A} = \mathbf{A}^{\mathrm{T}}$. This definition implies that **A** is a square matrix.

A matrix **A** is said to be *skew-symmetric* if $\mathbf{A}^{\mathrm{T}} = -\mathbf{A}$. It follows that **A** must be square and all of its diagonal elements zero.

Example 1.24

The matrix **A** is defined by

$$\mathbf{A} = \cos\theta\mathbf{I} + (1 - \cos\theta)\mathbf{B} + \sin\theta\mathbf{C}$$

where **B** and **C** are the matrices defined in Example 1.23 and θ may take any value. Derive the following properties of **A**:

(a) $\mathbf{A}\mathbf{A}^{\mathrm{T}} = \mathbf{I} = \mathbf{A}^{\mathrm{T}}\mathbf{A}$
(b) $\det\mathbf{A} = 1$
(c) $\det(\mathbf{A} - \mathbf{I}) = 0$
(d) $\mathbf{A}^n = \cos n\theta\mathbf{I} + (1 - \cos n\theta)\mathbf{B} + \sin n\theta\mathbf{C}$, where n is a positive integer.

▷ *Solution*

(a) It follows, since $\mathbf{B}^{\mathrm{T}} = \mathbf{B}$ and $\mathbf{C}^{\mathrm{T}} = -\mathbf{C}$, that

$$\mathbf{A}\mathbf{A}^{\mathrm{T}} = \{\cos\theta\mathbf{I} + (1 - \cos\theta)\mathbf{B} + \sin\theta\mathbf{C}\}\{\cos\theta\mathbf{I} + (1 - \cos\theta)\mathbf{B} - \sin\theta\mathbf{C}\}$$

Recalling the results $\mathbf{BC} = \mathbf{CB} = \mathbf{0}$, $\mathbf{B}^2 = \mathbf{B}$, $\mathbf{C}^2 = \mathbf{B} - \mathbf{I}$ we deduce, on expanding the right-hand side of the relation, that

$$\mathbf{A}\mathbf{A}^{\mathrm{T}} = \cos^2\theta\mathbf{I} + \{2\cos\theta(1 - \cos\theta) + (1 - \cos\theta)^2\}\mathbf{B} + \sin^2\theta(\mathbf{I} - \mathbf{B})$$

This relation reduces after some further simplification to $\mathbf{A}\mathbf{A}^{\mathrm{T}} = \mathbf{I}$.

Likewise $\mathbf{A}^{\mathrm{T}}\mathbf{A} = \mathbf{I}$ and hence $\mathbf{A}^{\mathrm{T}}$ is the inverse of **A**.

(b) Using the property $\det\mathbf{AB} = \det\mathbf{A}\det\mathbf{B}$ we have

$$1 = \det\mathbf{I} = \det\mathbf{A}\mathbf{A}^{\mathrm{T}} = \det\mathbf{A}\det\mathbf{A}^{\mathrm{T}} = (\det\mathbf{A})^2$$

as $\det\mathbf{A} = \det\mathbf{A}^{\mathrm{T}}$. Hence $\det\mathbf{A} = \pm 1$.

Now the elements of $\mathbf{A}$ are continuous functions of θ and therefore $\det\mathbf{A}$ must be a continuous function of θ. This continuity property of $\det\mathbf{A}$ implies that its value cannot jump from $+1$ to -1 as θ varies and thus $\det\mathbf{A}$ can take only one of these values. Now $\mathbf{A} = \mathbf{I}$ when $\theta = 0$ and therefore $\det\mathbf{A} = 1$ for all values of θ.

(c) It follows from property (a) of $\mathbf{A}$ that

$$\mathbf{A} - \mathbf{I} = \mathbf{A}(\mathbf{I} - \mathbf{A}^{\mathrm{T}}) = \mathbf{A}(\mathbf{I} - \mathbf{A})^{\mathrm{T}}$$

Hence

$$\det(\mathbf{A} - \mathbf{I}) = \det(\mathbf{A}(\mathbf{I} - \mathbf{A})^{\mathrm{T}}) = \det\mathbf{A}\det(\mathbf{I} - \mathbf{A})$$

using the properties of determinants. But $\det\mathbf{A} = 1$ by property (b), so that

$$\det(\mathbf{A} - \mathbf{I}) = \det(-\mathbf{I})\det(\mathbf{A} - \mathbf{I}) = -\det(\mathbf{A} - \mathbf{I})$$

Therefore $\det(\mathbf{A} - \mathbf{I}) = 0$ as required.

(d) This property is established by induction. Assume the property is true for $n = k$, that is

$$\mathbf{A}^k = \cos k\theta\mathbf{I} + (1 - \cos k\theta)\mathbf{B} + \sin k\theta\mathbf{C}$$

Hence

$$\begin{aligned}
\mathbf{A}^{k+1} &= \mathbf{A}\mathbf{A}^k \\
&= \{\cos\theta\mathbf{I} + (1 - \cos\theta)\mathbf{B} + \sin\theta\mathbf{C}\}\{\cos k\theta\mathbf{I} + (1 - \cos k\theta)\mathbf{B} \\
&\quad + \sin k\theta\mathbf{C}\} \\
&= \cos\theta\cos k\theta\mathbf{I} + (\cos\theta\sin k\theta + \sin\theta\cos k\theta)\mathbf{C} + \sin\theta\sin k\theta(\mathbf{B} - \mathbf{I}) \\
&\quad + \{\cos\theta(1 - \cos k\theta) + (1 - \cos\theta)\cos k\theta \\
&\qquad + (1 - \cos\theta)(1 - \cos k\theta)\}\mathbf{B}
\end{aligned}$$

using the various results contained in Example 1.23. Collecting up terms and using elementary trigonometrical identities leads to

$$\mathbf{A}^{k+1} = \cos(k+1)\theta\mathbf{I} + \{1 - \cos(k+1)\theta\}\mathbf{B} + \sin(k+1)\theta\mathbf{C}$$

By putting $k = 1$ in our working it is seen that the property is true for $n = 2$ and therefore true, by induction, for all n.

The matrix $\mathbf{A}$ defined in this example is known as the rotation matrix. It is discussed in detail in Chapter 5.

Example 1.25

The set of equations $\mathbf{A}\mathbf{x}^{\mathrm{T}} = \mathbf{0}$, where $\mathbf{A} = [a_{ij}]$ is a 3×3 matrix and $\mathbf{x}$ is the row vector $[x_1, x_2, x_3]$, has a solution such that x_1, x_2, x_3 are not all zero. Show that $\det\mathbf{A} = 0$.

> *Solution* The equations can be written as

$$a_{11}x_1 + a_{12}x_2 + a_{13}x_3 = a_{21}x_1 + a_{22}x_2 + a_{23}x_3 = a_{31}x_1 + a_{32}x_2 + a_{33}x_3 = 0$$

They are equivalent to the statement that the vector $\mathbf{x}$ is perpendicular to the vectors $\mathbf{a}_1$, $\mathbf{a}_2$, $\mathbf{a}_3$ whose components referred to the unit vectors $\mathbf{i}$, $\mathbf{j}$, $\mathbf{k}$ are respectively $[a_{11}, a_{12}, a_{13}]$, $[a_{21}, a_{22}, a_{23}]$, $[a_{31}, a_{32}, a_{33}]$. Alternatively we can write $\mathbf{x}\,.\,\mathbf{a}_1 = \mathbf{x}\,.\,\mathbf{a}_2 = \mathbf{x}\,.\,\mathbf{a}_3 = 0$.

We assume that the three vectors $\mathbf{a}_1$, $\mathbf{a}_2$, $\mathbf{a}_3$ are not all parallel. If they are all parallel then $\det\mathbf{A} = \mathbf{a}_1\,.\,\mathbf{a}_2 \times \mathbf{a}_3 = 0$ and $\mathbf{x}$ can be any vector perpendicular to $\mathbf{a}_1$, and the problem is therefore trivial. Taking $\mathbf{a}_2$ and $\mathbf{a}_3$ to be non-parallel we see that the solution to the equations $\mathbf{x}\,.\,\mathbf{a}_2 = \mathbf{x}\,.\,\mathbf{a}_3 = 0$ must be of the form $\mathbf{x} = \alpha\mathbf{a}_2 \times \mathbf{a}_3$ where α is any scalar. Substituting for $\mathbf{x}$ in the equation $\mathbf{x}\,.\,\mathbf{a}_1 = 0$ yields $\alpha\mathbf{a}_1\,.\,(\mathbf{a}_2 \times \mathbf{a}_3) = 0$. Since the elements of $\mathbf{x}$ are not all zero, α cannot be zero and therefore $\det\mathbf{A} = \mathbf{a}_1\,.\,\mathbf{a}_2 \times \mathbf{a}_3 = 0$.

We can conclude from this example that if $\det\mathbf{A} = 0$ then the equations $\mathbf{A}\mathbf{x}^{\mathrm{T}} = \mathbf{0}$ have a solution such that x_1, x_2, x_3 are not all zero. The equations $\mathbf{x}\,.\,\mathbf{a}_2 = \mathbf{x}\,.\,\mathbf{a}_3 = 0$ have the solution as before, $\mathbf{x} = \alpha\mathbf{a}_2 \times \mathbf{a}_3$. Now $\mathbf{x}\,.\,\mathbf{a}_1 = \alpha\mathbf{a}_1\,.\,\mathbf{a}_2 \times \mathbf{a}_3 = 0$ as $\det\mathbf{A} = 0$, and hence all three equations are satisfied.

1.5 DIFFERENTIATION OF MATRICES

We have not so far explicitly distinguished between matrices whose elements are numerical constants and those whose elements depend on one or more variables. Indeed the distinction has not been necessary since the matrix algebra contained in Section 1.4 applies equally well to both types of matrices. However, when we do wish to make the distinction we copy the notation used for vectors and write $\mathbf{A}(t)$ to denote for example that the elements of the matrix $\mathbf{A}$ are functions of the variable t. Again following the practice for vectors the argument t is invariably omitted or suppressed once the dependence of the elements on it has been noted. We can now define the derivative of a matrix.

Derivative of a matrix

Let

$$\mathbf{A}(t) = [a_{ij}(t)], \qquad i = 1, 2, \ldots, m, \qquad j = 1, 2, \ldots, n$$

The derivative of $\mathbf{A}(t)$ with respect to t is denoted by $\mathrm{d}\mathbf{A}/\mathrm{d}t$ and is defined to be the $m \times n$ matrix whose ijth element is equal to $\mathrm{d}a_{ij}/\mathrm{d}t$, that is

$$\frac{\mathrm{d}\mathbf{A}}{\mathrm{d}t} = \left[\frac{\mathrm{d}a_{ij}}{\mathrm{d}t}\right], \qquad i = 1, 2, \ldots, m, \qquad j = 1, 2, \ldots, n$$

Rules for differentiating the sums, differences and products of matrices are given below without proof. Proofs follow immediately from the definition of the derivative of a matrix and the formulae of the calculus of a single variable for differentiating sums and products. The rules are:

$$\frac{\mathrm{d}}{\mathrm{d}t}(\mathbf{A} \pm \mathbf{B}) = \frac{\mathrm{d}\mathbf{A}}{\mathrm{d}t} \pm \frac{\mathrm{d}\mathbf{B}}{\mathrm{d}t}$$

$$\frac{\mathrm{d}\mathbf{A}}{\mathrm{d}t} = \mathbf{0} \qquad \text{all elements of } \mathbf{A} \text{ constants}$$

$$\frac{\mathrm{d}}{\mathrm{d}t}(\alpha\mathbf{A}) = \alpha\frac{\mathrm{d}\mathbf{A}}{\mathrm{d}t} + \frac{\mathrm{d}\alpha}{\mathrm{d}t}\mathbf{A} \qquad \alpha \text{ is a scalar function of } t$$

$$\frac{\mathrm{d}}{\mathrm{d}t}(\mathbf{AB}) = \mathbf{A}\frac{\mathrm{d}\mathbf{B}}{\mathrm{d}t} + \frac{\mathrm{d}\mathbf{A}}{\mathrm{d}t}\mathbf{B}$$

We show now by means of examples how these rules can be extended to obtain the derivatives of more complex expressions. Whilst doing this we point out some possible pitfalls.

Example 1.26

The matrix $\mathbf{A}(t)$ is defined by

$$\mathbf{A}(t) = \begin{bmatrix} 0 & \sin t \\ \cos t & 0 \end{bmatrix}$$

Find

(a) $\mathrm{d}\mathbf{A}^2/\mathrm{d}t$
(b) $2\mathbf{A}(\mathrm{d}\mathbf{A}/\mathrm{d}t)$.

▷ *Solution*

(a) Firstly,

$$\mathbf{A}^2 = \begin{bmatrix} 0 & \sin t \\ \cos t & 0 \end{bmatrix}\begin{bmatrix} 0 & \sin t \\ \cos t & 0 \end{bmatrix} = \frac{1}{2}\begin{bmatrix} \sin 2t & 0 \\ 0 & \sin 2t \end{bmatrix}$$

Hence we have

$$\frac{\mathrm{d}\mathbf{A}^2}{\mathrm{d}t} = \begin{bmatrix} \cos 2t & 0 \\ 0 & \cos 2t \end{bmatrix}$$

(b) We obtain

$$2\mathbf{A}\frac{\mathrm{d}\mathbf{A}}{\mathrm{d}t} = 2\begin{bmatrix} 0 & \sin t \\ \cos t & 0 \end{bmatrix}\begin{bmatrix} 0 & \cos t \\ -\sin t & 0 \end{bmatrix} = \begin{bmatrix} -2\sin^2 t & 0 \\ 0 & 2\cos^2 t \end{bmatrix}$$

The reader should note that $d\mathbf{A}^2/dt \neq 2\mathbf{A}(d\mathbf{A}/dt)$. To find $d\mathbf{A}^2/dt$ without explicitly determing $\mathbf{A}^2$ we have to use the rule for differentiating the product $\mathbf{AB}$ with $\mathbf{B}$ put equal to $\mathbf{A}$. Thus

$$\frac{d\mathbf{A}^2}{dt} = \mathbf{A}\frac{d\mathbf{A}}{dt} + \frac{d\mathbf{A}}{dt}\mathbf{A}$$

and we therefore see that $d\mathbf{A}^2/dt$ is equal to $2\mathbf{A}(d\mathbf{A}/dt)$ only when $\mathbf{A}(d\mathbf{A}/dt) = (d\mathbf{A}/dt)\mathbf{A}$. Since

$$\mathbf{A}^2 = \frac{1}{2}\sin 2t\begin{bmatrix} 1 & 0 \\ 0 & 1 \end{bmatrix}$$

the matrices $\mathbf{A}^{2n-1}$ and $\mathbf{A}^{2n}$ are readily found, and readers can therefore verify for themselves that

$$\frac{d\mathbf{A}^n}{dt} \neq n\mathbf{A}^{n-1}\frac{d\mathbf{A}}{dt} \qquad \text{for any positive integer } n$$

Example 1.27

Show that

$$\frac{d\mathbf{A}^{-1}}{dt} = -\mathbf{A}^{-1}\frac{d\mathbf{A}}{dt}\mathbf{A}^{-1}$$

> *Solution* Differentiating both sides of the relation $\mathbf{AA}^{-1} = \mathbf{I}$ yields

$$\frac{d\mathbf{A}}{dt}\mathbf{A}^{-1} + \mathbf{A}\frac{d\mathbf{A}^{-1}}{dt} = \mathbf{0} \qquad \text{or} \qquad \mathbf{A}\frac{d\mathbf{A}^{-1}}{dt} = \frac{-d\mathbf{A}}{dt}\mathbf{A}^{-1}$$

Pre-multiplying both sides of this result by $\mathbf{A}^{-1}$ gives

$$\frac{d\mathbf{A}^{-1}}{dt} = -\mathbf{A}^{-1}\frac{d\mathbf{A}}{dt}\mathbf{A}^{-1}$$

as required.

In analogy with the rule of the calculus for differentiating a function of a function we might expect the answer to be $(-1)\mathbf{A}^{-2}(d\mathbf{A}/dt)$, where $\mathbf{A}^{-2}$ stands for $(\mathbf{A}^{-1})^2$. However, this will not be the case unless $\mathbf{A}^{-1}(d\mathbf{A}/dt)$ and $(d\mathbf{A}/dt)\mathbf{A}^{-1}$ are equal.

Example 1.28

The matrix $\mathbf{A}$ is defined by

$$\mathbf{A} = \begin{bmatrix} t & 1 \\ -t & 1 \end{bmatrix}$$

Show that when $t \neq 0$, $d\mathbf{A}^{-1}/dt$ exists but $(d\mathbf{A}/dt)^{-1}$ does not.

▷ *Solution* Det $\mathbf{A} = 2t$ and hence $\mathbf{A}$ has an inverse except when $t = 0$. The inverse of $\mathbf{A}$ can be found by inspection or by using the elementary results contained in Problem 1.40. It is given by

$$\mathbf{A}^{-1} = \frac{1}{2t}\begin{bmatrix} 1 & -1 \\ t & t \end{bmatrix}$$

so its derivative can be written down directly.

$$\frac{d\mathbf{A}}{dt} = \begin{bmatrix} 1 & 0 \\ -1 & 0 \end{bmatrix}$$

and therefore $\det(d\mathbf{A}/dt) = 0$. Thus the inverse of $d\mathbf{A}/dt$ does not exist.

Example 1.29

The matrix $\mathbf{A}$ is defined by

$$\mathbf{A} = \cos\theta\mathbf{I} + (1 - \cos\theta)\mathbf{B} + \sin\theta\mathbf{C}$$

where θ is a function of the variable t and the matrices $\mathbf{B}$ and $\mathbf{C}$ are as previously defined in Example 1.23. Given that the elements of $\mathbf{B}$ and $\mathbf{C}$ are constants, show that $\dot{\mathbf{A}}\mathbf{A}^{\mathrm{T}} = \dot{\theta}\mathbf{C}$, the dot notation being used, as for vector differentiation, to denote differentiation with respect to the parameter t.

▷ *Solution* Since $\mathbf{B}$ and $\mathbf{C}$ are matrices whose elements are constants, it follows that

$$\begin{aligned} \dot{\mathbf{A}} &= \dot{\theta}\{\sin\theta(\mathbf{B} - \mathbf{I}) + \cos\theta\mathbf{C}\} \\ &= \dot{\theta}\mathbf{C}(\cos\theta\mathbf{I} + \sin\theta\mathbf{C}) \end{aligned}$$

using the relation $\mathbf{C}^2 = \mathbf{B} - \mathbf{I}$ derived in Example 1.23. Noting that $\mathbf{B}^{\mathrm{T}} = \mathbf{B}$ and $\mathbf{C}^{\mathrm{T}} = -\mathbf{C}$ we see that

$$\begin{aligned} \dot{\mathbf{A}}\mathbf{A}^{\mathrm{T}} &= \dot{\theta}\mathbf{C}(\cos\theta\mathbf{I} + \sin\theta\mathbf{C})\{\cos\theta\mathbf{I} + (1 - \cos\theta)\mathbf{B} - \sin\theta\mathbf{C}\} \\ &= \dot{\theta}\mathbf{C}\{\cos^2\theta\mathbf{I} + \cos\theta(1 - \cos\theta)\mathbf{B} - \sin^2\theta(\mathbf{B} - \mathbf{I})\} \\ &= \dot{\theta}\mathbf{C} \end{aligned}$$

making repeated use of the result $\mathbf{CB} = \mathbf{0}$.

PROBLEMS

Section 1.2

1.1 The vectors **A**, **B**, **C** are given in terms of the unit vectors **i**, **j**, **k** by $\mathbf{A} = [2, -1, 1]$, $\mathbf{B} = [7, 4, 4]$, $\mathbf{C} = [8, 4, 1]$. Find
(a) The vectors $\mathbf{A} + \mathbf{B}$, $\mathbf{A} - \mathbf{B}$, $2\mathbf{A} + 3\mathbf{B} - \mathbf{C}$, $\mathbf{A} \times \mathbf{B}$
(b) The values of $|\mathbf{A}|$, $|\mathbf{B}|$, $\mathbf{A} \cdot \mathbf{B}$
(c) The angle between **A** and **B**
(d) A unit vector **n** parallel to **A**
(e) The angles between **n** and the vectors **i**, **j**, **k**
(f) The vectors $\mathbf{A} \times (\mathbf{B} \times \mathbf{C})$ and $(\mathbf{A} \times \mathbf{B}) \times \mathbf{C}$.

1.2 The coordinates of three sets of points referred to the axes Ox, Oy, Oz are given by
(a) [1, 1, 1], [0, 4, −1], [3, −5, 5]
(b) [2, 3, 7], [4, 5, 6], [0, 1, 8], [8, 7, −6]
(c) [1, 2, 9], [3, 4, 6], [9, 10, 3].
Determine vectorially which of the sets are collinear.

1.3 The coordinates of four sets of points referred to the axes Ox, Oy, Oz are given by
(a) [1, 2, 1], [4, 1, 6], [2, 3, 0], [4, 3, 2]
(b) [2, 0, 2], [3, 1, 3], [3, 2, 5], [6, 2, 2]
(c) [−1, 1, 6], [1, 3, 5], [−3, −1, 7], [0, 3, 9], [−3, −3, 0]
(d) [1, 2, 3], [2, 1, 3], [3, 1, 2], [3, 2, 1].
Determine vectorially which of the sets are coplanar.

1.4 Sets of vectors **a**, **b**, **c** are defined below in terms of the unit vectors **i**, **j**, **k**. Show that the members of each set are mutually perpendicular. Determine which of the sets are right-handed.
(a) $3\mathbf{a} = [2, 2, 1]$, $3\mathbf{b} = [2, -1, -2]$, $3\mathbf{c} = [1, -2, 2]$
(b) $7\mathbf{a} = [2, 3, 6]$, $7\mathbf{b} = [-6, -2, 3]$, $7\mathbf{c} = [3, -6, 2]$
(c) $11\mathbf{a} = [-9, -2, -6]$, $11\mathbf{b} = [-6, 6, 7]$, $11\mathbf{c} = [-2, -9, 6]$.
In each of (a), (b), (c), express **i**, **j**, **k** in terms of **a**, **b**, **c**.

Express the vector **A**, given by $\mathbf{A} = -7\mathbf{i} + 5\mathbf{j} + \mathbf{k}$, in terms of the vectors **a**, **b**, **c** defined in (a).

1.5 Prove the following identities:
(a) $(\mathbf{b} \times \mathbf{c}) \times (\mathbf{c} \times \mathbf{a}) \cdot \mathbf{a} \times \mathbf{b} = (\mathbf{a} \times \mathbf{b} \cdot \mathbf{c})^2$
(b) $\mathbf{a} \times (\mathbf{b} \times \mathbf{c}) + \mathbf{b} \times (\mathbf{c} \times \mathbf{a}) + \mathbf{c} \times (\mathbf{a} \times \mathbf{b}) = \mathbf{0}$
(c) $(\mathbf{a} \times \mathbf{b}) \times (\mathbf{b} \times \mathbf{c}) = (\mathbf{a} \times \mathbf{b} \cdot \mathbf{c})\mathbf{b}$.

1.6 The four vectors **A**, **B**, **C**, **D** are defined in terms of the unit vectors **i**, **j**, **k** by $\mathbf{A} = [1, 2, 3]$, $\mathbf{B} = [3, 6, 7]$, $\mathbf{C} = [4, -2, 2]$ and $\mathbf{D} = [7, -6, 3]$.
(a) Verify the identity

$$(\mathbf{A} \times \mathbf{B}) . (\mathbf{C} \times \mathbf{D}) = (\mathbf{A} . \mathbf{C})(\mathbf{B} . \mathbf{D}) - (\mathbf{A} . \mathbf{D})(\mathbf{B} . \mathbf{C})$$

(b) Find, using the results contained in Example 1.9, the relation that exists between the four vectors **A**, **B**, **C**, **D**.

(c) Express **C** in the form $\mathbf{C} = \alpha\mathbf{A} + \beta\mathbf{B} + \mu\mathbf{A} \times \mathbf{B}$.

1.7 The vectors $\underline{OA}$, $\underline{OB}$, $\underline{OC}$ are defined by $\underline{OA} = a\mathbf{i}$, $\underline{OB} = b\mathbf{j}$ and $\underline{OC} = c\mathbf{k}$. Show that the vector $bc\mathbf{i} + ca\mathbf{j} + ab\mathbf{k}$ is perpendicular to the plane defined by the lines AB and AC.

1.8 The points A, B, C, D are the vertices of a tetrahedron. Show, if E, F, G, H are respectively the midpoints of the lines AC, BD, AD, BC, that $2\underline{EF} = \underline{AB} + \underline{CD}$ and $2\underline{GH} = \underline{AB} - \underline{CD}$. Deduce, if the directions of the vectors $\underline{AB}$ and $\underline{CD}$ are perpendicular, that $|\underline{GH}| = |\underline{EF}|$.

Hint: use the relation $\underline{AE} + \underline{EF} + \underline{FB} = \underline{AB}$ to show that $2\underline{EF} = 2\underline{AB} - \underline{AC} - \underline{DB}$.

1.9 Using the identities (1.19) and (1.20) for triple-scalar products, show

(a) $$\begin{vmatrix} \alpha a_1 & \alpha a_2 & \alpha a_3 \\ b_1 & b_2 & b_3 \\ c_1 & c_2 & c_3 \end{vmatrix} = \alpha \begin{vmatrix} a_1 & a_2 & a_3 \\ b_1 & b_2 & b_3 \\ c_1 & c_2 & c_3 \end{vmatrix}$$

(b) $$\begin{vmatrix} a_1 & a_2 & a_3 \\ b_1 & b_2 & b_3 \\ c_1 & c_2 & c_3 \end{vmatrix} = - \begin{vmatrix} b_1 & b_2 & b_3 \\ a_1 & a_2 & a_3 \\ c_1 & c_2 & c_3 \end{vmatrix}$$

(c) $$\begin{vmatrix} a_1 & a_2 & a_3 \\ b_1 + \alpha c_1 & b_2 + \alpha c_2 & b_3 + \alpha c_3 \\ c_1 & c_2 & c_3 \end{vmatrix} = \begin{vmatrix} a_1 & a_2 & a_3 \\ b_1 & b_2 & b_3 \\ c_1 & c_2 & c_3 \end{vmatrix}.$$

Hence deduce, for third-order determinants, the following elementary properties of determinants:

(i) Interchanging any two rows of a determinant yields a second determinant whose value is equal to the value of the original determinant multiplied by minus one.

(ii) Adding a multiple of the elements of one row of a determinant to the corresponding elements of another row yields a determinant whose value is equal to that of the original determinant.

(iii) If two rows of a determinant are identical then the value of the determinant is zero.

1.10 Show that the three vectors $\underline{OA}$, $\underline{OB}$, $\underline{OC}$ are coplanar only if $\underline{OA} . \underline{OB} \times \underline{OC} = 0$. Deduce with the help of the first identity in Problem 1.5 that if the vectors $\underline{OA}$, $\underline{OB}$, $\underline{OC}$ are non-coplanar then so are the vectors $\underline{OB} \times \underline{OC}$, $\underline{OC} \times \underline{OA}$ and $\underline{OA} \times \underline{OB}$.

1.11 Referred to an origin O the position vectors of the four points A, B, C, D are given in terms of the unit vectors **i**, **j**, **k** by $\underline{OA} = [1, 1, 1]$,

$\underline{OB} = [3, 4, 2]$, $\underline{OC} = [2, 5, 3]$, $\underline{OD} = [4, 5, 6]$. Using the results contained in Example 1.10, determine the values of α, β, μ in the relation $\underline{AD} = \alpha\underline{AB} + \beta\underline{AC} + \mu\underline{AB} \times \underline{AC}$.

The point E is given by $\underline{OE} = [-4, 5, -3]$. Are the points D and E on the same or opposite sides of the plane defined by the vectors $\underline{AB}$ and $\underline{AC}$?

The point F is the mirror image of the point D in the plane. Show that the position of F is given by $\underline{OF} = \underline{OD} - 2\mu\underline{AB} \times \underline{AC}$. Hence determine the coordinates of the point F. Find the cartesian equation of the plane ABC and verify that the midpoint of the line DF lies on it.

1.12 (a) Explain why $\mathbf{A} \times (\mathbf{B} \times \mathbf{C})$ must be of the form $\alpha\mathbf{B} + \beta\mathbf{C}$ where α and β are scalars. Determine, by taking the scalar product of each side of the relation $\mathbf{A} \times (\mathbf{B} \times \mathbf{C}) = \alpha\mathbf{B} + \beta\mathbf{C}$ with $\mathbf{A}$, the ratio $\alpha : \beta$. Deduce that

$$\mathbf{A} \times (\mathbf{B} \times \mathbf{C}) = \mu((\mathbf{A} \cdot \mathbf{C})\mathbf{B} - (\mathbf{A} \cdot \mathbf{B})\mathbf{C})$$

where μ is a scalar to be found. Given that μ is independent of $\mathbf{A}$, $\mathbf{B}$, $\mathbf{C}$, obtain its value by choosing $\mathbf{A}$, $\mathbf{B}$, $\mathbf{C}$ appropriately.

(b) Show that if $\mathbf{A} \times (\mathbf{B} \times \mathbf{C}) = (\mathbf{A} \times \mathbf{B}) \times \mathbf{C}$ then either $\mathbf{A}$ and $\mathbf{C}$ are parallel or $\mathbf{B}$ and $\mathbf{A} \times \mathbf{C}$ are parallel.
Hint: expand $\mathbf{A} \times (\mathbf{B} \times \mathbf{C})$ and $(\mathbf{A} \times \mathbf{B}) \times \mathbf{C}$ and equate the results.

1.13 (a) Explain why the equation $\mathbf{X} \times \mathbf{A} = \mathbf{B}$, where $\mathbf{A} = [1, -2, 1]$ and $\mathbf{B} = [2, 1, 1]$ referred to the unit vectors $\mathbf{i}, \mathbf{j}, \mathbf{k}$, has no solution for the unknown vector $\mathbf{X}$.

(b) The vector $\mathbf{X}$ satisfies the equation $\mathbf{X} \times \mathbf{A} = \mathbf{B}$ and is perpendicular to the vector $\mathbf{C}$. Given that $\mathbf{A} = [1, -2, 1]$, $\mathbf{B} = [1, 1, 1]$, $\mathbf{C} = [2, 1, 1]$ referred to the unit vectors $\mathbf{i}, \mathbf{j}, \mathbf{k}$, find $\mathbf{X}$ using the following methods:

(i) Formulate and solve the equations for the components of $\mathbf{X}$.
(ii) Express $\mathbf{X}$ in the form $\alpha\mathbf{D} \times \mathbf{E}$, where $\mathbf{D}$ and $\mathbf{E}$ are known vectors, and solve for α.
(iii) Take the vector product of both sides of the equation $\mathbf{X} \times \mathbf{A} = \mathbf{B}$ with the vector $\mathbf{C}$.

1.14 Verify by direct substitution that $\mathbf{X} = \alpha\mathbf{A} + (\mathbf{A} \times \mathbf{B})/\mathbf{A}^2$, where α is an arbitrary scalar, is a solution of the vector equation $\mathbf{X} \times \mathbf{A} = \mathbf{B}$. Show that $\mathbf{X}^2 = \alpha^2\mathbf{A}^2 + \mathbf{C}^2$, where $\mathbf{A}^2\mathbf{C} = \mathbf{A} \times \mathbf{B}$. Given that $\mathbf{A} = [1, 3, -4]$ and $\mathbf{B} = [2, 6, 5]$ referred to the unit vectors $\mathbf{i}, \mathbf{j}, \mathbf{k}$ and $|\mathbf{X}| = 3$, find the two solutions for $\mathbf{X}$. If the two solutions are denoted by $\mathbf{X}_1$ and $\mathbf{X}_2$, verify that $\mathbf{X}_2 - \mathbf{X}_1$ is parallel to the vector $\mathbf{A}$. Obtain the two solutions for $\mathbf{X}$ by the alternative method of formulating and solving the equations satisfied by the components of $\mathbf{X}$.

1.15 Show that the equations $\mathbf{X} \times \mathbf{A} = \mathbf{B}$ and $\mathbf{X} \times \mathbf{C} = \mathbf{D}$, where $\mathbf{A} = [-7, -5, 10]$, $\mathbf{B} = [-5, -9, -8]$, $\mathbf{C} = [1, 5, -4]$, $\mathbf{D} = [1, 3, 4]$ referred to the unit vectors $\mathbf{i}$, $\mathbf{j}$, $\mathbf{k}$, have a solution for the unknown vector $\mathbf{X}$.

Find $\mathbf{X}$ using the following methods:

(a) Apply the formula derived in Example 1.14.

(b) Formulate and solve the six equations for the three (unknown) components of $\mathbf{X}$.

In (b), find the three (linear) relations that exist among the six equations.

1.16 (a) Verify that the vector $\mathbf{X}$ defined by $(\mathbf{A} \cdot \mathbf{C})\mathbf{X} = \alpha\mathbf{A} - \mathbf{B} \times \mathbf{C}$ satisfies the two vector equations $\mathbf{X} \times \mathbf{A} = \mathbf{B}$ and $\mathbf{X} \cdot \mathbf{C} = \alpha$, where $\alpha \neq 0$. The vectors $\mathbf{A}$ and $\mathbf{B}$ are such that $\mathbf{A} \cdot \mathbf{B} = 0$.

(b) Find the vector $\mathbf{X}$ that satisfies the equation $\mathbf{A} = \mathbf{X} \times \mathbf{B}$, where $\mathbf{A} \cdot \mathbf{B} = 0$, and is perpendicular to the vector $\mathbf{B}$.

1.17 The vector $\mathbf{X}$ satisfies the equation $\mathbf{X} - \mathbf{A} = \mathbf{B} \times (\mathbf{X} + \mathbf{A})$, where $\mathbf{A}$ and $\mathbf{B}$ are given vectors. By successively taking the scalar product and vector product of both sides of the equation with $\mathbf{B}$, show that:

(a) $\mathbf{A} \cdot \mathbf{B} = \mathbf{X} \cdot \mathbf{B}$

(b) $(1 + \mathbf{B}^2)\mathbf{X} = (1 - \mathbf{B}^2)\mathbf{A} + 2(\mathbf{A} \cdot \mathbf{B})\mathbf{B} + 2\mathbf{B} \times \mathbf{A}$.

Engineering graphics

In engineering graphics, problems are solved by drawing to give practice in orthographic projection (the basis of engineering drawing) and in three-dimensional visualization (an essential skill for a designer). Typical problems involve the determination of the angle between a straight line and a plane, the location of the line of intersection of two planes, and the direction of a line that makes specified angles with two given intersecting lines. These and many other problems in engineering graphics can be solved using vector algebra. Indeed vector algebra, together with matrix algebra, provide the basis of computer-aided drawing (CAD).

Problems 1.18 to 1.23 illustrate the application of vector algebra in engineering graphics. They can also serve as an introduction to the use of the computer in engineering graphics. Program writing should include procedures (or subroutines) for determining the magnitude of a vector, the components of a unit vector parallel to a given vector, and the scalar and vector product of two vectors. Input to a program will consist of one-dimensional arrays containing three elements corresponding to the co-ordinates of points used to define lines and planes.

For the purposes of Problems 1.18 to 1.23 it should be noted that:

(a) The angle between a line and a plane is equal to the angle between the line and its perpendicular projection onto the plane.

(b) The acute (or obtuse) angle between two planes is equal to the acute (or obtuse) angle between the normals to the planes.
(c) Two skew lines (previously defined as lines that do not intersect and arc not parallel) have a common perpendicular AB, where A is a point in one line and B a point in the other, and the length of AB is the shortest distance between any pair of points in the lines.

1.18 The projections of any two points C and D onto a given plane are E and F respectively. If $\mathbf{n}$ is a unit vector perpendicular to the plane and O is the origin for position vectors, show that:
(a) The angle between the line CD and the plane is given by α where $\cos\alpha|\underline{CD}| = |\underline{CD} \times \mathbf{n}|$.
(b) The length p of the perpendicular from any point P in the line CD to the plane is $|\underline{AP} \,.\, \mathbf{n}|$, where A is a given point in the plane.
(c) The position vector $\underline{OQ}$ of Q, the foot of the perpendicular from P to the plane, is given by $\underline{OQ} = \underline{OP} - (\underline{AP} \,.\, \mathbf{n})\mathbf{n}$.
(d) The position vector $\mathbf{r}$ of the point where the line CD meets the plane is given by $\mathbf{r} = \mathbf{c} + \alpha\mathbf{N}$, where $\underline{OA} = \mathbf{a}$, $\underline{OC} = \mathbf{c}$, $\mathbf{N}$ is a unit vector parallel to the line CD, and $(\mathbf{n} \,.\, \mathbf{N})\alpha = (\mathbf{a} - \mathbf{c}) \,.\, \mathbf{n}$.

1.19 The equations of two planes are given, in standard form, by $(\mathbf{r} - \mathbf{a}_1) \,.\, \mathbf{n}_1 = 0$ and $(\mathbf{r} - \mathbf{a}_2) \,.\, \mathbf{n}_2 = 0$. Explain why the equation of the line of intersection of the two planes has the form $\mathbf{r} = \mathbf{a} + \alpha\mathbf{n}_1 \times \mathbf{n}_2$, where $\mathbf{a}$ is the position vector of a point A in the line. Verify that the point with position vector $\beta\mathbf{n}_1 + \mu\mathbf{n}_2$, where β and μ are given by

$$|\mathbf{n}_1 \times \mathbf{n}_2|^2\beta = \mathbf{a}_1 \,.\, \mathbf{n}_1 - (\mathbf{a}_2 \,.\, \mathbf{n}_2)(\mathbf{n}_1 \,.\, \mathbf{n}_2)$$
$$|\mathbf{n}_1 \times \mathbf{n}_2|^2\mu = \mathbf{a}_2 \,.\, \mathbf{n}_2 - (\mathbf{a}_1 \,.\, \mathbf{n}_1)(\mathbf{n}_1 \,.\, \mathbf{n}_2)$$

lies in both planes and is therefore a possible position of the point A.

The coordinates of two sets of points referred to the axes Ox, Oy, Oz are given by [0, −1, 3], [3, −1, 0], [1, −3, 1] and [−2, −2, −2], [−3, −1, −2], [3, 0, 2]. Determine the equation of the line of intersection of the two planes defined by the two sets of points.

1.20 The equations of two intersecting straight lines are given by $\mathbf{r} = \mathbf{a} + \alpha_1\mathbf{n}_1$ and $\mathbf{r} = \mathbf{a} + \alpha_2\mathbf{n}_2$, where $\mathbf{a}$ is the position vector of their common point A and $\mathbf{n}_1$ and $\mathbf{n}_2$ are unit vectors. A third intersecting line, in the direction of the unit vector $\mathbf{n}$, is inclined at angles θ and ϕ respectively to the unit vectors $\mathbf{n}_1$ and $\mathbf{n}_2$. Show that $\mathbf{n}$ is given by

$$\mathbf{n} = \alpha\mathbf{n}_1 + \beta\mathbf{n}_2 + \mu\mathbf{n}_1 \times \mathbf{n}_2$$

where α, β are defined by $(1 - c^2)\alpha = a - bc$, $(1 - c^2)\beta = b - ac$, $a = \cos\theta$, $b = \cos\phi$, $c = \mathbf{n}_1 \,.\, \mathbf{n}_2$.

By taking the scalar product of the vector $\mathbf{n}$ with itself, show further that $\alpha^2 + \beta^2 + \mu^2(1 - c^2) + 2\alpha\beta c = 1$. Hence deduce, by substituting for α and β into this equation, that μ exists only if $a^2 + b^2 + c^2 \leq 1 + 2abc$.

Also deduce that when there are two possible values for μ the two expressions for the vector **n** define vectors that are the mirror images of each other in the plane that passes through the point A and has the vector $\mathbf{n}_1 \times \mathbf{n}_2$ as a normal. *Hint:* note that the two possible values of μ differ only in sign and use results contained in Problem 1.11.

Use the above results to solve the following problem. The points A and B have coordinates $[-77, 11, -11]$ and $[-54, -39, 56]$ with respect to the axes Ox, Oy, Oz. Determine the positions of the points D and E such that $\underline{\text{OD}}$ and $\underline{\text{OE}}$ are inclined at an angle of 50° to the vector $\underline{\text{OA}}$ and at 40° to the vector $\underline{\text{OB}}$ and $|\underline{\text{OD}}| = |\underline{\text{OE}}| = 85$.

1.21 (a) The equation of a straight line is $\mathbf{r} = \mathbf{a} + \alpha_1\mathbf{n}_1$ and that of a plane is $(\mathbf{r} - \mathbf{a}) \cdot \mathbf{n}_2 = 0$, where $\mathbf{n}_1, \mathbf{n}_2$ are unit vectors and **a** is the position vector of A, the point of intersection of the line and the plane. A second line passing through A is inclined at an angle θ to the first line and at an angle $(\pi/2) - \phi$ to the plane. Show that **n**, a unit vector in the direction of the second line, is given by $\mathbf{n} = \alpha\mathbf{n}_1 + \beta\mathbf{n}_2 + \mu\mathbf{n}_1 \times \mathbf{n}_2$, where α, β, μ satisfy the relations contained in Problem 1.20.

(b) Show that the problem of finding the direction of a line inclined at given angles to two planes is equivalent mathematically to finding the vector **n** of Problem 1.20.

1.22 Referred to the axes Ox, Oy, Oz the points A, B, C, D have coordinates $[2, 2, 0]$, $[4, 2, -2]$, $[3, 2, 2]$, $[5, 0, 2]$. Find the components of the vector products $\underline{\text{AB}} \times \underline{\text{AC}}$ and $\underline{\text{AB}} \times \underline{\text{AD}}$ referred to the unit vectors **i**, **j**, **k**. Deduce that the lines AB and CD are skew lines.

Show that the coordinates of any point E in the line AB can be expressed in the form $[2 + \alpha, 2, -\alpha]$, and those of any point F in the line CD in the form $[3 + \beta, 2 - \beta, 2]$, where α and β are scalars. Deduce that

$$\begin{aligned}\underline{\text{EF}}^2 &= 2\alpha^2 + 2\beta^2 - 2\alpha\beta + 2\alpha + 2\beta + 5 \\ &= (\alpha - \beta)^2 + (\alpha + 1)^2 + (\beta + 1)^2 + 3\end{aligned}$$

Write down the values of α and β that make the value of $\underline{\text{EF}}^2$ a minimum and hence determine the vector equation of the line that is the common perpendicular to the lines AB and CD.

1.23 Vectors $\mathbf{n}_1$, $\mathbf{n}_2$ are respectively parallel to the skew lines AB and CD, the line BD being the common perpendicular to the skew lines. The line through A parallel to BD meets the plane that passes through D and has BD as normal at E. Explain why $\underline{AE} = \underline{BD}$ and why $\underline{AE}$ is perpendicular to $\underline{CE}$. Hence deduce that

$$|\mathbf{n}_1 \times \mathbf{n}_2|\,|\underline{BD}| = |\underline{AC} \cdot \mathbf{n}_1 \times \mathbf{n}_2|$$

The points A and C have coordinates $[-1, 7, -2]$ and $[-7, -7, 6]$ referred to axes Ox, Oy, Oz, while $\mathbf{n}_1 = [3, -1, 1]$ and $\mathbf{n}_2 = [-3, 2, 4]$ referred to the unit vectors $\mathbf{i}$, $\mathbf{j}$, $\mathbf{k}$. By considering the vector relations $\underline{BD} + \underline{DC} = \underline{BC} = \underline{BA} + \underline{AC}$, determine the coordinates of B and D. *Hint:* use the fact that $\underline{BD}$, $\underline{BA}$ and $\underline{DC}$ must be of the form $r\mathbf{n}_1 \times \mathbf{n}_2$, $p\mathbf{n}_1$ and $q\mathbf{n}_2$, where p, q, r are scalars to be determined. Note that the values of p, q, r can be obtained with the help of results in Example 1.10.

Section 1.3

1.24 Sketch the curves defined by
(a) $\mathbf{r} = \cos t\mathbf{i} + \cos 2t\mathbf{j}$
(b) $\mathbf{r} = e^t(\cos t\mathbf{i} + \sin t\mathbf{j})$
(c) $\mathbf{r} = (1 + \cos t)(\cos t\mathbf{i} + \sin t\mathbf{j})$
(d) $\mathbf{r} = a\cos\omega t\mathbf{i} + b\sin\omega t\mathbf{j}$, where a, b, ω are constants.

1.25 Verify the rules

$$\frac{d}{dt}(\mathbf{A} \cdot \mathbf{B}) = \frac{d\mathbf{A}}{dt} \cdot \mathbf{B} + \mathbf{A} \cdot \frac{d\mathbf{B}}{dt}$$

$$\frac{d}{dt}(\mathbf{A} \times \mathbf{B}) = \frac{d\mathbf{A}}{dt} \times \mathbf{B} + \mathbf{A} \times \frac{d\mathbf{B}}{dt}$$

for $\mathbf{A} = \sin t\mathbf{i} + \cos t\mathbf{j}$ and $\mathbf{B} = \cos t\mathbf{i} - \sin t\mathbf{j}$.

1.26 The vector $\mathbf{r}$ is defined by $\mathbf{r} = \cos\alpha t\mathbf{i} - \sin\alpha t\mathbf{j} + \beta t\mathbf{k}$, where α and β are constants. Show that the vectors $\dot{\mathbf{r}}$ and $\ddot{\mathbf{r}}$ are connected by the relation $\ddot{\mathbf{r}} = \dot{\mathbf{r}} \times \mathbf{H}$, where $\mathbf{H} = \alpha\mathbf{k}$.

1.27 The vector $\mathbf{r}$ is given by $\mathbf{r} = e^t(\cos t\mathbf{i} + \sin t\mathbf{j})$. Show that the angle between the two vectors $\mathbf{r}$ and $\ddot{\mathbf{r}}$ is constant.

1.28 The vectors $\mathbf{a}$ and $\mathbf{b}$ are both functions of the variable t. Their derivatives with respect to t are given in terms of a third vector $\boldsymbol{\omega}$ by $\dot{\mathbf{a}} = \boldsymbol{\omega} \times \mathbf{a}$ and $\dot{\mathbf{b}} = \boldsymbol{\omega} \times \mathbf{b}$. Show that $\mathbf{a} \cdot \mathbf{b}$ is constant.

The vector $\mathbf{c}$ is defined by $\mathbf{c} = \mathbf{a} \times \mathbf{b}$. Find $\dot{\mathbf{c}}$ in terms of $\boldsymbol{\omega}$ and $\mathbf{c}$ and hence deduce that $|\mathbf{c}|$ is constant.

1.29 The vector $\mathbf{r}$ is given by $\mathbf{r} = (1 - e^{-t})\mathbf{v} + g(1 - t - e^{-t})\mathbf{k}$, where $\mathbf{v}$ and $\mathbf{k}$ are constant vectors. Show that $\mathbf{r}$ satisfies the differential equation $\ddot{\mathbf{r}} + \dot{\mathbf{r}} + g\mathbf{k} = \mathbf{0}$ and that the direction of the vector $\ddot{\mathbf{r}}$ is fixed.

1.30 The vector $\mathbf{r}$ is defined in terms of two non-parallel vectors $\mathbf{r}_1$, $\mathbf{r}_2$ and scalar constants a, b, c by $\mathbf{r} = a\mathbf{r}_1 + b\mathbf{r}_2 + c\mathbf{r}_1 \times \mathbf{r}_2$. If $\mathbf{r}_1$ and $\mathbf{r}_2$ are functions of the variable t such that $\dot{\mathbf{r}}_1 = \boldsymbol{\omega} \times \mathbf{r}_1$ and $\dot{\mathbf{r}}_2 = \boldsymbol{\omega} \times \mathbf{r}_2$, $\boldsymbol{\omega}$ being a further vector, show that $\dot{\mathbf{r}} = \boldsymbol{\omega} \times \mathbf{r}$.

When $t = 0$ the vectors $\mathbf{r}_1$, $\mathbf{r}_2$, $\mathbf{r}$ are given in terms of the unit vectors $\mathbf{i}$, $\mathbf{j}$, $\mathbf{k}$ by $\mathbf{r}_1 = [1, 2, 3]$, $\mathbf{r}_2 = [1, -1, -1]$ and $\mathbf{r} = [2, -4, 2]$. Obtain the values of a, b, c and hence determine the vector $\dot{\mathbf{r}}$ when $t = 0$, given that $\dot{\mathbf{r}}_1 = [1, -2, 1]$ and $\dot{\mathbf{r}}_2 = [0, 2, -2]$ when $t = 0$.

Explain why $\boldsymbol{\omega}$ is of the form $\alpha\dot{\mathbf{r}}_1 \times \dot{\mathbf{r}}_2$, where α is some scalar. Find $\boldsymbol{\omega}$ when $t = 0$ and hence verify the answer already obtained for $\dot{\mathbf{r}}$ when $t = 0$.

1.31 The vectors $\mathbf{r}_i(t)$ $(i = 1, 2, \ldots, m)$ are defined in terms of the parameter t by

$$\mathbf{r}_i(t) = \mathbf{r}_i = \cos\omega t\mathbf{a}_i + (1 - \cos\omega t)(\mathbf{n} \cdot \mathbf{a}_i)\mathbf{n} + \sin\omega t\mathbf{n} \times \mathbf{a}_i$$

where $\mathbf{a}_i$ $(i = 1, 2, \ldots, m)$ and $\mathbf{n}$ are fixed unit vectors and ω is a scalar constant. By differentiating the expression $\mathbf{r}_i \cdot \mathbf{r}_j$, where $i \neq j$, with respect to t, show that the angle between the vectors $\mathbf{r}_i$ and $\mathbf{r}_j$ is fixed and find an expression for this angle.

Hint: use the results contained in Example 1.19.

1.32 The vectors $\mathbf{a}$, $\mathbf{b}$, $\mathbf{c}$ are defined in terms of the parameters θ and ϕ by

$$\mathbf{a} = \sin\theta\cos\phi\mathbf{i} + \sin\theta\sin\phi\mathbf{j} + \cos\theta\mathbf{k}, \qquad \mathbf{b} = \frac{\partial\mathbf{a}}{\partial\theta}, \qquad \mathbf{c} = \mathbf{a} \times \mathbf{b}$$

Find the components of $\mathbf{b}$ and $\mathbf{c}$ and express $\partial\mathbf{a}/\partial\phi$ in terms of θ and $\mathbf{c}$. Show that $\mathbf{a}$ and $\mathbf{b}$ are perpendicular and deduce, by expanding $\mathbf{b} \times (\mathbf{a} \times \mathbf{b})$, that $\mathbf{b} \times \mathbf{c} = \mathbf{a}$.

The parameters θ and ϕ are both functions of the variable t. Show that $\dot{\mathbf{a}} = \dot{\theta}\mathbf{b} + \dot{\phi}\sin\theta\mathbf{c}$. Express $\dot{\mathbf{b}}$ in the form $\alpha\mathbf{a} + \beta\mathbf{c}$ and hence, by differentiating $\mathbf{a} \times \mathbf{b}$, obtain an expression for $\dot{\mathbf{c}}$ in terms of $\mathbf{a}$ and $\mathbf{b}$.

Verify that the derivatives of $\mathbf{a}$, $\mathbf{b}$, $\mathbf{c}$ are respectively equal to $\boldsymbol{\omega} \times \mathbf{a}$, $\boldsymbol{\omega} \times \mathbf{b}$, $\boldsymbol{\omega} \times \mathbf{c}$, where $\boldsymbol{\omega} = \dot{\phi}(\cos\theta\mathbf{a} - \sin\theta\mathbf{b}) + \dot{\theta}\mathbf{c}$. Obtain the components of $\boldsymbol{\omega}$ referred to $\mathbf{i}$, $\mathbf{j}$, $\mathbf{k}$ and express $\mathbf{i}$, $\mathbf{j}$, $\mathbf{k}$ in terms of $\mathbf{a}$, $\mathbf{b}$, $\mathbf{c}$.

1.33 The vector $\mathbf{a}$ is defined by

$$\mathbf{a} = \sin\alpha t\cos\beta t\mathbf{i} + \sin\alpha t\sin\beta t\mathbf{j} + \cos\alpha t\mathbf{k}$$

where α and β are constants. Find $\dot{\mathbf{a}}$ and deduce that $\mathbf{a}$ and $\dot{\mathbf{a}}$ are perpendicular.

Verify that the vector $\boldsymbol{\Omega}$, given by $\boldsymbol{\Omega} = [-\alpha\sin\beta t, \alpha\cos\beta t, \beta]$, is a solution of the vector equation $\dot{\mathbf{a}} = \boldsymbol{\omega} \times \mathbf{a}$, where $\boldsymbol{\omega}$ is the unknown vector. Explain why all other solutions to the equation can be expressed in the form $\boldsymbol{\omega} = \boldsymbol{\Omega} + \mu\mathbf{a}$, where μ is an arbitrary scalar. *Hint:* show if $\boldsymbol{\omega}_1$, $\boldsymbol{\omega}_2$ are two solutions of the equation then $(\boldsymbol{\omega}_1 - \boldsymbol{\omega}_2) \times \mathbf{a} = \mathbf{0}$.

Find the value of μ, in terms of t, when the two vectors $\mathbf{a}$ and $\boldsymbol{\Omega} + \mu\mathbf{a}$ are perpendicular.

1.34 The vectors $\mathbf{a}$ and $\mathbf{b}$ are functions of t; each is of unit magnitude, and $\mathbf{a} \times \mathbf{b} \neq \mathbf{0}$. The vectors $\boldsymbol{\omega}_1$ and $\boldsymbol{\omega}_2$ satisfy the relations $\dot{\mathbf{a}} = \boldsymbol{\omega}_1 \times \mathbf{a}$ and $\dot{\mathbf{b}} = \boldsymbol{\omega}_2 \times \mathbf{b}$. Show, if $\mathbf{a} \cdot \mathbf{b}$ has a constant value, that either $\boldsymbol{\omega}_1 = \boldsymbol{\omega}_2$ or $\boldsymbol{\omega}_1 - \boldsymbol{\omega}_2$ is perpendicular to $\mathbf{a} \times \mathbf{b}$.
Hint: differentiate $\mathbf{a} \cdot \mathbf{b}$ and substitute for $\dot{\mathbf{a}}$ and $\dot{\mathbf{b}}$ in the result.

1.35 The vector $\boldsymbol{\omega}$ satisfies the equations $\dot{\mathbf{a}} = \boldsymbol{\omega} \times \mathbf{a}$ and $\boldsymbol{\omega} \cdot \mathbf{i} = 0$, where $\mathbf{a} = [\sin t \cos t, \sin^2 t, \cos t]$ referred to $\mathbf{i}, \mathbf{j}, \mathbf{k}$. Determine the vector $\boldsymbol{\omega}$.
Hint: use the fact that $\boldsymbol{\omega}$ is perpendicular to both $\mathbf{i}$ and $\dot{\mathbf{a}}$.

Section 1.4

1.36 If

$$\mathbf{A} = \begin{bmatrix} 1 & 3 & 5 & 7 \\ 2 & 4 & 6 & 8 \end{bmatrix}, \quad \mathbf{B} = \begin{bmatrix} 1 & 0 & 1 & 0 \\ 0 & 1 & 1 & 0 \end{bmatrix}, \quad \mathbf{C} = \begin{bmatrix} 1 & -1 \\ 1 & 1 \end{bmatrix}$$

determine which of the following exist: $\mathbf{A} + \mathbf{B}$, $\mathbf{AB}$, $\mathbf{BA}$, $\mathbf{A}^{\mathrm{T}}\mathbf{B}$, $\mathbf{AB}^{\mathrm{T}}$, $\mathbf{AB}^{\mathrm{T}}\mathbf{C}$, $\mathbf{C}^{\mathrm{T}}\mathbf{BA}^{\mathrm{T}}$. Find those that do exist.

1.37 (a) Expand $(\mathbf{A} + \mathbf{B})^2$, stating when the expansion is possible.
(b) What relation must exist between $\mathbf{A}$ and $\mathbf{B}$ if $(\mathbf{A} + \mathbf{B})(\mathbf{A} - \mathbf{B}) = (\mathbf{A} - \mathbf{B})(\mathbf{A} + \mathbf{B})$?
(c) Simplify $(\mathbf{BAB}^{-1})^2$, $(\mathbf{BAB}^{-1})^3$ and $(\mathbf{BAB}^{-1})^n$, where n is a positive integer.

1.38 The matrices $\mathbf{A}$ and $\mathbf{B}$ are defined by

$$\mathbf{A} = \begin{bmatrix} 1 & 1 & 1 \\ 2 & 4 & -1 \\ 1 & 0 & 2 \end{bmatrix}, \quad \mathbf{B} = \begin{bmatrix} 2 & 2 & 1 \\ 1 & -1 & -2 \\ 3 & 4 & 3 \end{bmatrix}$$

Verify
(a) $\det\mathbf{A}^{\mathrm{T}} = \det\mathbf{A}$ and $\det\mathbf{B}^{\mathrm{T}} = \det\mathbf{B}$
(b) $\det(\mathbf{AB}) = \det\mathbf{A}\det\mathbf{B}$.

1.39 The matrices $\mathbf{A}$ and $\mathbf{B}$ are defined by

$$\mathbf{A} = \begin{bmatrix} 1 & 1 & 1 \\ 1 & 0 & 2 \\ 2 & 4 & -1 \end{bmatrix}, \quad \mathbf{B} = \begin{bmatrix} 2 & 2 & 1 \\ 3 & 4 & 3 \\ 9 & 7 & 2 \end{bmatrix}$$

Explain why $\mathbf{A}$ and $\mathbf{B}$ have inverses. Verify that the inverses are given by

$$\mathbf{A}^{-1} = \begin{bmatrix} -8 & 5 & 2 \\ 5 & -3 & -1 \\ 4 & -2 & -1 \end{bmatrix}, \qquad \mathbf{B} = \begin{bmatrix} -13 & 3 & 2 \\ 21 & -5 & -3 \\ -15 & 4 & 2 \end{bmatrix}$$

Determine the inverses of $\mathbf{A}^2$, $\mathbf{A}^{\mathrm{T}}$, $\mathbf{AB}$, $(\mathbf{AB})^{\mathrm{T}}$.

1.40 The matrix $\mathbf{A}$ is defined by

$$\mathbf{A} = \begin{bmatrix} a & b \\ c & d \end{bmatrix}$$

where $ad - bc \neq 0$. Verify that $\mathbf{A}^{-1}$ is given by

$$(ad - bc)\mathbf{A}^{-1} = \begin{bmatrix} d & -b \\ -c & a \end{bmatrix}$$

Find the inverses of the following 2×2 matrices where they exist:

(a) $\begin{bmatrix} 1 & -1 \\ 1 & 1 \end{bmatrix}$ (b) $\begin{bmatrix} 1 & 2 \\ 3 & 4 \end{bmatrix}$ (c) $\begin{bmatrix} 0 & 1 \\ -1 & 0 \end{bmatrix}$ (d) $\begin{bmatrix} 1 & 2 \\ 2 & 4 \end{bmatrix}$.

1.41 The cofactor of the element a_{ij} of the determinant

$$\begin{vmatrix} a_{11} & a_{12} & a_{13} \\ a_{21} & a_{22} & a_{23} \\ a_{31} & a_{32} & a_{33} \end{vmatrix}$$

is obtained by striking through the elements of the row and column that contain it and multiplying the resulting second-order determinant by $(-1)^{i+j}$. The cofactor of a_{ij} is denoted by A_{ij}.

Verify the following:

$$a_{11}A_{11} + a_{12}A_{12} + a_{13}A_{13} = \det\mathbf{A}$$
$$a_{21}A_{11} + a_{22}A_{12} + a_{23}A_{13} = 0$$
$$a_{31}A_{11} + a_{32}A_{12} + a_{33}A_{13} = 0$$

Write down six similar relations containing A_{ij} $(i = 2, 3, j = 1, 2, 3)$.

Deduce that the inverse of $\mathbf{A}$ (if it exists) is given by $(\det\mathbf{A})\mathbf{A}^{-1} = \mathbf{B}$, where $\mathbf{B}$ is the transpose of the matrix whose elements are the cofactors of a_{ij} $(i, j = 1, 2, 3)$.

Determine which of the following matrices possess an inverse and find those inverses that exist:

(a) $\begin{bmatrix} 0 & 1 & -1 \\ -1 & 0 & 1 \\ 1 & -1 & 0 \end{bmatrix}$ (b) $\begin{bmatrix} 1 & 3 & 7 \\ 4 & 2 & 1 \\ 6 & 1 & -4 \end{bmatrix}$ (c) $\begin{bmatrix} 1 & 1 & 1 \\ 1 & 3 & 2 \\ 2 & 1 & 3 \end{bmatrix}$.

1.42 Show, if $\mathbf{A}$ is an $n \times n$ skew-symmetric matrix, that $\det\mathbf{A} = (-1)^n \det\mathbf{A}$. Deduce that $\det\mathbf{A} = 0$ if n is odd.

Hint: $\det\mathbf{A} = \det\mathbf{A}^{\mathrm{T}} = \det(-\mathbf{A}) = \det((\mathbf{A})(-\mathbf{I}))$.

1.43 The matrix $\mathbf{D}$ defined by

$$\mathbf{D} = \begin{bmatrix} \alpha_1 & 0 & 0 \\ 0 & \alpha_2 & 0 \\ 0 & 0 & \alpha_3 \end{bmatrix}$$

is an example of a diagonal matrix, that is a matrix all of whose non-diagonal terms are zero. Find $\mathbf{D}^2$ and $\mathbf{D}^3$ and hence deduce an expression for $\mathbf{D}^n$, where n is a positive power. Is the result true when n is negative?

1.44 Assuming the result $(\mathbf{AB})^{\mathrm{T}} = \mathbf{B}^{\mathrm{T}}\mathbf{A}^{\mathrm{T}}$, prove the following:
(a) $(\mathbf{ABC})^{\mathrm{T}} = \mathbf{C}^{\mathrm{T}}\mathbf{B}^{\mathrm{T}}\mathbf{A}^{\mathrm{T}}$
(b) $\mathbf{AA}^{\mathrm{T}}$ is a symmetric matrix
(c) $\mathbf{A}^n$, where n is a positive integer, is symmetric if $\mathbf{A}$ is symmetric.
Hint: write $\mathbf{A}^{k+1} = \mathbf{A}^k\mathbf{A}$ and use induction.

1.45 The matrix $\mathbf{A}$ is defined by

$$3\mathbf{A} = \begin{bmatrix} -1 & 2 & -2 \\ -2 & 1 & 2 \\ 2 & 2 & 1 \end{bmatrix}$$

Verify
(a) $\mathbf{AA}^{\mathrm{T}} = \mathbf{A}^{\mathrm{T}}\mathbf{A} = \mathbf{I}$
(b) $\det\mathbf{A} = 1$
(c) $\det(\mathbf{A} - \mathbf{I}) = 0$.
Solve the set of equations $(\mathbf{A} - \mathbf{I})\mathbf{x}^{\mathrm{T}} = \mathbf{0}$, where $\mathbf{x} = [x_1, x_2, x_3]$.

1.46 The matrices $\mathbf{A}$ and $\mathbf{B}$ have the properties $\mathbf{AA}^{\mathrm{T}} = \mathbf{A}^{\mathrm{T}}\mathbf{A} = \mathbf{BB}^{\mathrm{T}} = \mathbf{B}^{\mathrm{T}}\mathbf{B} = \mathbf{I}$ and $\det\mathbf{A} = \det\mathbf{B} = 1$.
Show if $\mathbf{C} = \mathbf{AB}$ and $\mathbf{D} = \mathbf{BA}$ then
(a) $\mathbf{CC}^{\mathrm{T}} = \mathbf{C}^{\mathrm{T}}\mathbf{C} = \mathbf{DD}^{\mathrm{T}} = \mathbf{D}^{\mathrm{T}}\mathbf{D} = \mathbf{I}$
(b) $\det\mathbf{C} = \det\mathbf{D} = 1$.

1.47 The sum of the diagonal elements of a square matrix is called the trace of the matrix. Thus if $\mathbf{A}$ is a 3×3 matrix then the trace of $\mathbf{A}$, written as trace $\mathbf{A}$, is equal to $a_{11} + a_{22} + a_{33}$.
Show
(a) If $\mathbf{A}$ and $\mathbf{B}$ are both 3×3 matrices, trace $(\mathbf{AB})$ = trace $(\mathbf{BA})$.
(b) If $\mathbf{C}$ is a further 3×3 matrix, trace $(\mathbf{ABC})$ = trace $(\mathbf{BCA})$ = trace $(\mathbf{CAB})$.

1.48 The cubic polynomial $f(\alpha)$ defined by

$$f(\alpha) = \begin{vmatrix} a_{11} - \alpha & a_{12} & a_{13} \\ a_{21} & a_{22} - \alpha & a_{23} \\ a_{31} & a_{32} & a_{33} - \alpha \end{vmatrix}$$

is called the characteristic polynomial of the matrix $\mathbf{A} = [a_{ij}]$, $i, j = 1, 2, 3$.

Show by expanding the determinant that $f(\alpha) = -\alpha^3 + c_1\alpha^2 - c_2\alpha + c_3$, where $c_1 = \text{trace}\,\mathbf{A}$ and $c_3 = \det\mathbf{A}$.

Verify that the characteristic polynomial of the matrix

$$\mathbf{A} = \begin{bmatrix} 2 & -1 & 2 \\ -1 & 2 & 2 \\ -2 & -2 & 1 \end{bmatrix}$$

is given by $f(\alpha) = -\alpha^3 + 5\alpha^2 - 15\alpha + 27$. Also verify that $-\mathbf{A}^3 + 5\mathbf{A}^2 - 15\mathbf{A} + 27\mathbf{I} = \mathbf{0}$.

This is a particular example of the Cayley-Hamilton theorem: if $f(\mathbf{A}) = -\mathbf{A}^3 + c_1\mathbf{A}^2 - c_2\mathbf{A} + c_3\mathbf{I}$ then $f(\mathbf{A}) = \mathbf{0}$, that is a matrix satisfies its characteristic equation $f(\mathbf{A}) = \mathbf{0}$.

Section 1.5

1.49 The matrix **A** is defined by

$$\mathbf{A} = \begin{bmatrix} 1 & 0 & 0 \\ 0 & \cos t & -\sin t \\ 0 & \sin t & \cos t \end{bmatrix}$$

Verify that

(a) $\dfrac{\mathrm{d}}{\mathrm{d}t}(\mathbf{A}\mathbf{A}^{\mathrm{T}}) = \dfrac{\mathrm{d}\mathbf{A}}{\mathrm{d}t}\mathbf{A}^{\mathrm{T}} + \mathbf{A}\dfrac{\mathrm{d}\mathbf{A}^{\mathrm{T}}}{\mathrm{d}t}$

(b) $\dfrac{\mathrm{d}\mathbf{A}^{-1}}{\mathrm{d}t} = -\mathbf{A}^{-1}\dfrac{\mathrm{d}\mathbf{A}}{\mathrm{d}t}\mathbf{A}^{-1}$

(c) $\dfrac{\mathrm{d}\mathbf{A}^2}{\mathrm{d}t} = 2\mathbf{A}\dfrac{\mathrm{d}\mathbf{A}}{\mathrm{d}t}$

(d) $\left(\dfrac{\mathrm{d}\mathbf{A}}{\mathrm{d}t}\right)^{\mathrm{T}} = \dfrac{\mathrm{d}}{\mathrm{d}t}(\mathbf{A}^{\mathrm{T}})$

1.50 The matrices **D**, **E**, **F** are defined by

$$\mathbf{D} = \begin{bmatrix} 1 & 0 & 0 \\ 0 & \cos\theta & -\sin\theta \\ 0 & \sin\theta & \cos\theta \end{bmatrix}, \qquad \mathbf{E} = \begin{bmatrix} \cos\beta & 0 & \sin\beta \\ 0 & 1 & 0 \\ -\sin\beta & 0 & \cos\beta \end{bmatrix},$$

$$\mathbf{F} = \begin{bmatrix} \cos\alpha & -\sin\alpha & 0 \\ \sin\alpha & \cos\alpha & 0 \\ 0 & 0 & 1 \end{bmatrix}$$

where α, β, θ are functions of t. Find the matrix products $\dot{\mathbf{D}}\mathbf{D}^{\mathrm{T}}$, $\dot{\mathbf{E}}\mathbf{E}^{\mathrm{T}}$, $\dot{\mathbf{F}}\mathbf{F}^{\mathrm{T}}$. Show that these products could be obtained from the work of Example 1.29 by giving appropriate values to the components of **n**.

2 Concepts, laws and units

2.1 INTRODUCTION

Mechanics is the branch of physical science that deals with the state of rest or motion of bodies under the action of forces. It is of the utmost importance in the analysis essential in engineering design and manufacture. Before we can develop the methods used in analysis it is necessary to introduce and explain the laws on which mechanics is based, and to define the concepts that these laws embody. Further, in order to use the laws to obtain numerical answers we need to define units for the physical quantities involved. The first sections of this chapter, then, follow the sequence: concepts, laws and units. The concluding section briefly describes the process of mathematical modelling.

2.2 CONCEPTS

Body

The term 'body' in engineering has a wide interpretation. It may refer, for example, to a road bridge, an electron in a cathode ray tube, or a volume of liquid or gas. Whatever the material and form of the body the assumption is made that it may be represented by a number (which may range from one to infinity) of particles, of finite mass but negligible dimensions. The particles are acted upon by the surroundings of the body and interact with each other. This important assumption enables the laws of motion, which as we shall see are established for a particle, to be extended to apply to the whole body.

The nature of the interactions among the particles making up the body depends on the material of the body. Commonly three sets of assumptions about the behaviour of the material are made, and these lead to the main divisions of engineering mechanics: rigid body mechanics, deformable solid mechanics and fluid mechanics. This book will be mostly concerned with rigid body mechanics.

Mass

Mass is a scalar quantity that represents the amount of matter in a body. We shall see that it is the property of the body that determines its resistance to a change in velocity and its gravitational attraction to another body.

Length

Length is the quantity used to define the extent of a body and to locate a point within it. Length is a scalar quantity but when associated with direction and sense the outcome is a position vector (see Chapter 1).

Time

Time is the scalar measure of the sequence of events.

Force

Force is the action of one body on another that tends to move both bodies. Many experiments have established that force obeys the vector law of addition and, therefore, that force is a vector quantity. Forces are of two kinds; contact forces (between bodies that touch), and forces at a distance (between bodies that do not touch) such as gravity, electrostatic and magnetic forces. Contact forces act over a finite area even for hard bodies, but frequently the contact area is sufficiently small for it to be represented by a point. Forces at a distance act over the volume of the body, that is they act on every particle making up the body.

Absolute velocity

For terrestrial purposes the velocity of a particle is defined to be the time rate of change $\dot{\mathbf{r}}$ of its position vector $\mathbf{r}$ with respect to an observer (or origin) fixed relative to the earth. Now the earth rotates about its axis and at the same time moves in its orbit around the sun, and therefore the rate of change of the particle's position vector with respect to an observer fixed in a solar body (other than the earth) will be different.

The formulation of Newton's second law of motion (implicitly) requires a definition of the term 'velocity'. In order to provide this we introduce the concept of an inertial frame of reference. This is a frame of reference with a fixed origin (or origin moving with constant velocity) and whose associated reference axes possess fixed directions. The rate of change of a particle's position vector with respect to a point fixed in the inertial frame of reference is defined to be the particle's absolute velocity. In practice the concept of an inertial frame of reference is not very useful unless we can identify such a frame of reference. For terrestrial mechanics (but not spatial mechanics) it usually suffices to take, without serious error, axes fixed relative to the earth as inertial axes.

Absolute acceleration

Absolute acceleration is the time rate of change of absolute velocity.

These are the only concepts required for an understanding of the laws of mechanics: other concepts such as energy, work, impulse, and moment of a force will be introduced as the need for them arises in the rest of the book.

2.3 LAWS

The whole of engineering mechanics is based on the three laws of motion and the law of gravitation postulated by Newton (1642–1727). The validity of these laws has been tested by indirect means. Predictions, concerning the motion of bodies, that are based on the laws have been found to be consistent with experimental measurements.

Newton's laws of motion

The three laws of motion may be stated as follows:

I A particle remains at rest or continues to move in a straight line at constant velocity if there is no unbalanced force acting on it.

II The rate of change of the momentum of a particle acted upon by an unbalanced force is proportional to the magnitude of the force and is in the direction of the force.

III To the action of every force there is an equal and opposite (force) reaction.

Law II may be stated in mathematical terms as

$$\mathbf{F} = k\frac{\mathrm{d}(m\mathbf{v})}{\mathrm{d}t}$$

where

$\mathbf{F}$ is the unbalanced force on the particle
m is the mass of the particle
$\mathbf{v}$ is the absolute velocity of the particle
$m\mathbf{v}$ is the momentum of the particle
k is a factor of proportion.

If m is constant, the law becomes

$$\mathbf{F} = km\frac{\mathrm{d}\mathbf{v}}{\mathrm{d}t} = km\mathbf{f} \tag{2.1}$$

where $\mathbf{f}$ is the absolute acceleration of the particle. A suitable choice of units (see Section 2.4) enables us to take k to be equal to unity, so that

$$\mathbf{F} = m\mathbf{f} \tag{2.2}$$

If $\mathbf{F}$ is equal to the zero vector and m is constant then from law II, as given in (2.2), it follows that $\mathbf{f} = \mathbf{0}$ and the absolute velocity $\mathbf{v}$ of the particle must be constant. If $\mathbf{v} = \mathbf{0}$ then the particle remains at rest. Thus we see that law I follows from law II.

Law III states that forces always occur in equal and opposite pairs. In order to avoid confusion when analysing the behaviour of a body under the action of forces it is important to isolate the body and consider only the forces acting on it (see the use of free-body diagrams in Chapters 3 and 6).

Law II ceases to give reliable predictions when the magnitude of a particle's velocity approaches that of light ($2.998 \times 10^8\,\mathrm{m\,s^{-1}}$) or when a particle is of subatomic size: then it becomes necessary to use laws based on relativity. In engineering mechanics these limitations do not normally apply.

Newton's law of gravitation

Netwon's law of gravitation for the force exerted by one particle on another is expressed mathematically as

$$\mathbf{F} = -G\frac{m_1 m_2}{r^2}\mathbf{n} \tag{2.3}$$

where

F is the force exerted by particle m_1 on particle m_2
G is the universal constant of gravitation, determined by experiment
m_1, m_2 are the masses of the two particles
$\mathbf{r} = r\mathbf{n}$ is the position vector of m_2 relative to m_1; r is the scalar distance between the particles, so that **n** is a unit vector in the direction of **r**.

In the SI system of units (see Section 2.4) G has the value 6.673×10^{-11} $m^3\,kg^{-1}\,s^{-2}$.

The minus sign in (2.3) indicates that the gravitational force on particle m_2 is in the direction towards particle m_1, that is it is a force of attraction. By Newton's third law of motion there is an equal and opposite force exerted on particle m_1 by particle m_2.

Gravitational forces exist between any two particles, but for a particle located at or near the surface of the earth the only gravitational force of significant magnitude is that exerted by the earth, and this is called the weight of the particle. The direction of this gravitational force is inwards from the particle to the centre of the earth. If the weight of a particle, denoted by the vector **W**, is the only force acting on it, then from Newton's second law of motion (2.2) the acceleration **g** of the particle (called the acceleration due to gravity) is given by

$$\mathbf{W} = m\mathbf{g} \tag{2.4}$$

where m is the mass of the particle.

Experimental measurements of the magnitude of **g** (denoted by g) show that it varies over the earth's surface. This is because the earth is not a perfect sphere and also because its density varies in a non-symmetrical way. The standard value taken for g is $9.806\,651\,m\,s^{-2}$ ($9.81\,m\,s^{-2}$ in most engineering calculations): this is the value at sea level at a latitude of 45°. This value must not be taken when the distance of the particle from the surface of the earth is large, as it is, for example, for an earth satellite in orbit.

2.4 UNITS

The International System of Units (abbreviated to SI from the French Système International d'Unités) is a modern version of the metric system which has been internationally adopted to take the place, eventually, of all other systems. In the engineering applications in this book the SI system is generally used. Minor exceptions do occur; for example, engine speeds are sometimes given in terms of revolutions per minute.

The SI system defines base units for seven physical quantities, of which three – mass, length and time – are of importance in mechanics. The unit for any other physical quantity is obtained by combining base units according to some physical law (for an example see (2.5)). A unit formed in this way is referred to as a derived unit. Some derived units have special names such as newton, joule and watt (see later in this section for definitions). The definitions of the base units for mass, length and time are as follows:

Mass The unit is the kilogram, symbol kg. The kilogram is defined as the mass of the international prototype of the kilogram.
Time The unit is the second, symbol s. The second is defined as the duration of 9 192 631 770 periods of the radiation corresponding to the transition between the two hyperfine levels of the ground state of the caesium-133 atom.
Length The unit is the metre, symbol m. The metre is defined as the distance travelled by light in a vacuum in 1/299 792 458th part of a second.

The definition of mass was adopted in 1901, and the international prototype is a cylinder of platinum-iridium kept at the International Bureau of Weights and Measures in France.

Before 1967 the second was defined as 1/86 400 of the mean solar day: the change to an atomic standard gave much greater accuracy of realization and reproduction and the new definition was made consistent with the existing standard.

Prior to 1960 the unit for length was defined as the distance between two lines marked on the platinum-iridium bar which had been declared the prototype metre. An atomic standard for length was adopted in 1960. It was replaced by the present standard in 1983. This new definition of the metre is consistent with the previous standards.

A unit for angle is required in order to quantify direction. In the SI system the unit for plane angle is designated a dimensionless derived unit. The unit of plane angle is the radian, symbol rad. The radian is defined as the plane angle between two radii of a circle that cut off on the circumference an arc equal in length to the radius. The plane angle, measured in radians, between any two radii of a circle is therefore the ratio of the length of the arc cut off by the radii on the circumference to the length of the radius.

Direction is frequently specified in degrees, where the degree (symbol °) is $\pi/180$ rad.

The radian may be used to form derived units: two that are important in this book are the units for angular velocity, rad s^{-1}, and for angular acceleration, $\mathrm{rad\,s^{-2}}$. The running speed of a machine is frequently given in revolutions per minute (symbol $\mathrm{rev\,min^{-1}}$), where $1\,\mathrm{rev\,min^{-1}}$ is $2\pi/60\,\mathrm{rad\,s^{-1}}$ (based on the values $1\,\mathrm{rev} = 360° = 2\pi\,\mathrm{rad}$, and $1\,\mathrm{min} = 60\,\mathrm{s}$). The degree and the minute are not SI units but they have been retained for use with the SI system.

It follows from (2.2) as $m > 0$ that

$$|\mathbf{F}| = m|\mathbf{f}| \tag{2.5}$$

This relation leads to the definition of the newton, the unit of force in the SI system. The newton, symbol N, is defined as the magnitude of a force that will give a mass of 1 kg an acceleration of magnitude $1\,\mathrm{m\,s^{-2}}$, the unit for acceleration being derived from the base units of length and time. Putting SI units into (2.5) we have $1\,\mathrm{N} = 1\,\mathrm{kg} \times 1\,\mathrm{m\,s^{-2}}$, and if each of the physical quantities in (2.5) is divided by its SI unit to give its numerical value we obtain the numerical relation

$$\frac{|\mathbf{F}|}{\mathrm{N}} = \frac{m}{\mathrm{kg}}\frac{|\mathbf{f}|}{\mathrm{m\,s^{-2}}}$$

Since it is a derived unit no separate standard is required for the unit of force, but for convenience force-measuring instruments are calibrated against the weight of a known mass (determined by direct or indirect comparison with the prototype) at a location where g has been accurately measured.

Other examples of derived units, with special names, that are of importance in mechanics are as follows:

Quantity	*SI unit name*	*Unit symbol*	*Expression in terms of SI base units*
Energy	joule	J	$\mathrm{m^2\,kg\,s^{-2}}$
Frequency	hertz	Hz	$\mathrm{s^{-1}}$
Power	watt	W	$\mathrm{m^2\,kg\,s^{-3}}$
Pressure	pascal	Pa	$\mathrm{m^{-1}\,kg\,s^{-2}}$

Derived units can always, by definition, be expressed in terms of base units, but it is sometimes preferable to use a combination of a base unit and a derived unit with a special name. Thus the unit of momentum is equal to N s, while that for the moment of a force about a point (see (3.1) for the definition of the moment of a force) is equal to N m.

Prior to the adoption of the SI system of units it was common in engineering to use a system based on units for force, length and time rather than mass, length and time. The British unit of force was the pound force (symbol lbf) and was equal to the weight of a pound mass (symbol lb): the units of length and time were the foot (symbol ft) and second (symbol sec). Since, as we have seen, weight varies with local gravity it was necessary to adopt a standard magnitude for **g** of $32.2\,\mathrm{ft\,s^{-2}}$ to avoid variation of units with locality. The derived unit of mass was known as the slug, this being the mass that would have unit acceleration ($1\,\mathrm{ft\,s^{-2}}$) when acted upon by unit force (1 lbf). It follows from (2.4) that $|\mathbf{W}| = m|\mathbf{g}|$ and therefore $1\,\mathrm{lbf} = 1\,\mathrm{lb} \times 32.2\,\mathrm{ft\,s^{-2}}$, while from (2.5) we have $1\,\mathrm{lbf} = 1\,\mathrm{slug} \times 1\,\mathrm{ft\,s^{-2}}$ so that $1\,\mathrm{slug} = 1\,\mathrm{lbf\,ft^{-1}\,s^2} = 32.2\,\mathrm{lb}$. If each of the physical quantities in (2.5) is divided by its unit to give its numerical value we obtain the numerical relation

$$\frac{|\mathbf{F}|}{\mathrm{lbf}} = \frac{m}{\mathrm{slug}} \frac{|\mathbf{f}|}{\mathrm{ft\,s^{-2}}}$$

We mention this system of units because it, or a variation of it, is still in use.

Throughout this book we use unit vectors such as **i**, **j**, **k** or **a**, **b**, **c** as sets of reference vectors. These vectors are by implication dimensionless (see discussion prior to (1.6)). For example if a particle is moving parallel to the x axis then its velocity **v** can be expressed as $\mathbf{v} = v\mathbf{i}$, where the scalar v is equal to $\pm|\mathbf{v}|$ and therefore has the SI derived unit of $\mathrm{m\,s^{-1}}$. Thus we would write $\mathbf{v} = -2\,\mathrm{m\,s^{-1}}\,\mathbf{i}$ if the particle had a speed of $2\,\mathrm{m\,s^{-1}}$ in the negative direction of the x axis.

In the analytical applications of Chapters 5 and 7 no particular units have been specified. It is possible to use any coherent (that is consistent) set of units. Thus for example if we measure displacement in feet and time in seconds then in order to be consistent the magnitude of a velocity must be given in $\mathrm{ft\,s^{-1}}$.

2.5 MODELLING

The aim of engineering design and manufacture is to create a product that will meet the customer's needs as defined in the specification for the product. At the design stage (when the product does not exist and cannot be tested) the ability of the proposed design to meet the specification has to be predicted by analysing a model. This is not a scale physical model of the product, as the name might suggest (although such a model may be used in the design of a complex product) but an abstract model, on paper or in a computer, in which the product is represented by an assembly of elements having certain physical characteristics interconnected with one another in a certain way. The appropriate laws of physics are applied to the model to establish the governing mathematical equations, which are then manipulated and solved to give predictions of the performance of the product.

The model is always, to a greater or lesser extent, based on simplifying assumptions and approximations, and the aim is to use the simplest possible model that will give predictions at the required level of reliability.

Examples of mathematical modelling in mechanics are not given at this stage but are provided in the later chapters of this book.

3 Statics

3.1 INTRODUCTION

The part of mechanics called statics is concerned with bodies that are at rest. Many of the bodies we deal with in engineering, notably fixed structures and stationary bodies of liquid, are at rest and others, although moving, may be analysed as if they were at rest (see below). We need to be able to analyse the forces that act on and within a body in order to use another branch of engineering science (strength – or mechanics – of materials) to determine whether the body and its supports are strong enough to carry the forces satisfactorily (that is to carry them without breaking or deflecting to an unacceptable extent). If the body is at rest we assume that every one of the particles making up the body must also be at rest. Now we know from Newton's first law of motion (see Chapter 2) that if a particle is at rest it has no unbalanced force acting on it: the forces on the particle are said to be in equilibrium. In statics this application of Newton's first law of motion (used in conjunction with the third law of motion) is extended to give the conditions for the equilibrium of the whole body and parts of it, enabling the forces that act on and within the body to be determined.

Newton's first law of motion is not confined to a particle at rest but applies more generally to a particle that has a constant velocity in a straight line (of which rest is a special case with the velocity equal to zero). As a consequence the analysis developed in statics may be applied to bodies that have constant velocity (strictly as measured by an observer fixed in a Newtonian frame of reference, but for most engineering purposes as measured by an observer fixed to the earth; see Chapter 2). In practice the analysis is used without introducing significant error when the accelerations of a body are sufficiently small for the forces associated with the accelerations (see Chapter 6) to be negligible compared with the forces acting on the body.

The main assumptions made in developing the analysis are:

(a) The force between two bodies that are in contact acts over an area that is sufficiently small for it to be regarded as a point.
(b) The deformations caused by the forces acting on a body are small compared with the dimensions of the body, so that changes in the geometry of the body, and hence the geometry of the force system acting on the body, are negligible.
(c) Where friction is not negligible it may be modelled as dry (or Coulomb) friction (see Section 3.5).

The assumptions made in extending the analysis to hydrostatics (concerned with the forces within bodies of liquid at rest) are introduced in Section 3.4.

The work of this chapter is directly applicable to a wide range of engineering problems. At the same time it should develop confidence in dealing with force and moment of force and this will be valuable in Chapter 6 when we go on to deal with bodies undergoing accelerated motion.

3.2 FORCE SYSTEMS

We have already seen in Chapter 1 that force is a localized vector quantity in that it obeys the vector law of addition and is tied to a point, and thus has a particular line of action. The external effect of a force acting on a rigid body is independent of the point on the line of action at which it is applied. So, for example, in Fig. 3.1 the external forces at the supports of the rigid body at O and P are the same whether the force **F** is applied at point A, or point B, or any other point on the line of action AB. This result illustrates the principle of transmissibility, which enables a force to be treated as a *sliding* vector having magnitude, direction and line of action, so long as we are concerned with its *external* effect on a rigid body. The *internal* effect on the body, in terms of the forces and hence stresses within it, is critically influenced by the point of application of the force. This is illustrated in

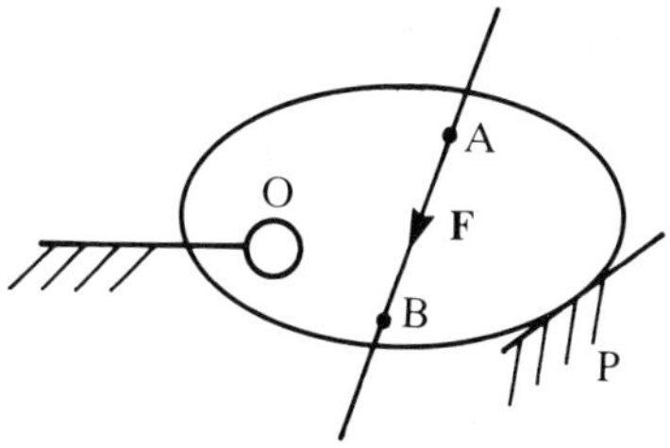

Fig. 3.1

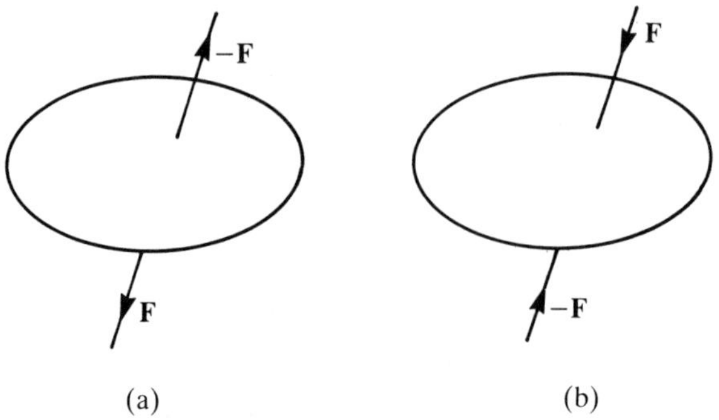

Fig. 3.2

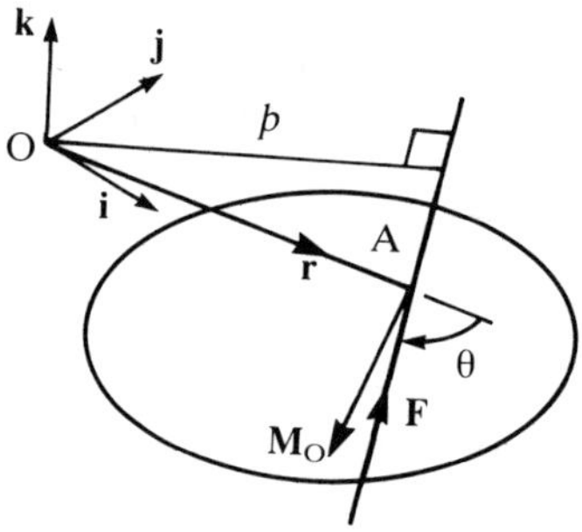

Fig. 3.3

Fig. 3.2(a) and (b), where forces that are externally equivalent give stresses of the opposite sign (tension in Fig. 3.2(a), compression in Fig. 3.2(b)) in the body on which they act. We will, apart from our brief consideration of liquids in Section 3.4, always be dealing with bodies assumed to be rigid and hence will make frequent use of the principle of transmissibility when analysing external force systems.

Moment of a force about a point O

In Fig. 3.3 the force $\mathbf{F}$ is acting at a point A of a body. If the position vector of A with respect to the point O is $\mathbf{r}$ then the moment of the force $\mathbf{F}$ about the point O is defined to be the vector $\mathbf{M}_O$, where

$$\mathbf{M}_O = \mathbf{r} \times \mathbf{F} \tag{3.1}$$

We observe, if B is any point on the line of action of $\mathbf{F}$, that

$$\underline{OB} \times \mathbf{F} = (\underline{OA} + \underline{AB}) \times \mathbf{F} = \mathbf{r} \times \mathbf{F} = \mathbf{M}_O$$

Thus the point A in the definition of $\mathbf{M}_O$ can be replaced by any point on the line of action of $\mathbf{F}$.

We note

$$|\mathbf{M}_O| = |\mathbf{r}| \times |\mathbf{F}| \sin\theta = p \times |\mathbf{F}|$$

where p (see Fig. 3.3) is the perpendicular distance from O to the line of action of $\mathbf{F}$.

If $\mathbf{r} = [x, y, z]$ and $\mathbf{F} = [X, Y, Z]$ referred to the unit vectors $\mathbf{i}, \mathbf{j}, \mathbf{k}$, then

$$\mathbf{M}_O = \begin{vmatrix} \mathbf{i} & \mathbf{j} & \mathbf{k} \\ x & y & z \\ X & Y & Z \end{vmatrix}$$
$$= [yZ - zY, zX - xZ, xY - yX] \tag{3.2}$$

The components of $\mathbf{M}_O$ are called the moments of the force $\mathbf{F}$ about the axes Ox, Oy, Oz.

When the line of action of $\mathbf{F}$ is located in the Ox, Oy plane we have $z = Z = 0$ and hence

$$\mathbf{M}_O = (xY - yX)\mathbf{k} \tag{3.3}$$

If the components of a force $\mathbf{F}$ and those of its moment $\mathbf{M}_O$ about O are given with respect to a set of unit vectors $\mathbf{i}, \mathbf{j}, \mathbf{k}$ (say), then the line of action of $\mathbf{F}$ is uniquely determined.

Recalling the solution of the vector equation $\mathbf{X} \times \mathbf{A} = \mathbf{B}$ contained in Example 1.13, namely

$$\mathbf{X} = \alpha\mathbf{A} + \frac{\mathbf{A} \times \mathbf{B}}{\mathbf{A}^2}$$

where α is arbitrary, we deduce that $\mathbf{r}$, which satisfies the equation $\mathbf{M}_O = \mathbf{r} \times \mathbf{F}$, is given by

$$\mathbf{r} = \frac{\mathbf{F} \times \mathbf{M}_O}{\mathbf{F}^2} + \alpha\mathbf{F} \tag{3.4}$$

This is the equation (see (1.22)) of a straight line that passes through the point whose position vector with respect to O is $(\mathbf{F} \times \mathbf{M}_O)/\mathbf{F}^2$ and is in the direction of the vector $\mathbf{F}$.

It is evident from the definition of the moment of a force about a point O that the sum of the moments about O of a number of forces acting at a (common) point is equal to the moment of their resultant about O.

Resultant of two parallel forces

It was stated in Chapter 1 that the resultant of forces acting at a (common) point is given by the parallelogram of forces and that this law has been

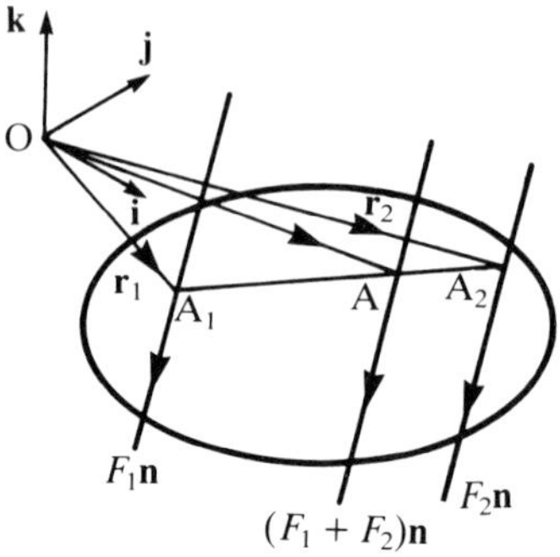

Fig. 3.4

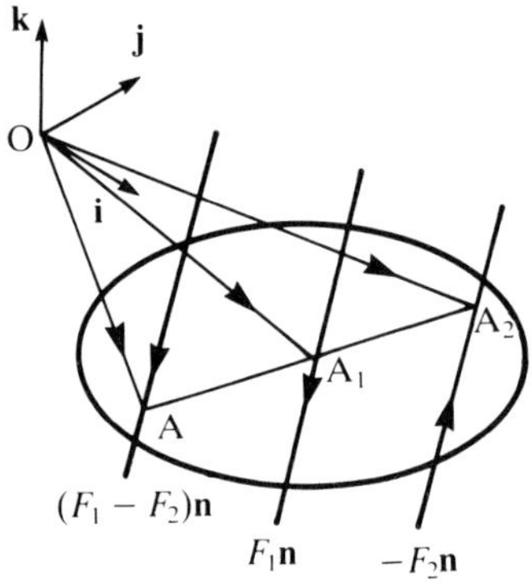

Fig. 3.5

verified by numerous experiments. Likewise it is possible to verify by means of elementary experiments (not described here) the truth of the law concerning the resultant of two parallel forces acting at different points. The law is in two parts:

(a) Like parallel forces (see Fig. 3.4): if $F_1, F_2 > 0$ and $\mathbf{n}$ is a unit vector, then the resultant of forces $F_1\mathbf{n}$ acting at A_1 and $F_2\mathbf{n}$ acting at A_2 is the single force $(F_1 + F_2)\mathbf{n}$ acting at the interior point A of the line A_1A_2, where

$$F_1|\underline{A_1A}| = F_2|\underline{A_2A}| \tag{3.5}$$

(b) Unlike parallel forces (see Figs 3.5 and 3.6): if $F_1, F_2 > 0$ and $\mathbf{n}$ is a unit vector, then the resultant of forces $F_1\mathbf{n}$ acting at A_1 and $-F_2\mathbf{n}$ acting at A_2 is the single force $(F_1 - F_2)\mathbf{n}$ acting at the exterior point A of the line A_1A_2, where

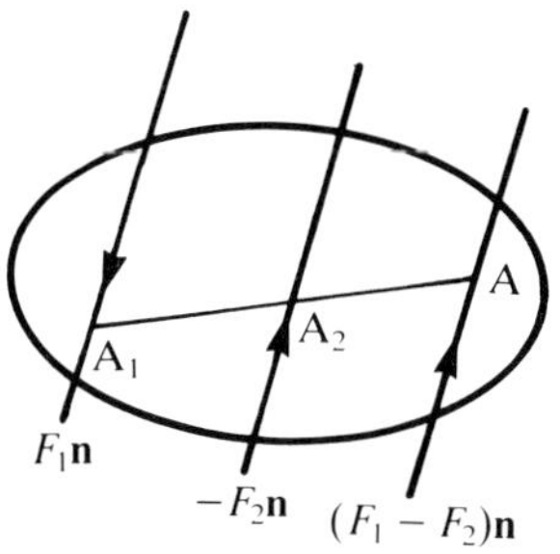

Fig. 3.6

$$F_1|\underline{A_1A}| = F_2|\underline{A_2A}| \tag{3.6}$$

If $F_1 > F_2$ (see Fig. 3.5) then A lies on A_2A_1 produced, but if $F_2 > F_1$ (see Fig. 3.6) then A lies on A_1A_2 produced. When $F_1 = F_2 = F$ there is no *exterior* point of A_1A_2 with the property (3.6) and a single resultant cannot be found. The pair of forces $F\mathbf{n}$ acting at A_1 and $-F\mathbf{n}$ at A_2 constitute an entity called a *couple*.

The truth of the statements in (a) and (b) can be demonstrated by invoking the device (see Problem 3.2) of introducing equal and opposite forces along the line A_1A_2 and then applying the parallelogram of forces.

Before considering the properties of couples we establish the following result for a pair of parallel forces that do not constitute a couple:

The sum of the moments of two parallel forces about any point O is equal to the moment of their resultant about O. (3.7)

The sum of the moments of the like parallel forces $F_1\mathbf{n}$ at A_1 and $F_2\mathbf{n}$ at A_2 is equal to

$$\begin{aligned}\underline{OA_1} \times F_1\mathbf{n} + \underline{OA_2} \times F_2\mathbf{n} &= (\underline{OA} + \underline{AA_1}) \times F_1\mathbf{n} + (\underline{OA} + \underline{AA_2}) \times F_2\mathbf{n} \\ &= \underline{OA} \times (F_1 + F_2)\mathbf{n} + (-F_1\underline{A_1A} + F_2\underline{AA_2}) \times \mathbf{n}\end{aligned}$$

The vectors $\underline{A_1A}$ and $\underline{AA_2}$ have the same sense and direction and therefore by virtue of (3.5) we have $-F_1\underline{A_1A} + F_2\underline{AA_2} = \mathbf{0}$. Thus the result holds for like parallel forces. A similar proof shows that (3.7) also holds for unlike parallel forces.

Properties of couples

The force pair that forms a couple has a number of properties. Firstly we show that the sum of the moments of the forces about any point O does not depend on O.

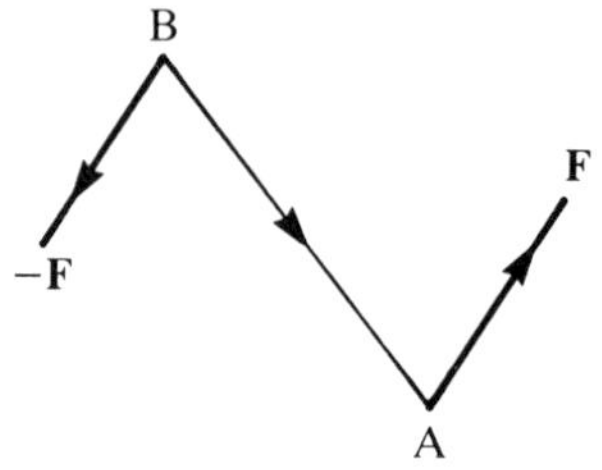

Fig. 3.7

For the forces **F** and −**F** acting at the points A, B (see Fig. 3.7) the sum of their moments about O is equal to

$$\underline{OA} \times \mathbf{F} + \underline{OB} \times -\mathbf{F} = (\underline{OB} + \underline{BA} - \underline{OB}) \times \mathbf{F} = \underline{BA} \times \mathbf{F}$$

The vector product $\underline{BA} \times \mathbf{F}$, which is clearly independent of O, is called the (vector) *moment* of the couple. We observe that it is equal to the moment of **F** at A about B and is *perpendicular* to the plane that contains the forces of the couple. The vector $\underline{BA}$ is called the *arm* of the couple.

The second property of couples that we establish relates to their moments. Any two couples may be combined to form a single couple whose moment is equal to the sum of the moments of the two couples. As a first step in the derivation of this property we consider couples lying in the same plane. In Fig. 3.8(a) the four forces $\mathbf{F}_1$, $-\mathbf{F}_1$, $\mathbf{F}_2$, $-\mathbf{F}_2$ are all parallel, whereas in Fig. 3.8(b) their lines of action intersect. The resultant of forces $\mathbf{F}_1$, $\mathbf{F}_2$ in Fig. 3.8(a) is the vector $\mathbf{F}_1 + \mathbf{F}_2$ acting at A, while the resultant of forces $-\mathbf{F}_1$, $-\mathbf{F}_2$ is $-(\mathbf{F}_1 + \mathbf{F}_2)$ acting at B. Thus the two couples combine to form a single couple. By taking moments about any point O and invoking (3.7) we deduce that the moment of the resultant couple is equal to the sum of the moments of the two couples.

We can in a similar manner establish the property for the couples in Fig. 3.8(b).

It follows at once that if we combine two couples (lying in the same plane) whose moments are equal but of opposite sense, the resulting (or resultant) couple has zero moment. Recalling the expression $\underline{BA} \times \mathbf{F}$ for the moment of the couple in Fig. 3.7 we see that the forces of this resultant couple must oppose each other or be of zero magnitude, that is the two couples are in equilibrium. Hence the effect of one couple on a body may be counterbalanced by applying to the body a second (coplanar) couple whose moment is equal and opposite to that of the first. The particular choice of forces and arm that produces this equal and opposite moment is immaterial provided we are concerned only with the external effect of the couples on the body. For this reason we may regard two (coplanar) couples with the same moment as equivalent.

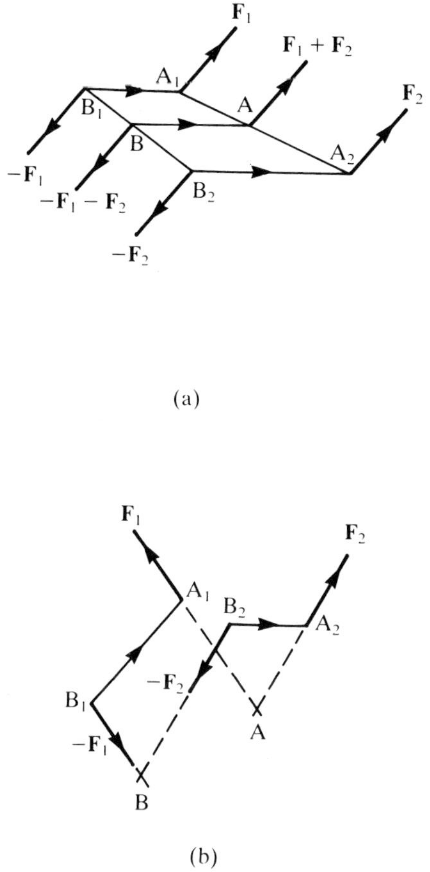

Fig. 3.8

We now extend our analysis to show how couples that lie in intersecting planes may be combined. Let the planes (labelled 1 and 2) of the couples intersect in the line CD, the moment of the couple lying in plane 1 being $\mathbf{M}_1$ and that of the couple in plane 2 being $\mathbf{M}_2$. Using the properties of couples just derived we see that the couple of moment $\mathbf{M}_1$ is equivalent to a force $F\mathbf{n}$ acting at the point A_1, where A_1 lies in CD and $\mathbf{n}$ is a unit vector in the direction of $\underline{CD}$, and a force $-F\mathbf{n}$ acting at the point B_1 whose position vector $\underline{A_1B_1}$ with respect to A_1 satisfies the equation $\mathbf{M}_1 = \underline{B_1A_1} \times (F\mathbf{n})$. Once the point A_1 and the value of F have been specified, $\underline{B_1A_1}$ can be found with the help of (3.4).

Likewise the couple of moment $\mathbf{M}_2$ is equivalent to the force pair $-F\mathbf{n}$ acting at B_2 and $F\mathbf{n}$ acting at A_2, where B_2 is taken to *coincide* with A_1 and A_2 is such that $\mathbf{M}_2 = \underline{B_2A_2} \times F\mathbf{n}$. The forces $F\mathbf{n}$ at A_1 and $-F\mathbf{n}$ at B_2 are equal and opposite so that the combined couples are equivalent to the force pair $F\mathbf{n}$ at A_2 and $-F\mathbf{n}$ at B_1. Since the moment of a couple is equal to the sum of the moments of the forces of it about any point we see, by taking moments about A_1, that the forces $F\mathbf{n}$ at A_2 and $-F\mathbf{n}$ at B_1 are equivalent to a couple of moment $\mathbf{M}_1 + \mathbf{M}_2$.

When the couples lie in parallel planes we proceed as follows. The couple of moment $\mathbf{M}_1$ is represented by a force pair $\mathbf{F}_1$, $-\mathbf{F}_1$, acting at the points A_1, B_1 (lying in the plane of the couple) chosen so that $\mathbf{M}_1 = \underline{B_1A_1} \times \mathbf{F}_1$. Likewise $\mathbf{M}_2$ is represented by the force pair $\mathbf{F}_2$, $-\mathbf{F}_2$ acting at A_2, B_2, where $\mathbf{F}_2$ is taken to be *parallel* to $\mathbf{F}_1$ and $\mathbf{M}_2 = \underline{B_2A_2} \times \mathbf{F}_2$. The forces $\mathbf{F}_1$ at A_1 and $\mathbf{F}_2$ at A_2 have a resultant $\mathbf{F}_1 + \mathbf{F}_2$ acting at the point A of the line A_1A_2 that can be determined using either (3.5) or (3.6) as appropriate. Similarly $-\mathbf{F}_1$ at B_1 and $-\mathbf{F}_2$ at B_2 have a resultant $-(\mathbf{F}_1 + \mathbf{F}_2)$ acting at the point B of the line B_1B_2, where AB is parallel to the planes of the couples. Thus the two couples combine to form a single (resultant) couple whose plane contains the line AB and is parallel to the planes of the couples. The moment of the couple is equal to the sum of the moments of the forces $\mathbf{F}_1 + \mathbf{F}_2$, $-(\mathbf{F}_1 + \mathbf{F}_2)$ acting at A, B (respectively) about any point. Recalling the result (3.7) we see that this sum is equal to $\mathbf{M}_1 + \mathbf{M}_2$. It is evident by taking appropriate (vector) values for $\mathbf{F}_1$, $\mathbf{F}_2$ that the resultant couple can be made to lie in any plane parallel to the planes of the given couples. We deduce by putting $\mathbf{M}_2 = 0$ that couples lying in parallel planes with the same moment are equivalent, so the earlier restriction that the couples must be coplanar to be equivalent is now removed.

Resultant of several like parallel forces

Let the forces be given by $\mathbf{F}_i = F_i\mathbf{n}$, where $\mathbf{n}$ is a unit vector, acting at the points A_i ($i = 1, 2, \ldots, N$). From (3.5) it is known that the resultant $(\mathbf{F}_1 + \mathbf{F}_2)$ of $\mathbf{F}_1$ acting at A_1 and $\mathbf{F}_2$ at A_2 acts at a point A that can be found. Similarly the point of application of the resultant $(\mathbf{F}_1 + \mathbf{F}_2 + \mathbf{F}_3)$ of $(\mathbf{F}_1 + \mathbf{F}_2)$ at A and $\mathbf{F}_3$ at A_3 can be found. It follows that the resultant of the N parallel forces is equal to the (force) vector $\Sigma_{i=1}^{N} \mathbf{F}_i$ acting at some point whose position can be determined.

In order to determine the position of the point we first need to extend to several parallel forces the result that the sum of the moments of two parallel forces about any point O is equal to the moment of their resultant about O. We know that the moment of $(\mathbf{F}_1 + \mathbf{F}_2)$ at A together with the moment of $\mathbf{F}_3$ at A_3 is equal to the moment of their resultant $\mathbf{F}_1 + \mathbf{F}_2 + \mathbf{F}_3$ about O. But the moment of $(\mathbf{F}_1 + \mathbf{F}_2)$ at A is equal to the sum of the moments of $\mathbf{F}_1$ at A_1 and $\mathbf{F}_2$ at A_2 about O. Hence we have established for three like parallel forces that the sum of their moments about O is equal to the moment of their resultant about O.

The extension to **N** like forces is immediate. If we denote the position vector of the point of application of the resultant of the N forces with respect to O by $\mathbf{r}$ then

$$\mathbf{r} \times \left(\sum_{i=1}^{N} \mathbf{F}_i \right) = \sum_{i=1}^{N} \mathbf{r}_i \times \mathbf{F}_i$$

where $\mathbf{r}_i = \underline{OA_i}$. Writing $\mathbf{R}$ for $\Sigma \mathbf{F}_i$ and $\mathbf{M}$ for $\Sigma \mathbf{r}_i \times \mathbf{F}_i$ yields the equation

$$\mathbf{r} \times \mathbf{R} = \mathbf{M} \tag{3.8}$$

so that, recalling (3.4), we have

$$\mathbf{r} = \frac{\mathbf{R} \times \mathbf{M}}{\mathbf{R}^2} + \alpha \mathbf{R}$$

where α is arbitrary.

Resultant of several parallel forces

Let the forces be parallel to the unit vector $\mathbf{n}$. We divide the forces into two groups, one group containing forces whose senses are the same as that of $\mathbf{n}$ and the second containing forces whose senses are opposite to that of $\mathbf{n}$. The forces of the first group are denoted by $\mathbf{F}_i$ $(i = 1, 2, \ldots, p)$ and those of the second by $\mathbf{F}_i$ $(i = p + 1, \ldots, N)$. The resultant of the first group is the force $\mathbf{R}_1$ where $\mathbf{R}_1 = \Sigma_{i=1}^{N} \mathbf{F}_i$, while the resultant of the second group is $\mathbf{R}_2$ where $\mathbf{R}_2 = \Sigma_{i=p+1}^{N} \mathbf{F}_i$. If $\mathbf{R}_1 + \mathbf{R}_2 = \mathbf{0}$ then the forces are equivalent to a couple, while if $\mathbf{R}_1 + \mathbf{R}_2 \neq \mathbf{0}$ then the forces have a resultant $\mathbf{R} = \mathbf{R}_1 + \mathbf{R}_2$ acting at the point whose position vector $\mathbf{r}$ with respect to a point O satisfies the vector equation (3.8).

Centre of gravity of a system of particles

The force exerted by the earth on a particle of mass m situated at the point P of its surface is $mg\mathbf{n}$, where $\mathbf{n}$ is a unit vector through P directed towards the earth's centre (see Chapter 2). It follows that the resultant gravitational force on N particles of masses m_s $(s = 1, 2, \ldots, N)$ whose mutual distances are small in comparison with the dimensions of the earth is $Mg\mathbf{n}$, where $M = \Sigma m_s$. This resultant acts at the point whose position vector $\mathbf{r}$ with respect to a point O satisfies (see (3.8)) the vector equation

$$\mathbf{r} \times Mg\mathbf{n} = \sum \mathbf{r}_s \times m_s g\mathbf{n}$$

where $\mathbf{r}_s$ is the position vector of the sth particle with respect to O. Hence

$$(M\mathbf{r} - \sum m_s \mathbf{r}_s) \times \mathbf{n} = \mathbf{0}$$

from which we deduce that

$$M\mathbf{r} = \sum m_s \mathbf{r}_s + \alpha \mathbf{n}$$

where α is arbitrary.

The point whose position vector with respect to O is equal to $(\Sigma m_s \mathbf{r}_s)/(\Sigma m_s)$ is called the *centre of gravity* of the particles. We take the resultant gravitational pull on the particles to act at this point, though we can if we wish assume it to act anywhere on a line that passes through the point and is in the direction of the unit vector $\mathbf{n}$.

Readers will see from the definition given in Section 6.4 that mathematically the centre of gravity and the centre of mass of a system of particles are the same point and therefore the former has all the mathematical properties of the latter. It should be appreciated, however, that the physical significance of one point is quite different from that of the other. Details of the positions of the centres of gravity of a variety of bodies are given in Appendix A.

The theory of statics considered so far enables us to combine forces acting at a point, to combine two couples lying in the same plane or in different planes, and also to determine the resultant of parallel forces. It remains to consider force systems in general and derive conditions of equilibrium. We do this by first showing that a system of forces is equivalent to a single force together with a couple.

Reduction of a system of forces to a single force acting at a point O and a couple

The force system consists of forces $\mathbf{F}_s$ acting at the points A_s $(s = 1, 2, \ldots, N)$ whose position vectors with respect to the point O are given by $\underline{OA_s} = \mathbf{r}_s$. We take it as axiomatic that the force $\mathbf{F}_s$ acting at A_s is equivalent to a force $\mathbf{F}_s$ whose line of action passes through the point O together with the couple of moment $\mathbf{r}_s \times \mathbf{F}_s$ that is formed by the force $\mathbf{F}_s$ acting at A_s and the force $-\mathbf{F}_s$ whose line of action passes through O (see Fig. 3.9). Hence the system of forces is equivalent to N forces $\mathbf{F}_s$ whose lines of action pass through O together with N couples whose moments are equal to $\mathbf{r}_s \times \mathbf{F}_s$ $(s = 1, 2, \ldots, N)$. The forces through O have a resultant equal to $\Sigma \mathbf{F}_s$, while the couples can be combined into a single couple whose moment is equal to $\Sigma \mathbf{r}_s \times \mathbf{F}_s$. Thus the system of forces is equivalent to a single force $\mathbf{R}$ whose line of action passes through O and a couple of moment $\mathbf{M}$, where

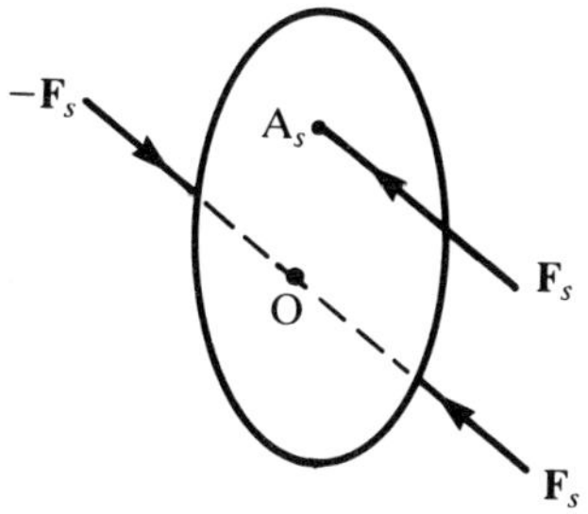

Fig. 3.9

$$\mathbf{R} = \sum \mathbf{F}_s, \qquad \mathbf{M} = \sum \mathbf{r}_s \times \mathbf{F}_s \tag{3.9}$$

We observe that $\mathbf{M}$ is equal to the sum of the moments about O of the forces $\mathbf{F}_s$ acting at the points A_s. Instead of reducing the system of forces to a single force $\mathbf{R}$ passing through the point O and a couple $\mathbf{M}$ we could have reduced it to a single force passing through the point O_1 and a couple. It follows from the analysis leading to (3.9) that the force will still be equal to $\mathbf{R}$ but the moment of the couple will, in general, be different. Denoting the couple by $\mathbf{M}_1$, we show that $\mathbf{M}$ and $\mathbf{M}_1$ are related by

$$\mathbf{M}_1 = \underline{O_1O} \times \mathbf{R} + \mathbf{M} \tag{3.10}$$

Using (3.9) it is seen that $\mathbf{M}_1$ is given by

$$\begin{aligned}\mathbf{M}_1 &= \sum \underline{O_1A_s} \times \mathbf{F}_s \\ &= \sum (\underline{O_1O} + \underline{OA_s}) \times \mathbf{F}_s \\ &= \underline{O_1O} \times \sum \mathbf{F}_s + \sum \underline{OA_s} \times \mathbf{F}_s \\ &= \underline{O_1O} \times \mathbf{R} + \mathbf{M}\end{aligned}$$

as required.

The relations (3.9) and (3.10) enable us to analyse force systems in some detail:

(a) $\mathbf{R} \neq \mathbf{0}$, $\mathbf{M} \neq \mathbf{0}$
If $\mathbf{R} \,.\, \mathbf{M} = 0$ the vector equation $\underline{O_1O} \times \mathbf{R} + \mathbf{M} = \mathbf{0}$, where $\underline{O_1O}$ is the unknown, has a solution. Hence by virtue of (3.10) the system of forces is equivalent to a single force acting at O_1.
If $\mathbf{R} \,.\, \mathbf{M} \neq 0$ then the system is equivalent to a force and couple.

(b) $\mathbf{R} \neq \mathbf{0}$, $\mathbf{M} = \mathbf{0}$
The system of forces reduces to a single force acting at the point O.

(c) $\mathbf{R} = \mathbf{0}$, $\mathbf{M} \neq \mathbf{0}$
The system of forces is equivalent to a couple. From (3.10) we deduce,

whatever the position of O_1, that $\mathbf{M}_1 = \mathbf{M}$, and hence the moment of the couple is equal to the sum of the moments of the forces of the system about any point.

(d) $\mathbf{R} = \mathbf{0}$, $\mathbf{M} = \mathbf{0}$
The system of forces is in equilibrium. It is evident from (3.10) if $\mathbf{R} = \mathbf{M} = \mathbf{0}$ that $\mathbf{M}_1 = \mathbf{0}$ whatever the position of O_1.

Example 3.1

The sums $\mathbf{M}$, $\mathbf{M}_1$, $\mathbf{M}_2$ of the moments of a system of forces about three non-collinear points O, O_1, O_2 are equal. Show that the system of forces is equivalent to a couple.

▷ *Solution* It follows from the relation (3.10) that

$$\mathbf{M}_1 = \underline{O_1O} \times \mathbf{R} + \mathbf{M}$$

$$\mathbf{M}_2 = \underline{O_2O} \times \mathbf{R} + \mathbf{M}$$

Thus $\underline{O_1O} \times \mathbf{R} = \underline{O_2O} \times \mathbf{R} = \mathbf{0}$. Since O, O_1, O_2 are non-collinear the vector $\mathbf{R}$ cannot be parallel to both $\underline{O_1O}$ and $\underline{O_2O}$ and must therefore be equal to the zero vector. Thus the system of forces reduces to a couple.

We note that if $\mathbf{M} = \mathbf{M}_1 = \mathbf{M}_2 = \mathbf{0}$ the system of forces is in equilibrium.

Example 3.2

Determine which of the following systems of forces reduces to a single force. All vectors are referred to the unit vectors **i**, **j**, **k** with O as origin.

(a) $\mathbf{F}_1 = [1, 2, 3]$, $\mathbf{F}_2 = [3, 4, 6]$, $\mathbf{F}_3 = [-1, 0, 2]$,
$\mathbf{r}_1 = [1, 0, 1]$, $\mathbf{r}_2 = [0, 0, 1]$, $\mathbf{r}_3 = [0, 1, 0]$.

(b) $\mathbf{F}_1 = [1, 2, 3]$, $\mathbf{F}_2 = [2, -4, -5]$, $\mathbf{F}_3 = [-2, 3, 3]$,
$\mathbf{r}_1 = [0, 1, -1]$, $\mathbf{r}_2 = [1, 0, 1]$, $\mathbf{r}_3 = [1, -1, 1]$.

Where the system reduces to a single force, find the position vector with respect to O of a point on the line of action of the force.

▷ *Solution*

(a) Recalling (3.9) we see that

$$\mathbf{R} = [1, 2, 3] + [3, 4, 6] + [-1, 0, 2] = [3, 6, 11]$$

while

$$\mathbf{M} = \begin{vmatrix} \mathbf{i} & \mathbf{j} & \mathbf{k} \\ 1 & 0 & 1 \\ 1 & 2 & 3 \end{vmatrix} + \begin{vmatrix} \mathbf{i} & \mathbf{j} & \mathbf{k} \\ 0 & 0 & 1 \\ 3 & 4 & 6 \end{vmatrix} + \begin{vmatrix} \mathbf{i} & \mathbf{j} & \mathbf{k} \\ 0 & 1 & 0 \\ -1 & 0 & 2 \end{vmatrix}$$

$$= [-2, -2, 2] + [-4, 3, 0] + [2, 0, 1] = [-4, 1, 3]$$

Thus

$$\mathbf{R} \cdot \mathbf{M} = -12 + 6 + 33 = 27$$

Hence the system of forces does not reduce to a single force.
(b) Here

$$\mathbf{R} = [1, 2, 3] + [2, -4, -5] + [-2, 3, 3] = [1, 1, 1]$$

while

$$\mathbf{M} = [5, -1, -1] + [4, 7, -4] + [-6, -5, 1] = [3, 1, -4]$$

Thus

$$\mathbf{R} \cdot \mathbf{M} = 3 + 1 - 4 = 0$$

Hence the system of forces does reduce to a single force.

If O_1 is the point of application of the single force then $\underline{O_1O}$ satisfies the vector equation (see 3.10) $\underline{OO_1} \times \mathbf{R} = \mathbf{M}$. Recalling (3.4) we see that $\underline{OO_1}$ can be taken equal to $(\mathbf{R} \times \mathbf{M})/\mathbf{R}^2$. Now

$$\mathbf{R} \times \mathbf{M} = \begin{vmatrix} \mathbf{i} & \mathbf{j} & \mathbf{k} \\ 1 & 1 & 1 \\ 3 & 1 & -4 \end{vmatrix} = [-5, 7, -2]$$

so that $3\,\underline{OO_1} = [-5, 7, -2]$.

Example 3.3

The force $\mathbf{F}_4$ is added to the system of forces defined in (a) of Example 3.2. Determine the line of action of $\mathbf{F}_4$ if the four forces reduce to a couple of moment [1, 4, 0].

Solution Let the position vector, with respect to O, of the point of application of $\mathbf{F}_4$ be $\mathbf{r}_4$. We require, using the conditions of (c), that

$$\mathbf{F}_4 = -[3, 6, 11], \qquad \mathbf{r}_4 \times \mathbf{F}_4 = -[-4, 1, 3] + [1, 4, 0] = [5, 3, -3]$$

Hence, recalling (3.4), we deduce that

$$166\mathbf{r}_4 = [51, -64, 21] + \alpha[3, 6, 11]$$

and thus the line of action of the force $\mathbf{F}_4$ is known.

Example 3.4

The density of a rod of length l and mass M at a point P of itself is proportional to the distance x of the point from the end O of the rod. Find the position of G, the centre of gravity of the rod.

▷ *Solution* We take the origin of coordinates to be at the end O of the rod and the axis of **x** to lie along the rod. The mass of a length δx of the rod whose distance from O is x is equal to $\alpha x\,\delta x$, where α is given by

$$\int_0^l \alpha x\,\mathrm{d}x = M \qquad \text{or} \qquad \alpha l^2 = 2M$$

The position of G, the centre of gravity of a body, is given by $M\underline{\mathrm{OG}} = \Sigma m_s \mathbf{r}_s$. Therefore, since the elemental mass $\alpha x\,\delta x$ of the rod has a position vector $x\mathbf{i}$,

$$M\underline{\mathrm{OG}} = \left(\int_0^l \alpha x^2\,\mathrm{d}x\right)\mathbf{i} = \frac{\alpha}{3}l^3\mathbf{i}$$

Hence $\underline{\mathrm{OG}} = (2/3)l\mathbf{i}$.

Two-dimensional force systems

Frequently in engineering practice we are dealing with a system of forces that lies in one plane. If we take this plane to be the Ox, Oy plane then the forces $\mathbf{F}_s$ and the position vectors $\mathbf{r}_s$ of their points of application with respect to O can be expressed in the form

$$\mathbf{F}_s = X_s\mathbf{i} + Y_s\mathbf{j}, \qquad \mathbf{r}_s = x_s\mathbf{i} + y_s\mathbf{j}$$

and hence

$$\mathbf{R} = \sum \mathbf{F}_s = (\sum X_s)\mathbf{i} + (\sum Y_s)\mathbf{j}$$
$$\mathbf{M} = \sum \mathbf{r}_s \times \mathbf{F}_s = \sum (x_s Y_s - y_s X_s)\mathbf{k}$$

If the system of forces is in equilibrium then $\mathbf{R} = \mathbf{M} = \mathbf{0}$ and therefore the components of the forces and the position vectors must satisfy three scalar relations called the conditions of equilibrium, namely

$$\sum X_s = \sum Y_s = \sum (x_s Y_s - y_s X_s) = 0 \qquad (3.11)$$

Example 3.5

Three coplanar forces are in equilibrium. Show that the lines of action of the forces must be either parallel or concurrent.

▷ *Solution* If two of the forces $\mathbf{F}_1$, $\mathbf{F}_2$ (say) are parallel then their resultant $\mathbf{F}_1 + \mathbf{F}_2$ is parallel to both of them. Now for equilibrium $\mathbf{R} = \mathbf{F}_1 + \mathbf{F}_2 + \mathbf{F}_3 = \mathbf{0}$ and therefore $\mathbf{F}_3$ is parallel to $\mathbf{F}_1 + \mathbf{F}_2$ and hence to $\mathbf{F}_1$ and $\mathbf{F}_2$.

If the two forces $\mathbf{F}_1$, $\mathbf{F}_2$ are not parallel we take as origin the point of intersection of their lines of action. Putting $x_1 = x_2 = y_1 = y_2 = 0$ we deduce, as $\mathbf{M} = \mathbf{0}$, that

$$x_3 Y_3 - y_3 X_3 = 0$$

If both of the coordinates x_3, y_3 are zero then $\mathbf{F}_3$ acts at O and the lines of action of the three forces are concurrent.

When one of the coordinates x_3 (say) is not zero we have $Y_3 = y_3 X_3 / x_3$, so that

$$\mathbf{F}_3 = X_3 \mathbf{i} + Y_3 \mathbf{j} = \frac{X_3}{x_3}(x_3 \mathbf{i} + y_3 \mathbf{j})$$

This implies that the force $\mathbf{F}_3$ is in the direction of $\mathbf{r}_3$ and therefore its line of action passes through O.

Free-body diagrams

So far we have been considering the equilibrium of the external force system acting on a body, where the external forces are reacted, as required by Newton's third law of motion, by the surroundings of the body. Within the body there will be internal forces exerted by one part of the body on another, which by Newton's third law of motion will be present in equal and opposite pairs. They will not, therefore, enter into the analysis of the external equilibrium of the body. Now if the body is in equilibrium we assume that any part of it, however large or small, must also be in equilibrium. When we wish to determine internal forces we exploit this assumption by isolating a part of the body and applying the conditions for equilibrium to the force system that acts on it. Some of the forces on the isolated part may be external to the body but in addition there will be forces exerted on the isolated part by the rest of the body. That is how we make the internal forces reveal themselves, as it were.

A drawing showing the forces acting on the isolated part of the body is called a free-body diagram. It is important that it should show all the forces that are *external to the isolated part of the body* and that Newton's third law of motion is applied in working from one isolated part of the body to another. Careful use of free-body diagrams is the only way, in our experience, of keeping proper control of the analysis. No example is given at this stage but extensive use is made of free-body diagrams in the examples of the next section.

3.3 APPLICATION OF STATICS

Beams

Beams are structural members that carry external forces (often called applied loads) over a gap between supports, predominantly by offering resistance to bending: they are normally at rest (relative to the earth). Typically they are long prismatic bars, and the lines of action of the external forces frequently lie in one plane passing through the axis of the bar, to give a two-dimensional force system. Analysis of the load-carrying capacity of a beam involves applying the principles of static equilibrium first to determine the actions on a given part of the beam, and then again, in combination with a knowledge of the strength characteristics of the material of the beam, to establish the internal resistance of the beam to the applied loads. The second part of the analysis is usually treated in the subject called mechanics (or strength) of materials and will not be our concern in this book.

The problem of determining the reactions (that is forces) at the points of support of a beam is said to be statically determinate if the reactions can be determined by the principles of statics. This means, as we have already seen in Section 3.2, that for a three-dimensional force system there are six scalar equations to be satisfied by the unknown forces at the supports, and for a two-dimensional system there are three equations to be satisfied. If the number of unknowns exceeds the number of equations then the forces at the supports can be determined only by combining the principles of static equilibrium with a knowledge of the force–deformation characteristics of the beam and the supports. The problem is said to be statically indeterminate and will not be dealt with in this book.

In Fig. 3.10(a) the magnitudes $|\mathbf{P}|$, $|\mathbf{Q}|$ of the vertical reactions provided by the two supports to the beam whose weight is $-W\mathbf{j}$ can be found by resolving vertically and taking moments, and therefore the problem is statically determinate. The quantities $|\mathbf{P}|$, $|\mathbf{Q}|$, $|\mathbf{R}|$ see Fig. 3.10(b) cannot be found in this way and therefore the problem is statically indeterminate.

Figure 3.11 shows the symbols for the three models commonly used to represent the supports of beams subjected to a two-dimensional system of forces (not shown), which is assumed to be in the plane of the page: the cross-hatching indicates a fixed body. The fixed support in Fig. 3.11(a) can react to forces in the x and y directions and to a couple whose moment is in

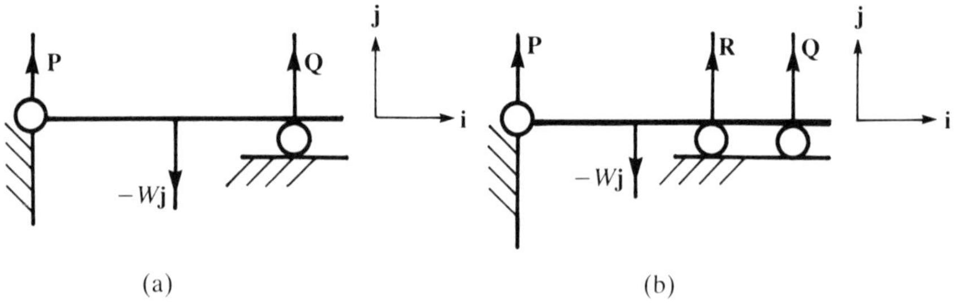

Fig. 3.10

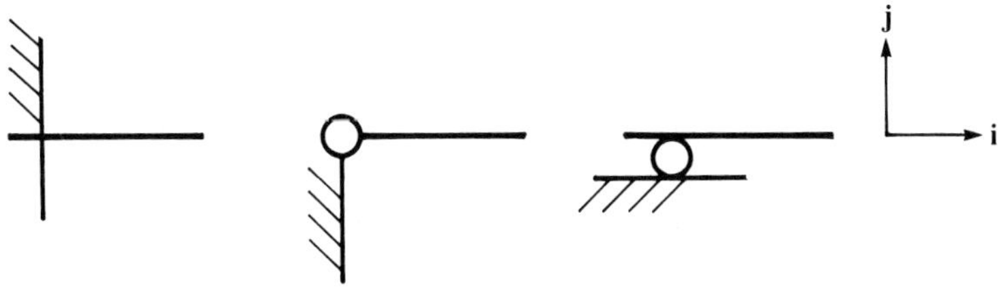

(a) (b) (c)

Fig. 3.11

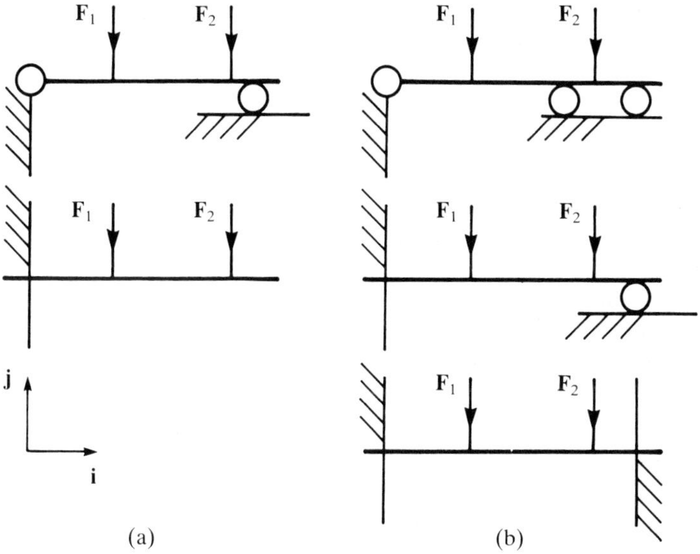

(a) (b)

Fig. 3.12

the z direction (perpendicular to the plane of the page); thus there are three unknowns associated with it. The simple support in Fig. 3.11(b) can react to forces in the x and y directions but not to a couple, and there are two unknowns associated with it. The roller support in Fig. 3.11(c) can react to a force in the y direction only (one unknown). We observe that the simple support corresponds to the revolute joint in planar mechanisms (see Chapter 4) and the roller support to a prismatic joint.

Figure 3.12(a) and (b) shows examples of force systems using the defined notation. The systems in Fig. 3.12(a) are statically determinate systems while those in Fig. 3.12(b) are statically indeterminate.

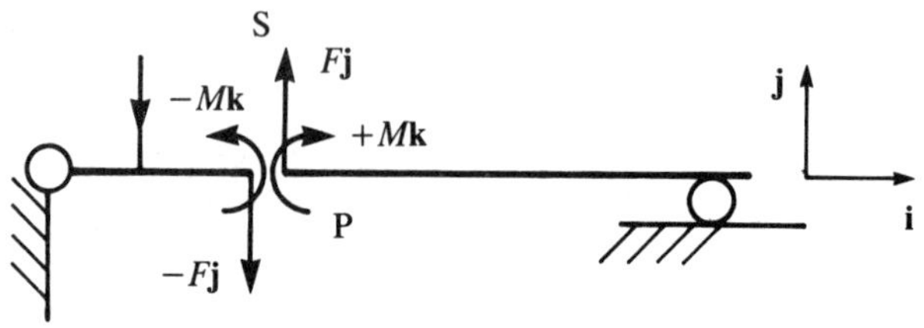

Fig. 3.13

Shear force and bending moment

We consider the equilibrium of a horizontal beam subject to vertical loads (see Fig. 3.13). If a vertical section S is drawn across the beam at P then it follows from the discussion earlier of free-body diagrams that the external forces acting on the part of the beam to the right of the section and the forces exerted on it (across the section S) by the part of the beam to the left of S must be in equilibrium. Since any system of forces is equivalent to a single force and a couple, the effect of the part of the beam to the left of the section on that to the right can be measured in terms of a single force **F** acting at the point P of S and a couple **M**. If we take the unit vector **i** to be along the beam and **j** vertically up we deduce, from the conditions of equilibrium and the fact that the external forces are vertical, that **F** must be of the form $F\mathbf{j}$ and **M** of the form $M\mathbf{k}$. The quantity **F** is called the shear force at P, while **M** is called the bending moment. The force $-F\mathbf{j}$ and the couple $-M\mathbf{k}$ represent the action of the part of the beam to the right of S on the part to the left and are in equilibrium with the external forces acting on the latter part (see Fig. 3.13).

We now show by means of two examples how the shear force and bending moment at a point of a beam can be determined. The solutions are based on the use of the scalar conditions of equilibrium given by (3.11). The condition $\Sigma X_s = 0$ does not directly enter into our considerations. However, we have used it earlier to infer that **F**, as the applied loads are vertical, must be of the form $F\mathbf{j}$.

Example 3.6

A uniform beam of length l and weight $-w\mathbf{j}$ per unit length is simply supported at its ends A and B. Find the shear force and bending moment at a point distance x from the end A.

▷ *Solution* By symmetry the reactions at A and B must be equal to $(w/2)l$ vertically upwards. The centre of gravity of the length AP of the beam,

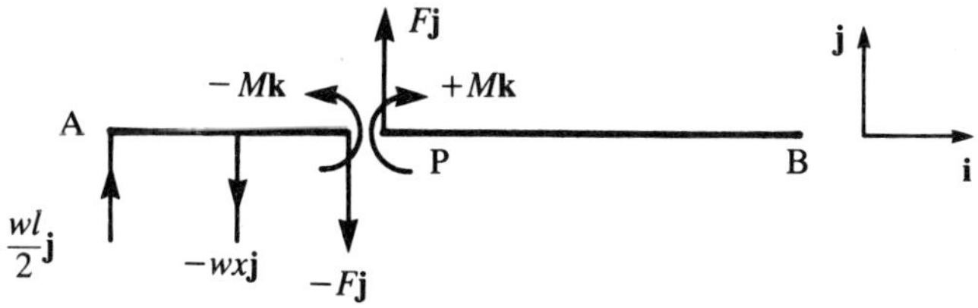

Fig. 3.14

where AP = x, is at the midpoint of AP. Hence we have, resolving vertically and using the notation of the free-body diagram in Fig. 3.14,

$$\frac{w}{2}l - wx - F = 0$$

so that

$$F = \frac{w}{2}(l - 2x)$$

Taking moments about A gives

$$\left(\frac{x}{2}\mathbf{i}\right) \times (-wx\mathbf{j}) - M\mathbf{k} + x\mathbf{i} \times (-F\mathbf{j}) = 0$$

$$\frac{w}{2}x^2 + Fx + M = 0$$

from which we deduce that

$$M = \frac{w}{2}x(x - l)$$

Figure 3.15(a) and (b) shows the variation of F and M with x. The plots are called the shear force and bending moment diagrams.

Example 3.7

A uniform beam of length l and weight $-wl\mathbf{j}$ is supported at two points equidistant from its ends. Find where these two points must be situated in order that the maximum magnitude of the bending moment be as small as possible.

Solution Let the supports be at a distance a from each end. Taking moments about P yields, with the help of the free-body diagram in Fig. 3.16(a),

$$-M + \frac{w}{2}x^2 = 0 \qquad \text{or} \qquad M = \frac{w}{2}x^2, \qquad 0 \leq x < a$$

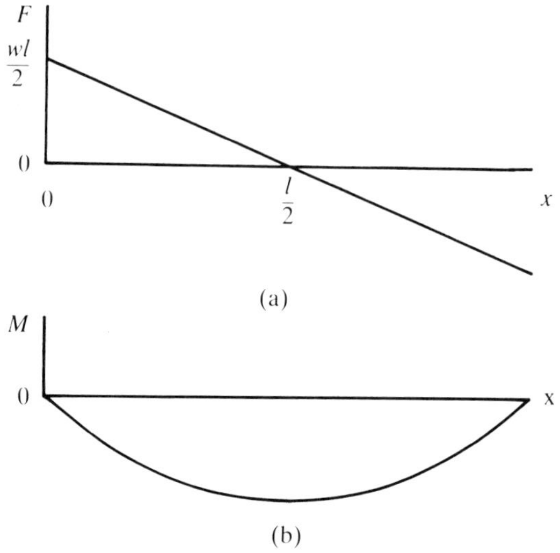

Fig. 3.15

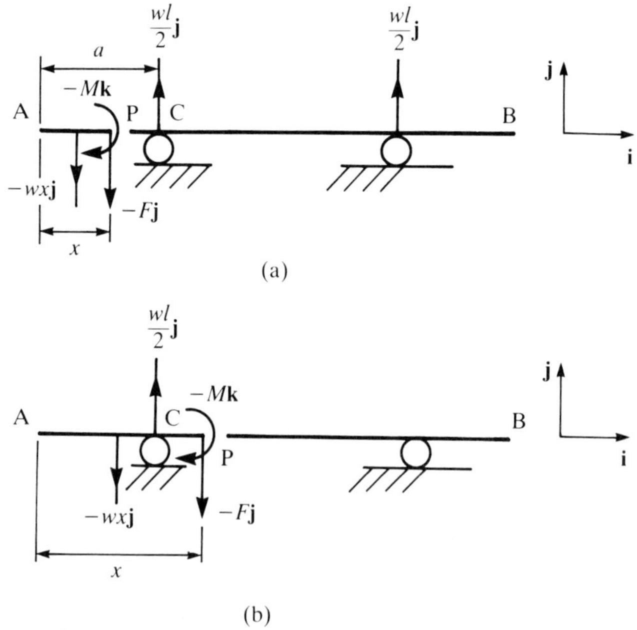

Fig. 3.16

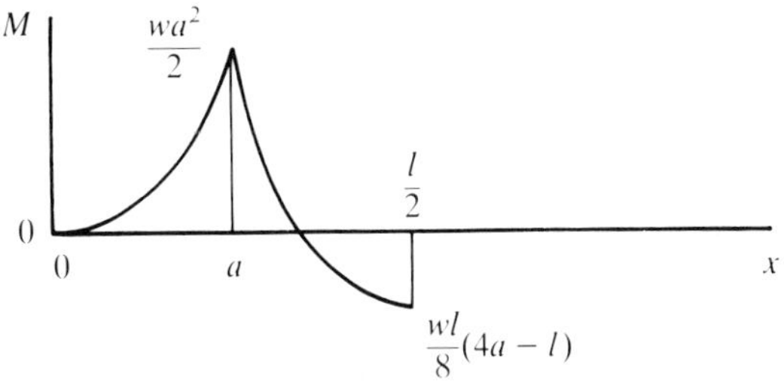

Fig. 3.17

From the free-body diagram in Fig. 3.16(b) we obtain the moment equation

$$-M + \frac{w}{2}x^2 - \frac{w}{2}l(x - a) = 0 \qquad \text{or}$$

$$M = \frac{w}{2}x^2 - \frac{w}{2}l(x - a), \qquad a < x < l - a$$

Figure 3.17 shows the bending moment diagram for $0 < x < l/2$. It is evident that the magnitude of the bending moment is greatest at $x = a$ where it has the value $(w/2)a^2$, or at $x = l/2$ where it has the value $|(wl/8)(4a - l)|$.

In order to choose the optimum value of a we sketch the graphs of the functions $(w/2)a^2$ and $|(wl/8)(4a - l)|$ for values of a between 0 and $l/2$. We see from Fig. 3.18 that the maximum magnitude of the bending moment is least for the value of a given by

$$\frac{w}{2}a^2 = \frac{w}{8}l(l - 4a) \qquad \text{or} \qquad 4a^2 + 4al - l^2 = 0$$

Hence $2a = (\surd 2 - 1)l$.

Frameworks

A framework is a structure that is made up of bars or members, each of which has two joints connecting it with other members. For the purposes of our analysis we assume, in a two-dimensional frame, that the connection between two members at a joint is effected by means of a small cylindrical pin fitting smoothly into holes drilled in the members. The assumption of perfect smoothness is an ideal one and in engineering practice is never fully realized. However, in the analysis that follows frictional forces at the joints are ignored, while the members of the frame are regarded as being rigid and sufficiently light that the effects of gravity may be ignored. The assumption of rigidity (see Section 3.1) means that members neither compress

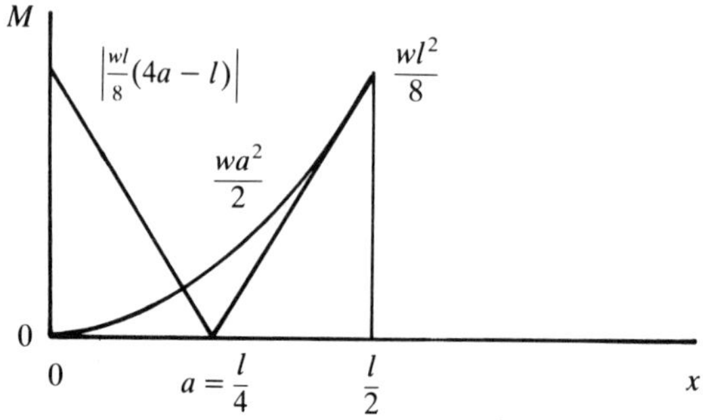

Fig. 3.18

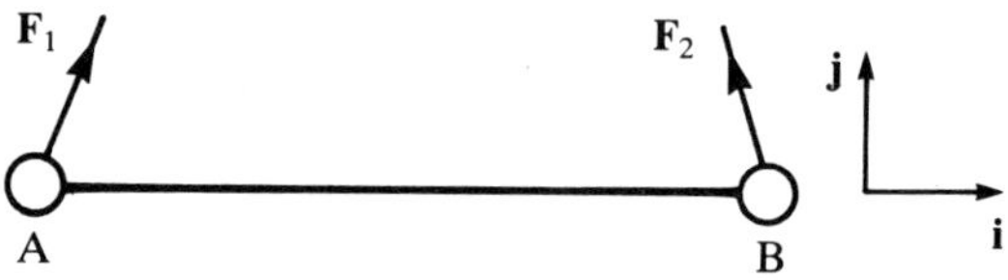

Fig. 3.19

nor extend when the framework is subject to a loading, so that its shape whether loaded or unloaded is fixed. In practice a framework deforms slightly when loaded but the extent of the deformation can only be determined (as we pointed out for beams) by taking into account the elastic properties of the material from which the framework is made. To the assumptions already made we add the further one that all external forces are applied to the pins and not to the members of the structure.

Since there are no frictional forces between a pin and a member, the forces of interaction between the two bodies must be normal to the surfaces in contact and therefore their lines of action must pass through the centre of the hole into which the pin fits. The forces thus have a single resultant, so that the action of the pins at the two joints of a single member AB is shown by the forces in the free-body diagram of Fig. 3.19. Since gravitational effects are ignored we deduce from the conditions of equilibrium that the lines of action of the forces are along the member and that these forces are equal and opposite. In Fig. 3.20(a) the forces exerted by the pins are tending to stretch the member. This tendency will be resisted by the member which is said to be in a state of tension. If the senses of all the forces are reversed, as in Fig. 3.20(b), the member is in compression. When we draw a free-body diagram to determine the magnitude of the thrust, that is of the

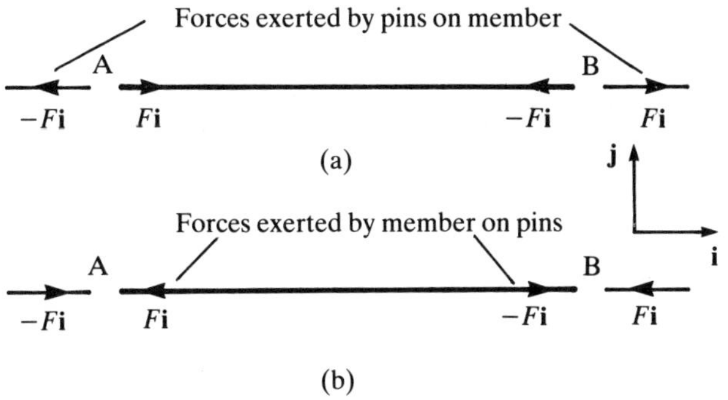

Fig. 3.20

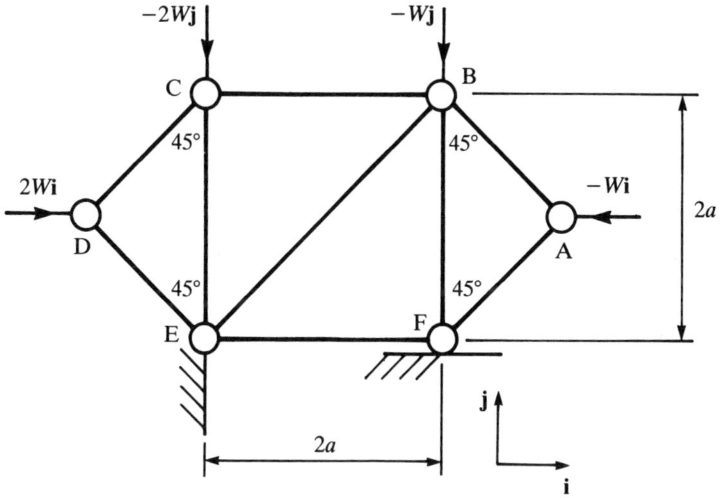

Fig. 3.21

tension or compression in a member, the forces shown at a joint are the forces exerted by the members on the pin together with any external forces. In short, we consider the equilibrium of the pin at a joint.

The procedure is illustrated by the example that now follows. Two solutions to the example are offered, their purpose being to highlight certain aspects of the theory of frameworks.

Example 3.8

Figure 3.21 represents a loaded framework of nine light smoothly jointed bars in a vertical plane; that is, the framework is smoothly pin-jointed to a fixed support at E and is supported by a smooth horizontal roller at F.

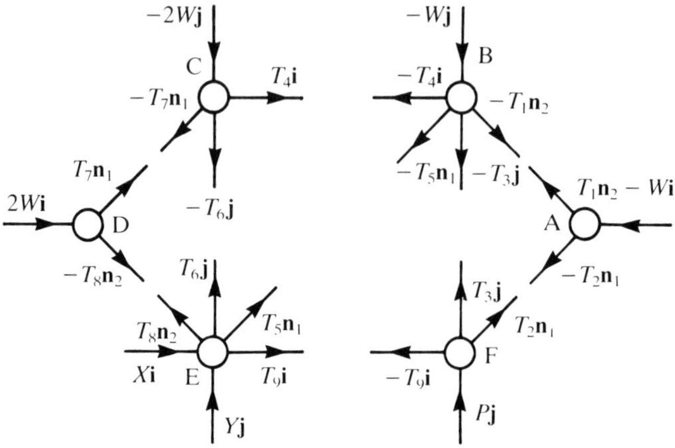

Fig. 3.22

The loads at B and C are vertical and those at A and D are horizontal, all loads acting in the plane of the framework. Calculate the forces in the bars, distinguishing between those bars that are in tension and those in compression.

▷ *Solution 1* Figure 3.22 shows the free-body diagrams for the forces in the bars at each joint in the framework. The effect of the support at E is represented by the force pair $X\mathbf{i}$, $Y\mathbf{j}$ and that of the support at F by the vertical force $P\mathbf{j}$. The reader should appreciate that certain assumptions are implicit in the figure. The forces in the members are unknown and hence we cannot strictly show the senses of the vectors that represent them. The senses shown apply when the members are in tension, that is the quantities T_i ($i = 1, 2, \ldots, 9$) are all positive. No difficulty arises in the analysis if we write down the single vector equation of equilibrium for the forces acting at a joint. Thus at A

$$-W\mathbf{i} - T_2\mathbf{n}_1 + T_1\mathbf{n}_2 = \mathbf{0}$$

where $\mathbf{n}_1$, $\mathbf{n}_2$ are unit vectors in the directions of $\underline{\mathrm{FA}}$ and $\underline{\mathrm{AB}}$. Expressing $\mathbf{n}_1$ and $\mathbf{n}_2$ in terms of $\mathbf{i}$ and $\mathbf{j}$ (that is resolving horizontally and vertically) yields the following two scalar equations and their force solutions:

At A: $\quad T_1 + T_2 + (\surd 2)W = 0, \qquad T_1 - T_2 = 0$

$$T_1 = T_2 = -W/\surd 2$$

Proceeding in a similar manner at other joints we obtain the following:

At D: $T_7 + T_8 + (2\sqrt{2})W = 0, \quad T_7 - T_8 = 0$
$T_7 = T_8 = -(\sqrt{2})W$

At C: $-(\sqrt{2})T_4 + T_7 = 0, \quad 2W + T_6 + T_7/\sqrt{2} = 0$
$T_4 = T_6 = -W$

At B: $-(2\sqrt{2})T_4 + T_1 - T_5 = 0, \quad W + T_3 + (T_1 + T_5)/\sqrt{2} = 0$
$T_3 = -W, \quad T_5 = W/\sqrt{2}$

At F: $(\sqrt{2})T_9 - T_2 = 0, \quad T_3 + P + T_2/\sqrt{2} = 0$
$T_9 = -W/2, \quad P = 3W/2$

At E: $X + T_9 - (T_8 - T_5)/\sqrt{2} = 0, \quad Y + T_6 + (T_8 + T_5)/\sqrt{2} = 0$
$X = -W, \quad Y = 3W/2$

All the bars of the framework are in compression except for the bar BE.

▷ *Solution 2* Since the framework is in equilibrium under the action of the applied loads at the points A, B, C, D, of the force pair X, Y at E, and of the single force P at F we can, by using the three conditions of equilibrium (see (3.11)) obtain the values of X, Y, P and then proceed to find the forces in the bars.

Taking moments about the point E yields

$$P2a + Wa - 2Wa - 2Wa = 0$$

giving $P = 3W/2$. Resolving horizontally and vertically we see $X = -W$, $Y = 3W/2$. We can can now write down the twelve equations obtained by resolving horizontally and vertically at the six joints. However, as there are now only nine unknowns, three of the equations will be redundant. It is evident from solution 1 that the scalar equations of equilibrium at the joint E and the equation obtained by resolving vertically at F are redundant equations.

Statical determinacy in frameworks

The stresses in the framework we have just considered are statically determinate, the framework being described as just rigid. However, if we had added another bar to Fig. 3.21, by joining C and F say, then there would have been insufficient (statical) equations to determine all the unknowns, although it would still be possible to find some of them, for example T_1, T_2, T_7, T_8. The structure with the additional bar is described as being over-rigid. If there had been one less bar, say CE were removed, then the number of equations would exceed the number of unknowns and a solution may or may not exist. In Example 3.8 it is evident that if CE were removed the part EDCB of the structure would collapse.

The question to be resolved is: what is the number of bars permitted in a framework if the forces within them are all to be statically determinate? The solutions to Example 3.8 suggest a way in which we might arrive at an answer to the question for a framework consisting of n joints. Vertical and horizontal resolution at each joint yield a total of $2n$ equations for the thrusts in the bars *and any unknown* forces at the external points of support. The latter, which are assumed to be statically determinate, cannot be more than three in number as there are only three equations of overall equilibrium to satisfy. Thus if we first use, as in solution 2, these equations to find the unknown forces at the external points of support, we argue that three of the $2n$ equations determined by resolution will be redundant. That leaves only $2n - 3$ independent equations, and these permit us to find $2n - 3$ unknowns. It is therefore concluded that the structure should contain $2n - 3$ bars if all the unknown forces are to be statically determinate. The conclusion is somewhat glib, as the following examples show.

If in Example 3.8 the loads B and C are interchanged the reader will find, using either of the two solution methods (or the method of sections to be described shortly), that the force in CE is zero. Thus the bar CE could be removed and the structure still remain in equilibrium. The number of bars is now $2n - 4$ where $n = 6$.

The framework in Fig. 3.23(a) has six joints at A, B, C, D, E, F and nine bars (that is $2n - 3$ where $n = 6$), the joints being such that they form the vertices of a regular hexagon. The loads are all of magnitude W and act as shown along the lines of the diagonals. By symmetry the free-body diagram for the forces at a joint will take the form shown in Fig. 3.23(b). Resolving in the direction of a diagonal we deduce $T_1 + T_2 = W$ and this is the only equation available to us to determine T_1 and T_2. Thus frameworks containing n joints and $2n - 3$ bars can be statically indeterminate structures.

We hasten to reassure the reader that frameworks with n joints and $2n - 3$ bars will for the most part be statically determinate. He or she may be further reassured that any unusual property of a framework (see Fig. 3.23(b)) will always be revealed by a study of the $2n$ equations of equilibrium derived by resolving in two directions at each of the n joints.

Method of sections

The method of sections provides an alternative means of finding the forces in the bars of a framework. It is particularly useful when we require some but not all of the forces. In order to describe the method we return once more to Example 3.8.

Figure 3.24 shows the framework of Example 3.8 divided into two parts

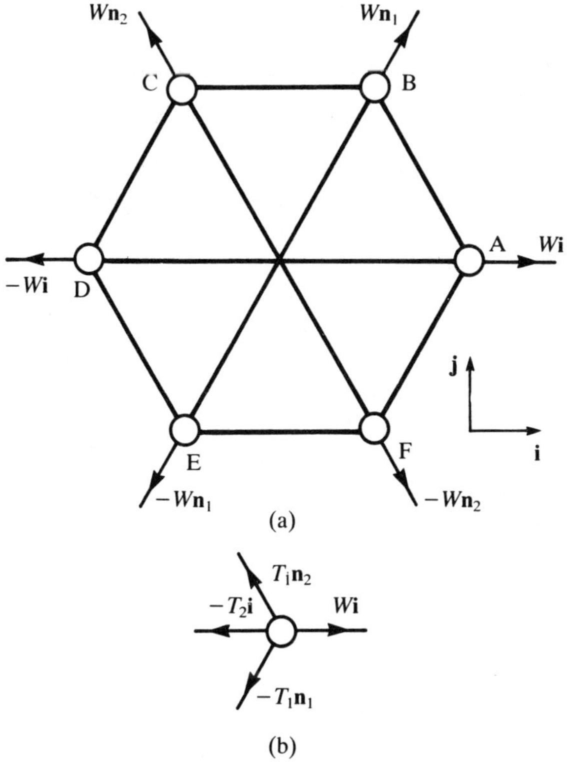

Fig. 3.23

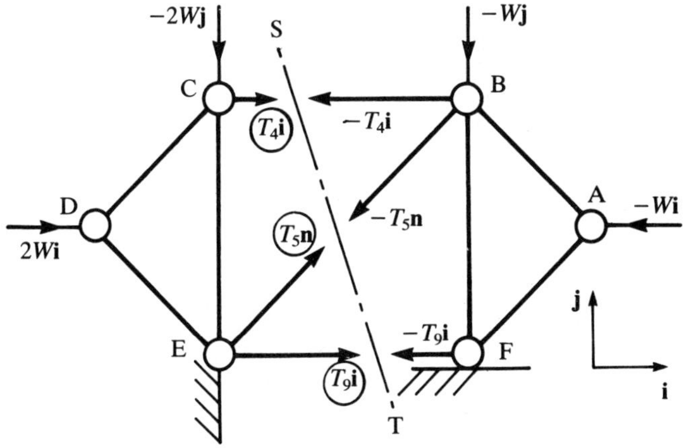

Fig. 3.24

by the broken line ST. To the left and right of this line we show equal and opposite pairs of forces $T_4\mathbf{i}$, $-T_4\mathbf{i}$, $T_5\mathbf{n}$, $-T_5\mathbf{n}$, $T_9\mathbf{i}$, $-T_9\mathbf{i}$. It is essential to understand what they represent. We begin by reminding the reader that in solutions 1 and 2 we took each member of the framework to be in tension, and if the value of the tension is found to be negative the member is in compression. We recall with the help of Fig. 3.20(a) that when a bar is in tension the force exerted by the pin is outwards from the line of the bar. Hence if we consider the equilibrium of the part of the bar CB to the left of the line ST we see that the force exerted by the pin at C on this part must be equal and opposite to the force exerted on it by the part of the bar CB to the *right* of ST. This (latter) force is represented by the ringed symbol $T_4\mathbf{i}$. It follows that the part of the framework to the left of ST is in equilibrium under the action of the ringed forces $T_4\mathbf{i}$, $T_5\mathbf{n}$, $T_9\mathbf{i}$, the loadings at C and D and the support at E. Thus if we have already determined (see solution 2 of Example 3.8) the force pair X, Y at E, we can apply the conditions of equilibrium to find T_4, T_5, T_9. Taking moments about the point E gives T_4 immediately; resolving vertically yields T_5; and finally resolving horizontally gives T_9.

3.4 HYDROSTATICS

Hydrostatics is the study of the conditions under which systems of forces maintain masses of liquid in a state of equilibrium, together with the determination of the reactions of liquids upon the bodies with which they are in contact.

Shear stress

Figure 3.25 shows a section of a material substance or body that is in equilibrium under the action of external forces. As we have argued else-

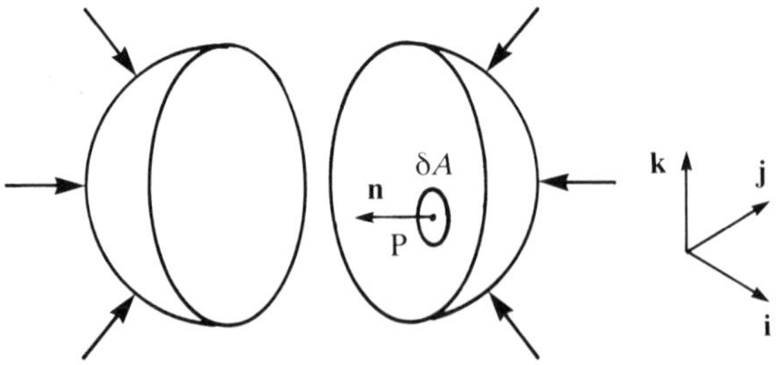

Fig. 3.25

where in this chapter, the part of the substance to the left of the section exerts forces, across the section, on the part to the right of it. These forces, which are reacted in an equal and opposite manner by the part of the body to the right of the section, are equivalent to a single force together with a couple.

Let us now consider an elemental area δA of the section containing the point P. The action of one part of the body on the other is transmitted in part across this elemental area. We assume that the action across the elemental area can be represented by a single force that depends on the dimensions of the elemental area, the position of the point P, and the direction of the unit vector $\mathbf{n}$ normal to the elemental area. The force is expressed as $\delta A \times \mathbf{R_n}(\mathrm{P})$, where the vector $\mathbf{R_n}(\mathrm{P})$, as the symbols indicate, depends on P and $\mathbf{n}$. The vector $\mathbf{R_n}(\mathrm{P})$ represents force per unit area and is called the stress vector at P for the elemental area. It can be resolved into two components, one normal to the area and the other lying in its plane. The former is called the normal stress and the latter the shear stress. We assume when a liquid is in equilibrium that the *shear stress is zero*. It follows therefore that if the material substance we are considering is a liquid then the force exerted across the elemental area by the part of the liquid to the *left* of the section on the part to the *right* can be expressed in the form $-p_{\mathbf{n}}\delta A\mathbf{n}$, where $p_{\mathbf{n}}$ is a scalar that depends on $\mathbf{n}$. The reader should (see Fig. 3.25) note the sense of the vector $\mathbf{n}$ in relation to the section and the two parts of the fluid. The quantity $p_{\mathbf{n}}$ is called the liquid pressure at P on the elemental area. It is, as we will see, shortly, a positive scalar. Before establishing this property it is necessary to derive another property concerning $p_{\mathbf{n}}$.

Pressure at a point

We now prove that the pressure at a point P of a liquid in equilibrium is the same for all elemental areas, that is it is independent of the unit vector $\mathbf{n}$.

We consider the equilibrium of a part of the liquid that is in the shape of the tetrahedron OABC (volume V) where the plane ABC is taken to be horizontal (Fig. 3.26). The resultant force exerted on the face OBC of the tetrahedron by the rest of the liquid is in the direction of the (inward) normal to the face (see Fig. 3.26). It can therefore be expressed in the form $(1/2)p_1\underline{\mathrm{OC}} \times \underline{\mathrm{OB}}$, where p_1 is some scalar. Similar expressions (containing scalars p_2, p_3, p_4) apply for the forces on the other faces. If ϱ is the liquid density and the unit vector $\mathbf{k}$ is vertically upwards, then using the condition $\mathbf{R} = \mathbf{0}$ (see (3.9)) of equilibrium we have

$$p_1\underline{\mathrm{OC}} \times \underline{\mathrm{OB}} + p_2\underline{\mathrm{OA}} \times \underline{\mathrm{OC}} + p_3\underline{\mathrm{OB}} \times \underline{\mathrm{OA}} + p_4\underline{\mathrm{AB}} \times \underline{\mathrm{AC}} - 2\varrho gV\mathbf{k} = \mathbf{0} \qquad (3.12)$$

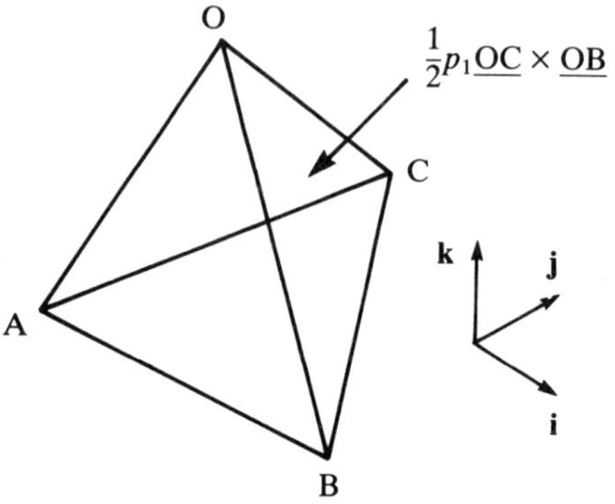

Fig. 3.26

Now $V = (1/3)h(\text{area ABC})$, where h is the length of the perpendicular from the vertex O to the plane ABC. Hence

$$2\varrho g V\mathbf{k} = (1/3)\varrho g h\,\underline{AB} \times \underline{AC}$$

But $\underline{AB} \times \underline{AC} = (\underline{AO} + \underline{OB}) \times (\underline{AO} + \underline{OC})$, so that after some manipulation the relation (3.12) can be expressed in the form

$$\{p_1 - p_4 - (1/3)\varrho g h\}\underline{OC} \times \underline{OB} + \{p_2 - p_4 - (1/3)\varrho g h\}\underline{OA} \times \underline{OC} + \{p_3 - p_4 - (1/3)\varrho g h\}\underline{OB} \times \underline{OA} = \mathbf{0} \qquad (3.13)$$

The directions of the vectors $\underline{OA}$, $\underline{OB}$, $\underline{OC}$ are non-coplanar and so therefore are those of $\underline{OC} \times \underline{OB}$, $\underline{OA} \times \underline{OC}$, $\underline{OB} \times \underline{OA}$. Thus for any value of h the coefficients of these vectors in (3.13) must vanish. Further if we let h tend to zero the quantities p_1, p_2, p_3, p_4 become the pressures at O with respect to the elemental faces of the tetrahedron. In the limit $p_1 = p_2 = p_3 = p_4$, so the pressure at P is the same in all directions. Since the pressure no longer depends on $\mathbf{n}$ we shall, henceforth, denote it simply by p or by p with a suffix as in p_A, p_B, ... the latter notation being used when we are considering the pressure at a number of points A, B, ...

It remains for us to determine an expression for the pressure at a point in a given liquid. As a preliminary to deriving this expression we show that the pressure is the same at all the points of a liquid that lie in a given horizontal plane.

Figure 3.27(a) shows an elemental volume (within a liquid) that has the shape of a right circular cylinder, the generators of the cylinder being horizontal. This elemental volume of liquid is in equilibrium under the action of the other parts of the liquid on it and gravity. Resolving (see Fig. 3.27(a)) in the direction perpendicular to the end faces of the cylinder yields the relation

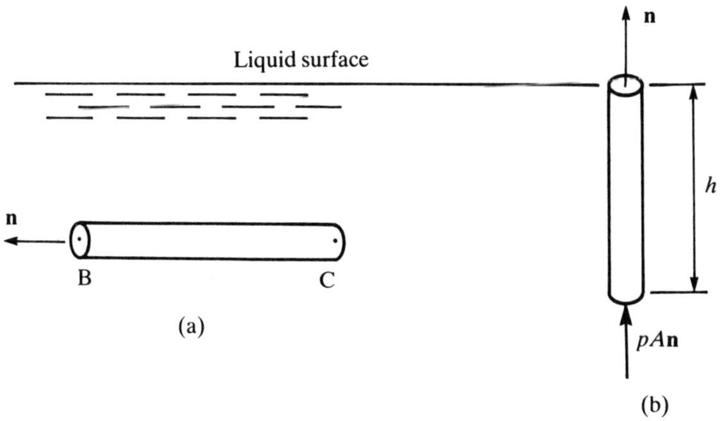

Fig. 3.27

$$-p_B \delta A\mathbf{n} + p_C \delta A\mathbf{n} = \mathbf{0}$$

where δA is the area of an end face. Thus $p_B = p_C$ and the result follows.

To find an expression for the pressure we consider a second cylindrical volume of liquid (see Fig. 3.27(b)), the generators of the cylinder now being vertical and one of the end faces of the cylinder lying in the (free) surface of the liquid. If this surface is subject to atmospheric pressure, denoted by p_a, and h is the height of the cylinder, then resolving vertically leads to

$$-p_a \delta A\mathbf{n} - \varrho g h \delta A\mathbf{n} + p\delta A\mathbf{n} = \mathbf{0}$$

Hence

$$p = p_a + \varrho g h \tag{3.14}$$

Centre of pressure

In order to define the term 'centre of pressure' we consider the action of a liquid on a flat plate immersed in it. The forces exerted by the liquid on the elemental areas that form a surface of the plate are all normal to the plate and in the same sense and therefore constitute a system of like parallel forces. These parallel forces have a single resultant whose line of action can be determined by means of the theory developed earlier in the chapter. The point where this line of action meets the surface of the plate is called the centre of pressure.

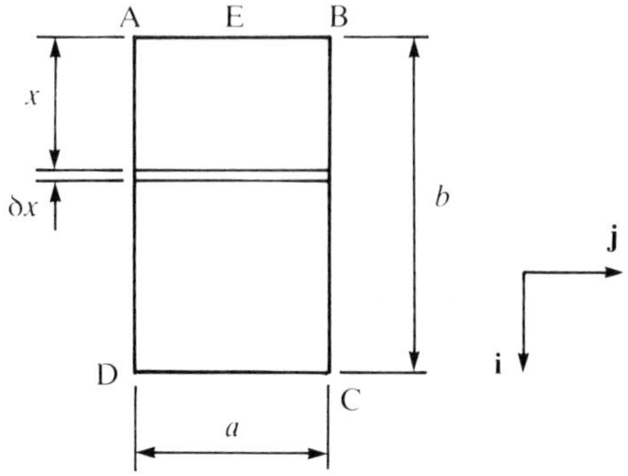

Fig. 3.28

Example 3.9

A rectangular plate whose sides are of lengths a and b is immersed in a liquid, with the sides of length b vertical and a side of length a in the surface of the liquid. Determine the position of the centre of pressure. Neglect the effects of atmospheric pressure.

▷ *Solution* Consider a strip of the plate of thickness δx and distance x from the side AB of the plate that is in the surface of the liquid (see Fig. 3.28). The resultant force on this strip is $\varrho gxa\,\delta x$ and acts at the midpoint of the strip. Hence the resultant force on one side of the plate is of magnitude

$$\int_0^b \varrho gax\,\mathrm{d}x = \varrho ga\frac{b^2}{2}$$

Let the centre of pressure be at the point P. This point lies on the line that passes through the midpoint E of AB and is parallel to the sides AD and BC of the rectangle ABCD. Now the moment of the resultant force about E is equal to the sum of the moments about E of the forces on the elemental areas that make up the rectangular areas, and therefore

$$\frac{\varrho}{2}ab^2\,\mathrm{EP} = \int_0^b \varrho gax^2\,\mathrm{d}x$$

from which we deduce EP $= 2b/3$.

It is left as an exercise to the reader to show that:

(a) The position of the centre of pressure relative to the rectangle ABCD is unaltered if the plate is inclined at any angle $\theta(\neq 0)$ to the horizontal.

(b) If the plate (assumed vertical) is lowered a distance d then the depth of the centre of pressure is equal to $h + b^2/12h$, where $h = (b/2) + d$.

The result in (b) is a particular cxample of a more general result concerning centres of pressure. The position of the centre of pressure of a plane area immersed in a liquid can be expressed (see Problem 3.18) in terms of quantities known as first and second moments of area. The principles of statics used in the course of Example 3.9 form the basis of the work in Problem 3.18.

3.5 FRICTION

The force of gravity on a body of mass M at rest on an inclined plane (angle α) can be resolved into two components: $Mg\cos\alpha$ perpendicular to the plane and $Mg\sin\alpha$ down a line of greatest slope of the plane. Since the body is in equilibrium the plane must exert on the body a force perpendicular to the plane of magnitude $Mg\cos\alpha$. This force is called the normal reaction of the plane. In addition the plane must exert on the body a force of magnitude $Mg\sin\alpha$ up the plane. This latter force is provided by the action of friction. From experience it is known that if α is greater than a certain value the body will slide down the plane, so that there is a limit to the frictional force the plane can exert on the body. This maximum frictional force is called static (or limiting) friction. Coulomb (1736–1806), to whom the laws of friction are attributed, observed that when a body is in motion the frictional force on it is less than static friction, and used the term 'sliding friction' to describe this frictional force.

Frictional forces come into play when a train ascends a gradient. Provided the wheels of the train are not slipping, that is there is no relative motion between each wheel and the railway line at the point of contact, a force of friction is exerted on the train which is less than sliding friction. This frictional force, which enables the train to ascend the gradient, is called rolling friction. Frictional forces are essential for many types of motion. Their absence, as drivers of vehicles in icy conditions are aware, inhibits motion.

Coulomb formulated the following laws of friction:

1 Friction is proportional to the magnitude of the normal reaction between the surfaces in contact, that is it is independent of the areas of the surfaces.
2 Friction is independent of the relative speed between surfaces.
3 If R is the magnitude of the normal reaction between the surfaces of two bodies in contact and F is the magnitude of static friction then $F/R = \mu$, where μ is called the coefficient of static friction. Its value depends on

the nature of the surfaces in contact. A similar law applies for sliding friction, the coefficient of sliding friction being less than that of static friction.

It is important, when considering the sliding frictional force between two bodies B_1 and B_2, to identify its direction correctly. If $\mathbf{v}_1$ denotes the velocity of the point of B_1 momentarily in contact with body B_2, and $\mathbf{v}_2$ denotes that of the point of B_2 in contact with B_1, then the sliding frictional force on B_2 is in the direction of the unit vector $\mathbf{t}$ defined by

$$\mathbf{v}_2 - \mathbf{v}_1 = -\alpha\mathbf{t}$$

where α is a positive scalar. By Newton's third law the frictional force on B_1 is in the direction of $-\mathbf{t}$.

The laws of friction just described are sufficiently accurate for most purposes. However, surfaces in contact deform slightly and this deformation gives rise to a couple. The magnitude of the moment of this couple is small and will only be significant when sliding friction is small.

PROBLEMS

Section 3.2

3.1 The components of a force $\mathbf{F}$ referred to the unit vectors $\mathbf{i}$, $\mathbf{j}$, $\mathbf{k}$ are [2, 1, 2]. Given that the moment $\mathbf{M}_O$ of the force about the origin O is equal to [2, 6, −5], find the equation of the line of action of the force $\mathbf{F}$.

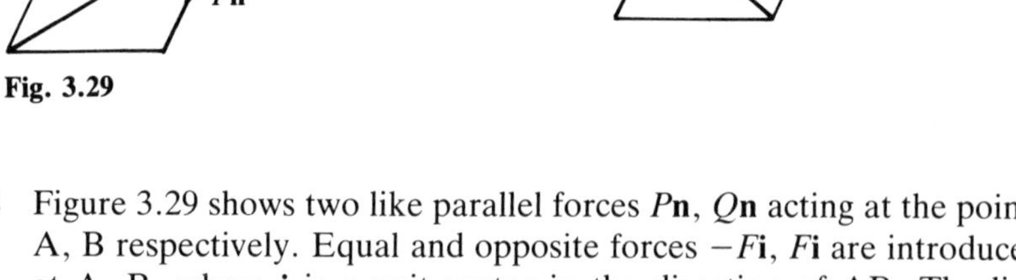

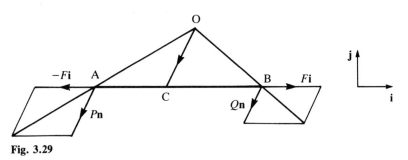

Fig. 3.29

3.2 Figure 3.29 shows two like parallel forces $P\mathbf{n}$, $Q\mathbf{n}$ acting at the points A, B respectively. Equal and opposite forces $-F\mathbf{i}$, $F\mathbf{i}$ are introduced at A, B, where $\mathbf{i}$ is a unit vector in the direction of $\underline{AB}$. The line of action of the force $-F\mathbf{i} + P\mathbf{n}$ acting at A meets that of the force $F\mathbf{i} + Q\mathbf{n}$ acting at B in O, while the line through O parallel to the unit vector $\mathbf{n}$ meets AB in C. Use the relations

$$\alpha(\underline{\mathrm{OC}} + \underline{\mathrm{CA}}) = \alpha\underline{\mathrm{OA}} = -F\mathbf{i} + P\mathbf{n}$$
$$\beta(\underline{\mathrm{OC}} + \underline{\mathrm{CB}}) = \beta\underline{\mathrm{OB}} = F\mathbf{i} + Q\mathbf{n}$$

to show that the resultant of the forces $P\mathbf{n}$, $Q\mathbf{n}$ is $(P + Q)\mathbf{n}$ acting along $\underline{\mathrm{OC}}$, where $P\underline{\mathrm{CA}} = Q\underline{\mathrm{BC}}$.

Does the proof have to be amended if P, Q are of different sign and not numerically equal?

3.3 Forces $4\mathbf{n}$, $2\mathbf{n}$, $-2\mathbf{n}$, $-\mathbf{n}$ act at the points A, B, C, D respectively. If $\underline{\mathrm{OA}} = [1, 3, 2]$, $\underline{\mathrm{OB}} = [1, 2, 1]$, $\underline{\mathrm{OC}} = [2, 3, 2]$, $\underline{\mathrm{OD}} = [-2, 4, 1]$ and $3\mathbf{n} = [1, 2, 2]$ referred to the unit vectors $\mathbf{i}$, $\mathbf{j}$, $\mathbf{k}$, determine the equation of the line of action of the resultant of the four forces. What is the length of the perpendicular from O to the line of action?

3.4 Forces $\mathbf{F}_1$, $\mathbf{F}_2$, $\mathbf{F}_3$ act at points whose position vectors with respect to a point O are $\mathbf{r}_1$, $\mathbf{r}_2$, $\mathbf{r}_3$. Determine which of the following reduce to a single force and obtain, where appropriate, the equation of the line of action of this single force:

(a) $\mathbf{F}_1 = [2, 3, -5]$, $\mathbf{F}_2 = [-3, -2, 4]$, $\mathbf{F}_3 = [1, -1, 1]$,
$\mathbf{r}_1 = [1, 2, 3]$, $\mathbf{r}_2 = [2, 1, 3]$, $\mathbf{r}_3 = [1, 3, 2]$

(b) $\mathbf{F}_1 = [1, 2, 3]$, $\mathbf{F}_2 = [2, 3, 5]$, $\mathbf{F}_3 = [4, 9, 1]$,
$\mathbf{r}_1 = [1, 0, 1]$, $\mathbf{r}_2 = [-1, 0, 1]$, $\mathbf{r}_3 = [2, 3, 0]$

(c) $\mathbf{F}_1 = [2, 3, 6]$, $\mathbf{F}_2 = [1, 1, 1]$, $\mathbf{F}_3 = [-3, -3, -6]$
$\mathbf{r}_1 = [1, 3, 4]$, $\mathbf{r}_2 = [1, -1, -3]$, $\mathbf{r}_3 = [-1, 2, 0]$.

3.5 A system of forces reduces to a single force $\mathbf{R}$ acting at the point O together with a couple of moment $\mathbf{M}$. If the system also reduces to a single force acting at the point O_1 and a couple of a moment $\mathbf{M}_1$, show that $\mathbf{R} \cdot \mathbf{M} = \mathbf{R} \cdot \mathbf{M}_1$.

Determine the point O_1 if $\mathbf{M}_1$ is parallel to $\mathbf{R}$, that is $\mathbf{M}_1 = \alpha\mathbf{R}$, where α is some scalar. Obtain the value of α. Show, for the system of forces $\alpha\underline{\mathrm{AB}}$, $\beta\underline{\mathrm{BC}}$, $\mu\underline{\mathrm{CD}}$ acting at the points A, B, C respectively, that $\mathbf{R} \cdot \mathbf{M} = 6\alpha\mu V$, where V is the volume of the tetrahedron ABCD.

3.6 Forces $X\mathbf{i}$, $Y\mathbf{j}$, $-Z\mathbf{k}$ act along three of the edges of a rectangular parallelepiped (see Fig. 3.30), the lengths of the edges being respectively a, b, c. Show that if

$$\frac{a}{X} + \frac{b}{Y} + \frac{c}{Z} = 0$$

the system of forces reduces to a single force, and obtain the equation of the line of action of the force.

3.7 The four points O, A, B, C are the vertices of a tetrahedron. If $\mathbf{n}_1$, $\mathbf{n}_2$, $\mathbf{n}_3$, $\mathbf{a}$, $\mathbf{b}$, $\mathbf{c}$ are unit vectors in the directions of the vectors $\underline{\mathrm{OA}}$, $\underline{\mathrm{OB}}$, $\underline{\mathrm{OC}}$, $\underline{\mathrm{BC}}$, $\underline{\mathrm{CA}}$, $\underline{\mathrm{AB}}$, show that the system of forces $P\mathbf{n}_1$, $Q\mathbf{n}_2$, $R\mathbf{n}_3$, $X\mathbf{a}$, $Y\mathbf{b}$, $Z\mathbf{c}$ located respectively in these vectors reduces to a single force provided that

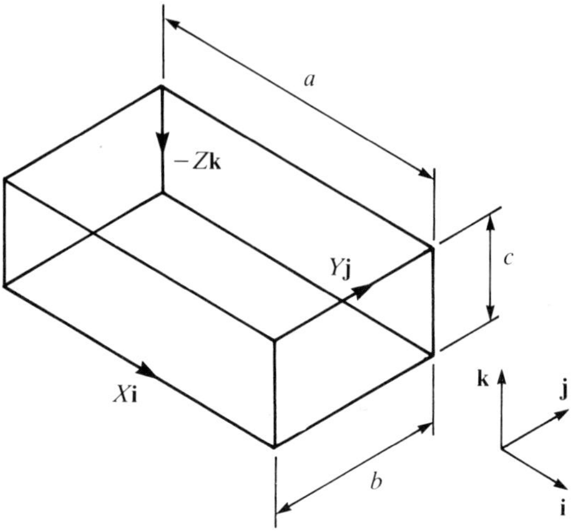

Fig. 3.30

$$\frac{PX}{|\underline{\mathrm{OA}}|\,|\underline{\mathrm{BC}}|} + \frac{QY}{|\underline{\mathrm{OB}}|\,|\underline{\mathrm{CA}}|} + \frac{RZ}{|\underline{\mathrm{OC}}|\,|\underline{\mathrm{AB}}|} = 0$$

Hint: express the vectors **a**, **b**, **c** in terms of $\mathbf{n}_1$, $\mathbf{n}_2$, $\mathbf{n}_3$ and the lengths of the edges of the tetrahedron and then obtain an expression for $\mathbf{R} \cdot \mathbf{M}$ in the form $\alpha(\mathbf{n}_1 \times \mathbf{n}_2) \cdot \mathbf{n}_3$.

3.8 A thin plate of weight $-W\mathbf{k}$ and centre of gravity G is freely pivoted about a point O of itself and is kept in equilibrium by strings AC, BC attached to the plate at the points A, B on opposite sides of OG and to a point C vertically above O. Derive the following equations:

$$T_1\mathbf{n}_1 + T_2\mathbf{n}_2 - W\mathbf{k} + \mathbf{F} = \mathbf{0}$$
$$\underline{\mathrm{CO}} \times \mathbf{F} + W\mathbf{k} \times \underline{\mathrm{OG}} = \mathbf{0}$$

where T_1, T_2 are the tensions in the strings AC, BC, the vector **F** is the (unknown) reaction of the support at O, $\mathbf{n}_1$, $\mathbf{n}_2$ are unit vectors in the directions of $\underline{\mathrm{AC}}$ and $\underline{\mathrm{BC}}$, and **k** is vertically upwards. Deduce that T_1, T_2 satisfy the vector equation

$$(T_1\mathbf{n}_1 + T_2\mathbf{n}_2) \times \underline{\mathrm{CO}} + W\mathbf{k} \times \underline{\mathrm{OG}} = \mathbf{0}$$

Solve this equation to show that

$$T_1\mathbf{n}_1 = \frac{W\,\mathrm{OG}\,\mathrm{BD}}{\mathrm{OC}\,\mathrm{OD}\,\mathrm{AB}}\underline{\mathrm{AC}}, \qquad T_2\mathbf{n}_2 = \frac{W\,\mathrm{OG}\,\mathrm{AD}}{\mathrm{OC}\,\mathrm{OD}\,\mathrm{AB}}\underline{\mathrm{BC}}$$

where D is the point of intersection of the lines OG and AB, and OC for example denotes $|\underline{\mathrm{OC}}|$. *Hint:* Take the scalar product of the equation with $\mathbf{n}_2$, noting that

$$\underline{OG} = \frac{OG}{OD}\underline{OD} \qquad \text{and} \qquad \underline{CO} = \frac{AD}{AB}\underline{CB} + \frac{BD}{AB}\underline{CA}$$

(see Example 1.2).

3.9 Referred to the unit vectors **i**, **j**, **k**, two couples have moments $\mathbf{M}_1$ and $\mathbf{M}_2$ given by $\mathbf{M}_1 = [1, 2, 1]$, $\mathbf{M}_2 = [2, 3, 1]$. If the couple of moment $\mathbf{M}_1$ is equivalent to the force pair $\mathbf{F}_1$, $-\mathbf{F}_1$ acting at the points P_1, Q_1 respectively, show that a possible expression for $\mathbf{F}_1$ is $[1, -1, 1]$. Referred to the origin O the point P_1 is given by $\underline{OP_1} = [1, 1, 1]$. Determine $\underline{OQ_1}$ if the arm $\underline{P_1Q_1}$ is perpendicular to $\mathbf{F}_1$, which is taken to be equal to $[1, -1, 1]$. Show that the couple of moment M_2 can be represented by a force pair $[-1, 1, -1]$ acting at P_1 and $[1, -1, 1]$ acting at some point Q_2. Find $\underline{OQ_2}$ if $\underline{P_1Q_2}$ is perpendicular to $[1, -1, 1]$.

Verify that the force pair $[1, -1, 1]$ acting at Q_2 and $[-1, 1, -1]$ at Q_1 have a moment equal to $\mathbf{M}_1 + \mathbf{M}_2$.

3.10 The components of three forces $\mathbf{F}_1$, $\mathbf{F}_2$, $\mathbf{F}_3$ and the position vectors $\mathbf{r}_1$, $\mathbf{r}_2$, $\mathbf{r}_3$ of their points of application with respect to the origin O are given by

$$\begin{aligned} &\mathbf{F}_1 = [1, -1, -1], \qquad &&\mathbf{F}_2 = [-1, 2, 3], \qquad &&\mathbf{F}_3 = [0, -1, -2] \\ &\mathbf{r}_1 = [1, 0, 1], &&\mathbf{r}_2 = [0, 1, 1], &&\mathbf{r}_3 = [-1, -1, 0] \end{aligned}$$

Show that the forces are equivalent to a couple. Show also that the couple can be represented by a pair of forces $\mathbf{Q}$, $-\mathbf{Q}$, where $\mathbf{Q} = [1, 2, -2]$, provided the perpendicular distance between the forces is $\sqrt{2}$.

Section 3.3

3.11 A light beam of length l is simply supported horizontally at its ends A and B. It carries a vertical load $-W\mathbf{j}$ at a point C, where $AC = b$. Obtain expressions for the bending moment and shear force at a point P of the beam whose distance from the end A is x, distinguishing between the cases $0 \le x < b$ and $b < x \le l$. What can be said about the shear force at the point $x = b$?

Determine the point of the beam where the magnitude of the bending moment is greatest. Show that whatever the value of b this magnitude cannot exceed $Wl/4$.

A second vertical load $-W_1\mathbf{j}$ is applied to the beam at D, where $AD = c > b$. Using the results derived for the case of the single load, write down expressions for the bending moment according as $0 \le x \le b$, $b \le x \le c$, $c \le x \le l$.

3.12 A uniform beam AB of weight $-w\mathbf{j}$ per unit length is clamped horizontally at the end A while the other end B is free. If the length of the

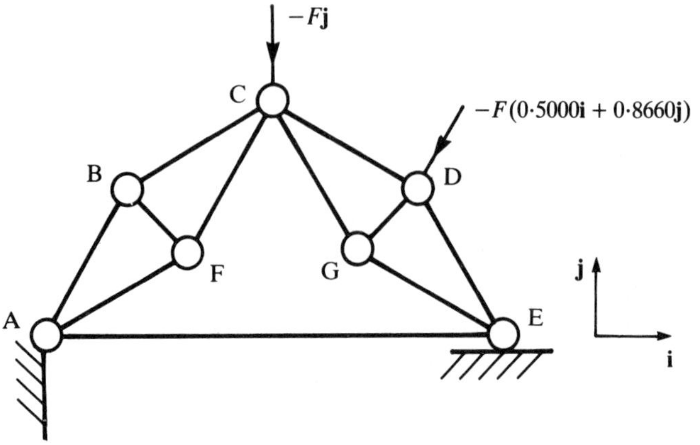

Fig. 3.31

beam is l find the force and couple which the clamp exerts on the beam. Determine the bending moment and shear force at a point P distance x from the clamped end, firstly by considering the equilibrium of the length AP of the beam and secondly by considering that of the length PB.

3.13 A light horizontal beam simply supported at two points A, B carries a concentrated vertical load $-W\mathbf{j}$ at the point C. Find an expression for the bending moment at D, a point of the beam, and show that it is equal to the bending moment at C when the concentrated load is applied at D instead of C.

3.14 Two uniform horizontal beams AB and BC of lengths a and b are smoothly hinged together at B. The beam AB is simply supported at A and D, where AD $= c\ (<a)$, while the beam BC is simply supported at C. If each beam is of weight $-w\mathbf{j}$ per unit length determine the shear force and bending moment at the point P of the beams, where AP $= x$ and $0 \leq x \leq a + b$.

3.15 Figure 3.31 shows a symmetrical pin-jointed framework of light members. All members apart from BF, GD, AE are of length l. The angles BCF and FCG are respectively equal to 30° and 60°. The framework is pivoted at A and supported by a vertical reaction at E, and is subjected (see Fig. 3.31) to a vertical load at C and a load at D perpendicular to CD. Find the reactions at A and E. Using the method of sections find also the force in the member AE and hence obtain the forces in the members AF and AB.

3.16 Figure 3.32 shows a pin-jointed framework of light members which is supported at B and E. The frame carries loads of $-3\text{kN}\mathbf{j}$ and $-6\text{kN}\mathbf{j}$

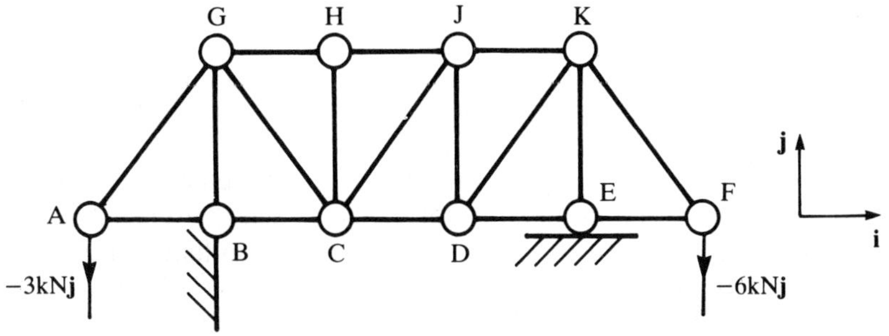

Fig. 3.32

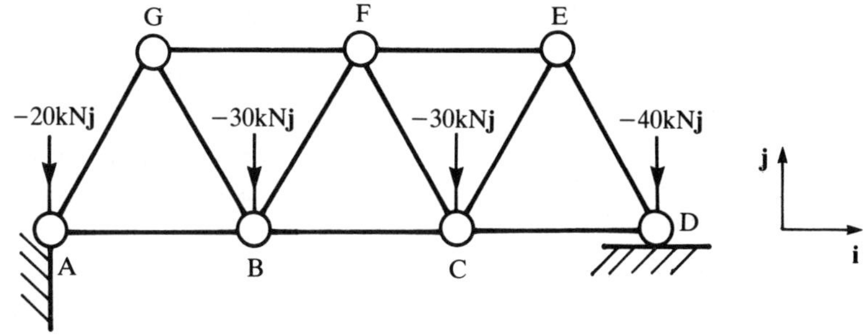

Fig. 3.33

at A and F respectively. The members AB, BC, CD, DE and EF are each 3 m long while BG, CH, DJ, EK are each 4 m long.

Determine the reactions of the supports at B and E. By combining the method of sections and the procedure of resolving in two perpendicular directions at a joint, determine also the forces in the members DE, GH, CH, CJ and DK.

3.17 Figure 3.33 shows a pin-jointed framework which is supported at A and D. Loads are applied at A, B, C, D as indicated in the figure. If the members of the framework are all of equal length, find the forces in the members BC, CD, CE, CF.

Section 3.4

3.18 The centroid of a plane area is defined to be the point G whose coordinates $\bar{x}$, $\bar{y}$ referred to axes Ox, Oy taken in the plane of the area are given by

$$A\bar{x} = \Sigma x\,\delta A, \qquad A\bar{y} = \Sigma y\,\delta A$$

where δA is an elemental area containing the point of the area with coordinates x, y and the summation is taken over all the elemental areas that make up the total area A.

A plane area immersed in a liquid has its plane vertical. Axes Ox, Oy are taken in the plane of the area such that Ox is horizontal and Oy is vertically upwards. Show that the resultant thrust (neglecting atmospheric pressure) on each side of the area is of magnitude $\varrho g(Ad - \Sigma y\,\delta A)$, where d is the depth of O and ϱ is the fluid density. Deduce that the resultant thrust is equal to ϱgAh, where h is the depth of the centroid of the area. Deduce also that if Ox is chosen so as to pass through G then the depth of the centre of pressure is equal to $h + (\Sigma y^2\,\delta A)/Ah$.

Note: the quantity $\Sigma y^2\,\delta A$ is called the second moment of the area A about Ox.

3.19 A rectangular door in the vertical side of a reservoir can turn freely about its lower (horizontal) edge, and is fastened at its two upper corners. The door is 1 metre wide and 2 metres high and its upper edge is 1.5 metres below the water level. Determine the reactions at the upper corners, assuming them to be equal. Neglect atmospheric pressure. Take the mass of a cubic metre of water to be equal to 10^3 kg.

4 Planar kinematics of particles and rigid bodies

4.1 INTRODUCTION

Kinematics is the part of dynamics that is concerned with the analysis of the motion of bodies without consideration of the forces causing or influencing that motion. The central problem in kinematics is the determination of the absolute velocity and acceleration (both linear and angular) of a body, where the term 'absolute' implies measurement by an observer fixed in a Newtonian frame of reference (or for most engineering purposes fixed to the earth; see Chapters 2 and 6). We need absolute velocity and acceleration because usually, in solving engineering problems, we have to go on to apply Newton's laws of motion in order to relate forces and motion (see Chapter 6).

If a particle moves in a fixed plane it is said to have two-dimensional or planar motion. The motion of a body is described as planar when all the particles that compose the body move in planes parallel to some fixed plane. The fixed plane is known as the (*fixed*) plane of the motion. Planar motion occurs frequently in engineering practice and is concerned with single particles and rigid bodies, that is bodies that do not deform in their motion. The analysis provides an introduction to the more complicated problems of motion in three dimensions that are dealt with in Chapters 5, 6 and 7.

We start by considering the planar kinematics of a particle. This is directly applicable to engineering problems in which the rotation of the body may be ignored: as it may, for example, in the analysis of the orbital motion of a satellite. The work is then extended to deal with the planar motion of a rigid body. This is followed by a number of applications that illustrate the scope of the theory.

In the work that follows Ox, Oy, Oz are described as fixed axes. By this

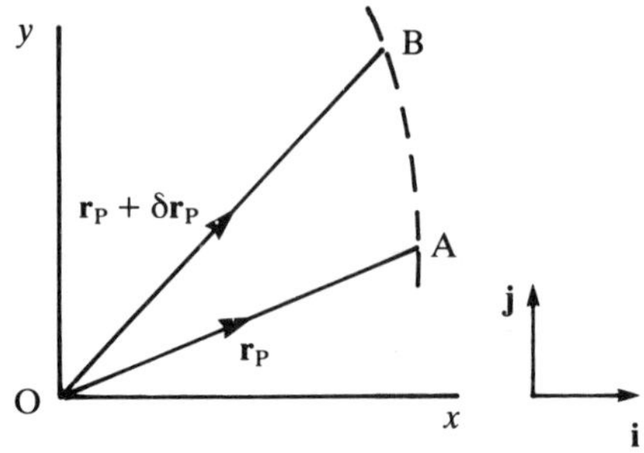

Fig. 4.1

we mean that they are fixed to a Newtonian frame of reference so that velocity and acceleration measured relative to these axes are absolute quantities.

4.2 PLANAR KINEMATICS OF A PARTICLE

In Fig. 4.1 the axes Ox, Oy, Oz are fixed and the plane Ox, Oy is the plane of the motion. This lies in the plane of the page so that for right-handed axes the axis Oz is out of the page. The particle (referred to by the single letter P) follows a path in the plane of the motion (shown by the broken line in Fig. 4.1) which is determined by the forces that act on it. At time t the particle is at the point A and its absolute position vector $\underline{\mathrm{OA}} = \mathbf{r}_P$. During the time interval δt the particle moves along its path to the point B whose absolute position vector $\underline{\mathrm{OB}} = \mathbf{r}_P + \delta\mathbf{r}_P$. The absolute velocity $\mathbf{v}_P$ of the particle P is defined by

$$\mathbf{v}_P = \lim_{\delta t \to 0} \frac{\delta \mathbf{r}_P}{\delta t} = \frac{d\mathbf{r}_P}{dt} = \dot{\mathbf{r}}_P \tag{4.1}$$

The absolute acceleration of the particle, denoted by $\mathbf{f}_P$, is defined by

$$\mathbf{f}_P = \frac{d\mathbf{v}_P}{dt} = \frac{d}{dt}\left(\frac{d\mathbf{r}_P}{dt}\right) = \frac{d^2\mathbf{r}_P}{dt^2} = \ddot{\mathbf{r}}_P \tag{4.2}$$

The definitions of $\mathbf{v}_P$ and $\mathbf{f}_P$ in (4.1) and (4.2) apply also when the path of the particle is not a plane curve, that is when its motion is three-dimensional.

We shall adopt the convention of using a single subscript as in $\mathbf{v}_P$ and $\mathbf{f}_P$ to indicate absolute velocity and acceleration, but a single subscript as in $\mathbf{r}_P$ should *not* be taken to imply an absolute position vector. The double subscript notation for relative velocity and acceleration is introduced later in the analysis of the planar motion of a rigid body.

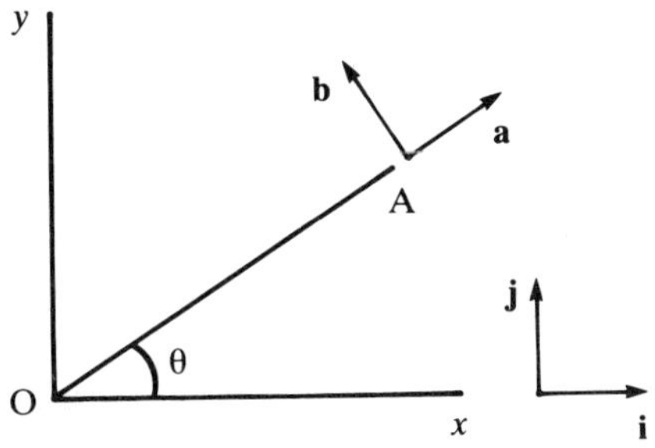

Fig. 4.2

There are three ways in which the scalar components of the vector quantities $\mathbf{v}_P$ and $\mathbf{f}_P$ are commonly expressed; these are now described.

Cartesian components

If the point A occupied by the particle P has coordinates $[x, y, 0]$ referred to the axes Ox, Oy, Oz then $\underline{OA} = \mathbf{r}_P = [x, y, 0]$ referred to the unit vectors **i**, **j**, **k**. Hence using the rules for differentiating a vector we have

$$\mathbf{v}_P = \dot{\mathbf{r}}_P = [\dot{x}, \dot{y}, 0] \tag{4.3}$$

$$\mathbf{f}_P = \ddot{\mathbf{r}}_P = [\ddot{x}, \ddot{y}, 0] \tag{4.4}$$

Polar components

We denote the unit vectors taken along and perpendicular to the position vector $\underline{OA}$ (see Fig. 4.2) by **a** and **b**, while θ is the angle between the direction of $\underline{OA}$ and the axis Ox.

Using the results contained in Example 1.17, namely

$$\frac{d\mathbf{a}}{d\theta} = \mathbf{b}, \qquad \frac{d\mathbf{b}}{d\theta} = -\mathbf{a}$$

and the rule for differentiating a function of a function, it follows that

$$\dot{\mathbf{a}} = \dot{\theta}\mathbf{b}, \qquad \dot{\mathbf{b}} = -\dot{\theta}\mathbf{a} \tag{4.5}$$

Now $\mathbf{r}_P = r\mathbf{a}$, where $r = |\underline{OA}| = |\mathbf{r}_P|$, and therefore differentiating $\mathbf{r}_P$ with respect to time we obtain, with the help of (4.5),

$$\mathbf{v}_P = \dot{\mathbf{r}}_P = \dot{r}\mathbf{a} + r\dot{\mathbf{a}} = \dot{r}\mathbf{a} + r\dot{\theta}\mathbf{b} \tag{4.6}$$

Differentiating a second time, again noting (4.5), yields

$$\begin{aligned}\mathbf{f}_P = \ddot{\mathbf{r}}_P &= \ddot{r}\mathbf{a} + \dot{r}\dot{\mathbf{a}} + (\dot{r}\dot{\theta} + r\ddot{\theta})\mathbf{b} + r\dot{\theta}\dot{\mathbf{b}} \\ &= (\ddot{r} - r\dot{\theta}^2)\mathbf{a} + (2\dot{r}\dot{\theta} + r\ddot{\theta})\mathbf{b}\end{aligned} \tag{4.7}$$

The components of the vectors $\mathbf{v}_P$ and $\mathbf{f}_P$ in the direction of the unit vector $\mathbf{a}$ are called the radial components of velocity and acceleration, while those in the direction of the unit vector $\mathbf{b}$ are called the transverse components. If $|\mathbf{r}_P| = \text{constant} = c$ then the particle describes a circle at an angular rate $\dot{\theta}$ with components of velocity and acceleration given by

$$\mathbf{v}_P = c\dot{\theta}\mathbf{b} \tag{4.8}$$

$$\mathbf{f}_P = -c\dot{\theta}^2\mathbf{a} + c\ddot{\theta}\mathbf{b} \tag{4.9}$$

The expression for $\mathbf{v}_P$ in (4.8) can be put in an alternative form that is of use later:

$$\mathbf{v}_P = c\dot{\theta}\mathbf{b} = c\dot{\theta}\mathbf{k} \times \mathbf{a} = \dot{\theta}\mathbf{k} \times c\mathbf{a} = \dot{\theta}\mathbf{k} \times \mathbf{r}_P \tag{4.10}$$

Normal and tangential components

Figure 4.3(a) and (b) shows the planar curve described by the particle P in its motion. The vector $\mathbf{t}$ is a unit vector located in the tangent at A and in the sense of s increasing, where s denotes arc length measured from a fixed point on the curve. By virtue of the results contained in Example 1.18 we have $\mathbf{t} = \mathrm{d}\mathbf{r}_P/\mathrm{d}s$, and hence

$$\mathbf{v}_P = \dot{\mathbf{r}}_P = \frac{\mathrm{d}s}{\mathrm{d}t}\frac{\mathrm{d}\mathbf{r}_P}{\mathrm{d}s} = \dot{s}\mathbf{t} = v\mathbf{t} \tag{4.11}$$

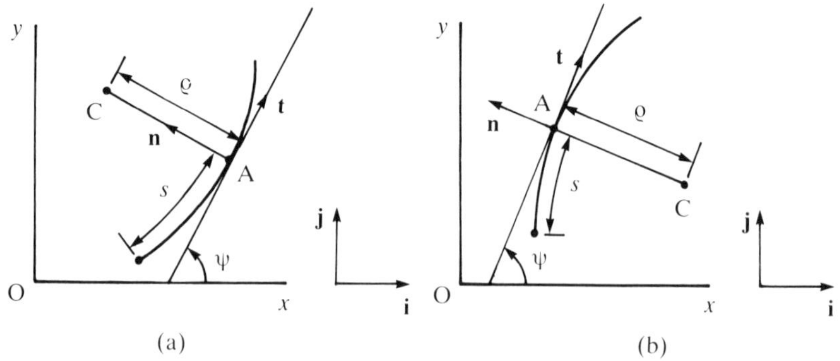

Fig. 4.3

the scalar quantity $\dot{s}$ being denoted by v.

A second differentiation with respect to time yields

$$\begin{aligned}\mathbf{f}_P = \dot{\mathbf{v}}_P &= \ddot{s}\mathbf{t} + \dot{s}\frac{d\mathbf{t}}{dt}\\ &= \ddot{s}\mathbf{t} + \dot{s}\frac{d\psi}{dt}\frac{d\mathbf{t}}{d\psi}\\ &= \ddot{s}\mathbf{t} + \frac{\dot{s}^2}{ds/d\psi}\frac{d\mathbf{t}}{d\psi}\end{aligned}$$

where ψ is the angle (see Fig. 4.3(a) and (b)) between the direction of $\mathbf{t}$ and Ox. Recalling once more the results in Example 1.18, it is seen that the expression for $\mathbf{f}_P$ can be written as

$$\mathbf{f}_P = \ddot{s}\mathbf{t} + \frac{v^2}{ds/d\psi}\mathbf{n} \tag{4.12}$$

where $\mathbf{n}$ is a unit vector perpendicular to $\mathbf{t}$ and such that $\mathbf{n} = \mathbf{k} \times \mathbf{t}$. In Fig. 4.3(a) the quantity $ds/d\psi$ is positive while in Fig. 4.3(b) it is negative. Thus in both cases the sense of the vector component of $\mathbf{f}_P$ parallel to $\mathbf{n}$ is that of the inward normal. If we write $\varrho = |ds/d\psi|$ then the magnitude of the component of acceleration along the inward normal is v^2/ϱ. The quantity ϱ is called the radius of curvature at A, while the point C on the inward normal such that $|\overline{CA}| = \varrho$ is called the centre of curvature.

Example 4.1

The radial component of acceleration of a particle P describing a plane curve is zero. If $\dot{\theta} = \omega$ (a constant), find the polar equation of the curve given that $\theta = 0$, $r = c$, $\dot{r} = 0$ when $t = 0$.

▷ *Solution* Integrating the equation $\dot{\theta} = \omega$ gives $\theta = \omega t +$ constant; the constant of integration is zero as $\theta = 0$ when $t = 0$. Since the radial component of acceleration is zero, it follows from (4.7) that $\ddot{r} - r\dot{\theta}^2 = 0$ and hence that

$$\ddot{r} - \omega^2 r = 0$$

The general solution of this differential equation is given by

$$r = A\cosh \omega t + B\sinh \omega t$$

where A and B are arbitrary constants. Using the initial conditions, namely $r = c$, $\dot{r} = 0$ when $t = 0$, the values of A and B are found to be $A = c$, $B = 0$. Thus the polar equation of the curve is $r = c\cosh\theta$.

4.3 PLANAR KINEMATICS OF A RIGID BODY

Definition of a rigid body

Every real body deforms when a force is applied to it, and the extent of the deformation for a given force is determined by the stiffness of the material from which the body is made. Bodies made from the materials (particularly metals) commonly used for engineering products deform very little, even when substantial loads are applied, and the deformations can often be ignored in the formulation of a mathematical model of a problem. The term 'rigid body' is used to describe a body whose deformation can be ignored. Before giving a formal mathematical definition of this term, some preliminary discussion of the relevant ideas is necessary.

Any body, rigid or otherwise, is assumed to be composed of a number (normally large) of material particles. The number of particles is taken as $N + 1$ and the particles are designated $P_0, P_1, P_2, \ldots, P_N$. We shall take P_0, P_1, P_2 as reference particles and assume them to be non-collinear. Thus we can write

$$\mathbf{r}_i = \alpha_i \mathbf{r}_1 + \beta_i \mathbf{r}_2 + \mu_i(\mathbf{r}_1 \times \mathbf{r}_2) \tag{4.13}$$

where $\mathbf{r}_i$ $(i = 1, 2, \ldots, N)$ are the position vectors of the particles P_i with respect to P_0 and α_i, β_i, μ_i are scalars. Using the results contained in Example 1.10 leads to

$$\begin{aligned} |\mathbf{r}_1 \times \mathbf{r}_2|^2 \alpha_i &= (\mathbf{r}_i \cdot \mathbf{r}_1)\mathbf{r}_2^2 - (\mathbf{r}_1 \cdot \mathbf{r}_2)(\mathbf{r}_i \cdot \mathbf{r}_2) \\ |\mathbf{r}_1 \times \mathbf{r}_2|^2 \beta_i &= (\mathbf{r}_i \cdot \mathbf{r}_2)\mathbf{r}_1^2 - (\mathbf{r}_1 \cdot \mathbf{r}_2)(\mathbf{r}_i \cdot \mathbf{r}_1) \\ |\mathbf{r}_1 \times \mathbf{r}_2|^2 \mu_i &= \mathbf{r}_i \cdot (\mathbf{r}_1 \times \mathbf{r}_2) \end{aligned} \tag{4.14}$$

If the body is rigid, distances between pairs of particles must be fixed and so must angles between pairs of lines joining particles. Thus the quantities $\mathbf{r}_i^2$, $\mathbf{r}_i \cdot \mathbf{r}_j$ and $|\mathbf{r}_i \times \mathbf{r}_j|^2$ $(i, j = 1, 2, \ldots, N)$ will all have fixed values that are independent of the position of the body in space. It follows immediately that α_i and β_i have fixed values. Taking the scalar product of $\mathbf{r}_i$ with itself and noting that $\mathbf{r}_1 \cdot \mathbf{r}_1 \times \mathbf{r}_2$ and $\mathbf{r}_2 \cdot \mathbf{r}_1 \times \mathbf{r}_2$ are both zero yields

$$\mathbf{r}_i^2 = \alpha_i^2 \mathbf{r}_1^2 + \beta_i^2 \mathbf{r}_2^2 + 2\alpha_i \beta_i \mathbf{r}_1 \cdot \mathbf{r}_2 + \mu_i^2 |\mathbf{r}_1 \times \mathbf{r}_2|^2$$

Since the quantities in this relation are known, apart from μ_i^2, to be constant, it follows that μ_i^2 must be constant. Writing $\mu_i^2 = \psi_i^2$ yields:

$$\mathbf{r}_i \cdot (\mathbf{r}_1 \times \mathbf{r}_2)/|\mathbf{r}_1 \times \mathbf{r}_2|^2 = \mu_i = \pm \psi_i$$

Hence P_i is situated either at the point A whose position vector with respect to the particle P_0 is

$$\alpha_i \mathbf{r}_1 + \beta_i \mathbf{r}_2 + \psi_i \mathbf{r}_1 \times \mathbf{r}_2$$

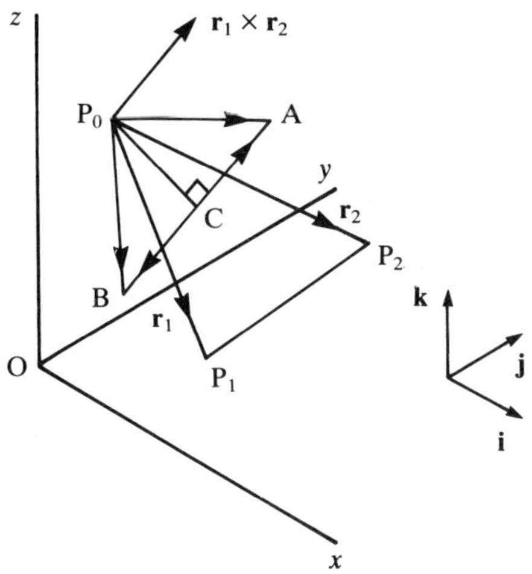

Fig. 4.4

or at the point B whose position vector is

$$\alpha_i \mathbf{r}_1 + \beta_i \mathbf{r}_2 - \psi_i \mathbf{r}_1 \times \mathbf{r}_2$$

B is the mirror image of A in the plane defined by the particles P_0, P_1, P_2 (see Fig. 4.4 and Problem 1.11). Now the vector $\mathbf{r}_1 \times \mathbf{r}_2$ is fixed relative to the particles P_0, P_1, P_2 and its sense (see Fig. 4.4 again) can therefore be used to distinguish one side of the plane from the other. If the body does not deform the particle P_i will always remain on the same side of the plane of P_0, P_1, P_2, and consequently the position of P_i is obtained by choosing appropriately from the points A, B. Thus μ_i has a constant value equal to one of the values $\pm\psi_i$. The three quantities α_i, β_i, μ_i are called the body coordinates of the particle P_i with respect to the reference particles P_0, P_1, P_2. We can now provide a formal definition of the term 'rigid body':

If in any displacement the distances between the three reference particles P_0, P_1, P_2 and the body coordinates of each particle P_i ($i = 3, 4, \ldots, N$) remain unaltered, then the body is described as a rigid body.

The body coordinates are determined from a given configuration of the particles using the formulae given in (4.14).

We now proceed to determine expressions for the absolute velocities and accelerations of the particles of a rigid body whose motion is two-dimensional.

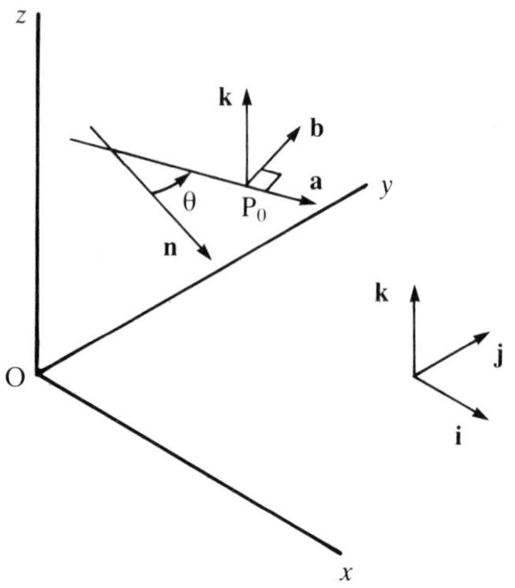

Fig. 4.5

Determination of absolute velocity and acceleration of a particle of a rigid body

We choose the Ox, Oy plane to coincide with the (fixed) plane of motion of the rigid body and the three reference particles P_0, P_1, P_2 so that $\mathbf{r}_1$, $\mathbf{r}_2$, $\mathbf{k}$ form a right-handed set of mutually perpendicular unit vectors. Thus P_0, P_1, P_2 all lie in a plane parallel to the fixed plane of motion. In order to emphasize the nature of $\mathbf{r}_1$, $\mathbf{r}_2$ we relabel them as $\mathbf{a}$, $\mathbf{b}$ respectively (see Fig. 4.5). The angle between the direction of $\mathbf{a}$ and that of the unit vector $\mathbf{n}$ lying in the plane of the reference particles and having a fixed direction in space is taken to be θ (see Fig. 4.5 again).

It follows from the relation (4.13) that the position vector $\mathbf{r}_i$ of any particle P_i with respect to the particle P_0 can be written as

$$\mathbf{r}_i = \alpha_i \mathbf{a} + \beta_i \mathbf{b} + \mu_i \mathbf{k} \tag{4.15}$$

where α_i, β_i, μ_i are all scalar constants. Differentiating both sides of the relation (4.15) with respect to time, and using relations (4.5), leads to

$$\begin{aligned}\dot{\mathbf{r}}_i &= \alpha_i \dot{\mathbf{a}} + \beta_i \dot{\mathbf{b}} \\ &= \dot{\theta}(\alpha_i \mathbf{b} - \beta_i \mathbf{a}) \\ &= \dot{\theta}\mathbf{k} \times (\alpha_i \mathbf{a} + \beta_i \mathbf{b}) \\ &= \dot{\theta}\mathbf{k} \times (\alpha_i \mathbf{a} + \beta_i \mathbf{b} + \mu_i \mathbf{k}) = \dot{\theta}\mathbf{k} \times \mathbf{r}_i\end{aligned} \tag{4.16}$$

Denoting the position vectors of the particles P_i ($i = 0, 1, 2, \ldots, N$) with respect to the fixed origin O by $\mathbf{R}_i$ we have $\mathbf{R}_i = \mathbf{R}_0 + \mathbf{r}_i$, and therefore differentiation with respect to time yields

$$\dot{\mathbf{r}}_i = \dot{\mathbf{R}}_i - \dot{\mathbf{R}}_0 = \mathbf{v}_{P_i} - \mathbf{v}_{P_0} \tag{4.17}$$

The quantity $\mathbf{v}_{P_i} - \mathbf{v}_{P_0}$ is the velocity of P_i relative to the velocity of P_0. We frequently write $\mathbf{v}_{P_iP_0}$ for it, the order of the letters in the double subscript indicating that we are considering the velocity of P_i relative to P_0. Thus $\mathbf{v}_{P_0P_i} = -\mathbf{v}_{P_iP_0}$.

For simplicity of notation $\mathbf{v}_{P_i}$ is replaced by $\mathbf{v}_i$ ($i = 0, 1, 2, \ldots, N$), and hence combining (4.16) and (4.17) we obtain

$$\begin{aligned} \mathbf{v}_i &= \mathbf{v}_0 + \dot{\theta}\mathbf{k} \times \mathbf{r}_i \\ &= \mathbf{v}_0 + \boldsymbol{\omega} \times \mathbf{r}_i \qquad \text{where } \boldsymbol{\omega} = \dot{\theta}\mathbf{k} \end{aligned} \tag{4.18}$$

The vector $\boldsymbol{\omega}$ is called the angular velocity vector of the rigid body, while $\dot{\theta}$ is referred to as the angular speed. The direction of $\boldsymbol{\omega}$ in planar motion is always perpendicular to the plane of the motion. It is important to appreciate how the angular speed, frequently denoted by ω, is measured. Referring to Fig. 4.5 we see that $\dot{\theta}$ is the rate of change of the angle between the direction of the unit vector $\mathbf{a}$ (fixed in the body) and the direction of the unit vector $\mathbf{n}$ (fixed in space), both vectors being *parallel* to the plane of motion. Using (4.18) we have for any two particles P_i, P_j

$$\mathbf{v}_j - \boldsymbol{\omega} \times \mathbf{r}_j = \mathbf{v}_0 = \mathbf{v}_i - \boldsymbol{\omega} \times \mathbf{r}_i$$

and hence

$$\mathbf{v}_j = \mathbf{v}_i + \boldsymbol{\omega} \times (\mathbf{r}_j - \mathbf{r}_i) \tag{4.19}$$

The relation (4.19) is particularly important since it shows how the absolute velocities of any two particles of a rigid body are connected. It also holds for the three-dimensional motion of a rigid body except that the vector $\boldsymbol{\omega}$ has (as will be seen in Chapter 5) components in the directions of all three unit vectors $\mathbf{i}, \mathbf{j}, \mathbf{k}$ instead of only $\mathbf{k}$ for planar motion. Recalling (4.10) and (4.18) we see that the relation (4.19) may be interpreted as follows:

The velocity of P_j is equal to the velocity of P_i plus the velocity of a particle describing a circle centre P_i, radius $|\overline{P_iP_j}|$, at an angular rate $\dot{\theta}$, the plane of the circle having $\mathbf{k}$ as normal.

Differentiating both sides of the relation (4.19) with respect to time we obtain

$$\mathbf{f}_j = \mathbf{f}_i + \dot{\boldsymbol{\omega}} \times (\mathbf{r}_j - \mathbf{r}_i) + \boldsymbol{\omega} \times (\dot{\mathbf{r}}_j - \dot{\mathbf{r}}_i)$$

or

$$\mathbf{f}_j = \mathbf{f}_i + \dot{\boldsymbol{\omega}} \times (\mathbf{r}_j - \mathbf{r}_i) + \boldsymbol{\omega} \times (\mathbf{v}_j - \mathbf{v}_i)$$

and therefore using (4.19) we obtain

$$\mathbf{f}_j = \mathbf{f}_i + \dot{\boldsymbol{\omega}} \times (\mathbf{r}_j - \mathbf{r}_i) + \boldsymbol{\omega} \times (\boldsymbol{\omega} \times (\mathbf{r}_j - \mathbf{r}_i)) \tag{4.20}$$

where $\mathbf{f}_i$ $(i = 0, 1, 2, \ldots, N)$ is the absolute acceleration of the particle P_i.

The expression on the right-hand side of (4.20) shows how the absolute acceleration of the particle P_j is related to the absolute acceleration of the particle P_i.

We observe that if Q_i is the foot of the perpendicular from the position of the particle P_i to the plane defined by the particles P_0, P_1, P_2, and Q_0 is the point in space occupied by P_0 then, by virtue of (4.15),

$$\underline{Q_0Q_i} = \alpha_i\mathbf{a} + \beta_i\mathbf{b}$$

Hence noting (4.16) and (4.18) we have

$$\dot{\mathbf{r}}_i = \dot{\theta}\mathbf{k} \times (\alpha_i\mathbf{a} + \beta_i\mathbf{b}) = \boldsymbol{\omega} \times \underline{Q_0Q_i}$$

so that with the help of (4.17) we deduce

$$\mathbf{v}_i = \mathbf{v}_0 + \boldsymbol{\omega} \times \underline{Q_0Q_i}, \qquad \mathbf{v}_j = \mathbf{v}_i + \boldsymbol{\omega} \times \underline{Q_iQ_j} \tag{4.21}$$

Differentiating both sides of the last relation with respect to time gives an alternative form of (4.20), namely

$$\mathbf{f}_j = \mathbf{f}_i + \dot{\boldsymbol{\omega}} \times \underline{Q_iQ_j} + \boldsymbol{\omega} \times (\boldsymbol{\omega} \times \underline{Q_iQ_j}) \tag{4.22}$$

Expanding the triple-vector product and noting that $\boldsymbol{\omega}$ is perpendicular to $\underline{Q_iQ_j}$ we deduce, as $\dot{\boldsymbol{\omega}} = \ddot{\theta}\mathbf{k}$,

$$\mathbf{f}_j = \mathbf{f}_i + \ddot{\theta}\mathbf{k} \times \underline{Q_iQ_j} - \dot{\theta}^2\underline{Q_iQ_j} \tag{4.23}$$

The components of the relative acceleration $\mathbf{f}_{P_jP_i}$ are shown in Fig. 4.6. The quantity $\dot{\boldsymbol{\omega}}$ is called the absolute angular acceleration of the body.

It is evident from the relations (4.21) and (4.22) that the motion of the particles of the rigid body can be determined from the motion of the points Q_i $(i = 0, 1, 2, \ldots, N)$ lying in the plane of the reference particles P_0, P_1, P_2. Accordingly we can, if we wish, restrict our analysis in any problem to the motion of these points.

We now consider some particular examples of planar motion of a rigid body.

Rectilinear and curvilinear translation of a rigid body

If the angular speed $\dot{\theta} = 0$ then the direction of the unit vector $\mathbf{a}$ is fixed and $\mathbf{v}_i = \mathbf{v}_0$ for all the particles P_i $(i = 1, 2, \ldots, N)$. If, in addition, $\mathbf{v}_0$ is a constant vector then all the particles move in parallel straight lines and the motion is said to be rectilinear translation. When $\mathbf{v}_0$ is not a constant vector

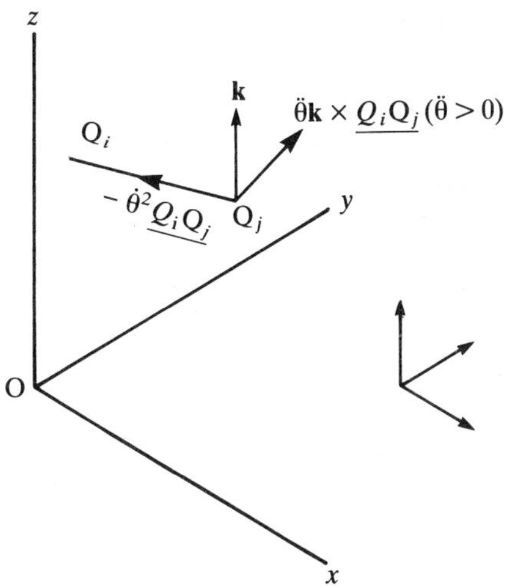

Fig. 4.6

the particles move in parallel curved paths, the motion being described as a curvilinear translation. The paths of the particles can be determined from the three linear differential equations

$$\dot{x} = f(t), \qquad \dot{y} = g(t), \qquad \dot{z} = 0$$

where $[x, y, z]$ are the coordinates of a typical particle P_i referred to the fixed axes Ox, Oy, Oz and $\mathbf{v}_0 = [f(t), g(t), 0]$ referred to the unit vectors $\mathbf{i}$, $\mathbf{j}$, $\mathbf{k}$, the functions $f(t)$ and $g(t)$ being known functions of time.

Rotation of a rigid body about a fixed axis

If the velocity $\mathbf{v}_0$ of the point Q_0 is zero then, by virtue of (4.21),

$$\mathbf{v}_i = \boldsymbol{\omega} \times \underline{Q_0Q_i} \qquad \text{where } \underline{Q_0Q_i} = \alpha_i\mathbf{a} + \beta_i\mathbf{b} \tag{4.24}$$

Thus the points Q_i are describing circles centre Q_0 radii $\sqrt{(\alpha_i^2 + \beta_i^2)}$ with angular speed $\dot{\theta}$. Likewise the particles P_i will also be describing circles (in planes parallel to the plane of the motion) whose centres lie on the fixed line through Q_0 in the direction of the unit vector $\mathbf{k}$. We say that the body is rotating about a fixed axis which is called the axis of rotation. A more general situation in which the direction of the axis of rotation is not fixed but varies continuously is considered in Chapter 5.

Putting $i = 0$ in (4.23) we see that the acceleration of the particle P_j is given by

$$\mathbf{f}_j = \ddot{\theta}\mathbf{k} \times \underline{Q_0Q_j} - \dot{\theta}^2\underline{Q_0Q_j} \tag{4.25}$$

Relative motion between bodies

In problems in mechanics we are frequently concerned with the relative motion of bodies whose surfaces come into contact. The contact may be at a single point, along a line or a curved surface, as for example in a cylindrical or spherical joint. A common but unsatisfactory practice is to label a point of contact by a single letter. The practice is unsatisfactory because it is not clear whether the point is to be regarded as belonging to one body or the other. A preferable alternative is to use a single letter A with subscripts as in A_1, A_2, the subscript 1 indicating that the point of contact is to be taken as a point of one body and the subscript 2 that it is to be taken as a point of the other. The notation is readily extended if more than two bodies are under consideration (see for example the motion of linkages). With the help of the relations (4.21) and (4.23) we can derive expressions for the velocities $\mathbf{v}_{A_1}$, $\mathbf{v}_{A_2}$ and the accelerations $\mathbf{f}_{A_1}$, $\mathbf{f}_{A_2}$. If there is no relative motion at the point of contact then $\mathbf{v}_{A_1} = \mathbf{v}_{A_2}$ while if there is relative motion $\mathbf{v}_{A_1} - \mathbf{v}_{A_2} = \mathbf{v}$ (say), where $\mathbf{v}$ is assumed given. The velocity $\mathbf{v}$ is a measure of the slipping between the two surfaces in contact.

Example 4.2

A circular disc centre C and radius a moves so that its plane always remains in the same (fixed) vertical plane. During the motion the disc maintains contact with a fixed horizontal plane. Given that at time t the relative velocity of slip between the disc and the horizontal plane at the point of contact is $\mathbf{v}(t)$ and the velocity of the centre C of the disc is $\mathbf{u}(t)$, obtain expressions for the quantities $\dot{\theta}$ and $\ddot{\theta}$, where $\dot{\theta}$ is the angular speed of the disc.

▷ *Solution* We take the disc to be body 1, the horizontal plane to be body 2, and A_1, A_2 to be labels for the point of contact between the disc and the plane (see Fig. 4.7). Noting the given information we have $\mathbf{v}_{A_1} - \mathbf{v}_{A_2} = \mathbf{v}$ and $\mathbf{v}_{A_2} = \mathbf{0}$, so that

$$\mathbf{v}_{A_1} = \mathbf{v} \tag{4.26}$$

Using (4.19) gives

$$\mathbf{v} = \mathbf{v}_{A_1} = \mathbf{u} + \boldsymbol{\omega} \times \underline{CA_1}$$

Taking the directions of the unit vectors $\mathbf{i}$, $\mathbf{j}$ to be as shown in Fig. 4.7 we see that $\underline{CA_1} = -a\mathbf{j}$, and hence

$$\mathbf{v} - \mathbf{u} = \dot{\theta}\mathbf{k} \times (-a\mathbf{j}) = a\dot{\theta}\mathbf{i} \tag{4.27}$$

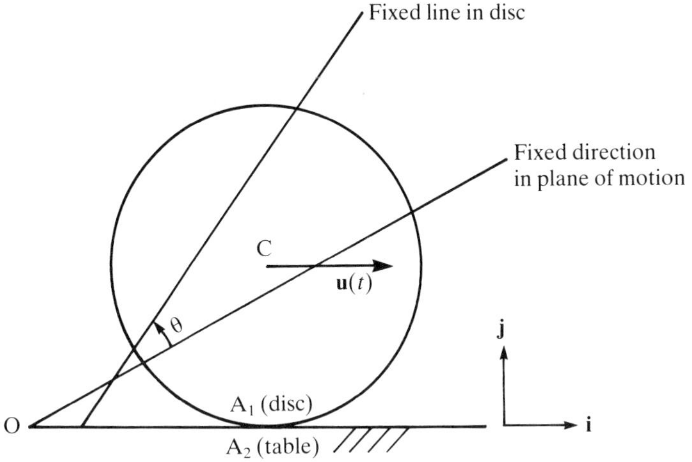

Fig. 4.7

where θ is the angle between a line fixed in the plane of the disc and moving with the disc, and a fixed line in the vertical plane of motion.

The relation (4.27) leads to the determination of $\dot{\theta}$ for given $\mathbf{v}(t)$ and $\mathbf{u}(t)$. If $\mathbf{v} = \mathbf{u}$ then $\dot{\theta} = 0$ so that the disc slips along the plane but does not rotate. When $\mathbf{u} \neq \mathbf{v} \neq \mathbf{0}$ the disc still slips but also rotates. The motion of the disc is then frequently described as a combination of rolling and slipping. When $\mathbf{v} = \mathbf{0}$ there is no slipping at the point of contact and the motion is often said to be one of pure rolling. The angular speed is then given by $a\dot{\theta}\mathbf{i} = -\mathbf{u}$.

To determine $\ddot{\theta}$ we differentiate the relation (4.27) with respect to time and obtain

$$\dot{\mathbf{v}} - \dot{\mathbf{u}} = a\ddot{\theta}\mathbf{i} \tag{4.28}$$

It is important to realize why this differentiation is possible. The derivation of (4.27) makes use of the relation (4.26), which holds at only one particular instant (the instant of contact) and therefore cannot be differentiated with respect to time. In general $\dot{\mathbf{v}}_{A_1} \neq \dot{\mathbf{v}}$. During any time interval the point of contact will change and hence it must be given different labels B_1, B_2 (say). However, replacing A_1, A_2, by B_1, B_2 in the analysis leading to the relation (4.27) produces an equation of precisely the same form but holding for a different value of the time. Thus (4.27) holds for all the relevant values of time.

Using the relation (4.20) we have

$$\begin{aligned}\mathbf{f}_{A_1} &= \dot{\mathbf{u}} + \dot{\boldsymbol{\omega}} \times \underline{CA_1} + \boldsymbol{\omega} \times (\boldsymbol{\omega} \times \underline{CA_1}) \\ &= \dot{\mathbf{u}} + \ddot{\theta}\mathbf{k} \times (-a\mathbf{j}) - \omega^2\underline{CA_1} \\ &= \dot{\mathbf{u}} + a\ddot{\theta}\mathbf{i} - \omega^2\underline{CA_1} \\ &= \dot{\mathbf{v}} - \dot{\theta}^2\underline{CA_1}\end{aligned}$$

noting (4.28). We deduce that $\mathbf{f}_{A_1} = \dot{\mathbf{v}}$ only when the angular velocity of the disc is zero.

There are many practical problems, notably in the kinematics of mechanisms, that are concerned with the relative motion of a particle and a rigid body that are in contact with each other, the plane of motion of the particle being parallel to the fixed plane of motion of the rigid body (see Section 4.1 for the definition of this plane). Generally the motion of the particle (which itself may be part of another rigid body) relative to the body and the absolute motion of the body are known, and we are thus faced with the task of deriving expressions for the absolute velocity and acceleration of the particle. We show how this can be done in the course of the next section.

Determination of the absolute velocity and acceleration of a particle P moving relative to a rigid body, the plane of motion of the particle being parallel to the fixed plane of motion of the rigid body

As we have explained already, the motion of any particle P_i of a rigid body can be determined from the motion of the point of projection Q_i of the particle's position onto the plane of the reference particles P_0, P_1, P_2. Since the velocity of the particle P is parallel to the fixed plane of motion of the rigid body, we can assume, without loss of generality, that the particle moves in the plane of P_0, P_1, P_2. We denote the position of the particle P in this plane at any instant of time by Q (no suffix). Accordingly we can write as before (see relations preceding (4.21))

$$\underline{OQ_i} = \underline{OQ_0} + \underline{Q_0Q_i} = \underline{OQ_0} + \alpha_i\mathbf{a} + \beta_i\mathbf{b} \qquad (i = 1, 2, \ldots, N)$$

and

$$\underline{OQ} = \underline{OQ_0} + \underline{Q_oQ} = \underline{OQ_0} + \alpha\mathbf{a} + \beta\mathbf{b} \tag{4.29}$$

where O is a fixed origin, α_i, β_i are fixed quantities, α, β are variables, and **a**, **b** are the unit vectors defined just prior to and used in the relation (4.15).

Differentiating both sides of the relation (4.29) with respect to time yields

$$\begin{aligned}\mathbf{v}_P = \underline{\dot{OQ}} &= \mathbf{v}_0 + \alpha\dot{\mathbf{a}} + \beta\dot{\mathbf{b}} + \dot{\alpha}\mathbf{a} + \dot{\beta}\mathbf{b} \\ &= \mathbf{v}_0 + \alpha\dot{\theta}\mathbf{b} - \beta\dot{\theta}\mathbf{a} + \dot{\alpha}\mathbf{a} + \dot{\beta}\mathbf{b} \\ &= \mathbf{v}_0 + \dot{\theta}\mathbf{k} \times (\alpha\mathbf{a} + \beta\mathbf{b}) + \dot{\alpha}\mathbf{a} + \dot{\beta}\mathbf{b} \end{aligned} \tag{4.30}$$

using (4.5). Noting (4.21) we have

$$\mathbf{v}_i = \mathbf{v}_0 + \dot{\theta}\mathbf{k} \times (\alpha_i\mathbf{a} + \beta_i\mathbf{b})$$

When the position Q of the particle P coincides with the point Q_i fixed in the rigid body, $\alpha = \alpha_i$ and $\beta = \beta_i$, and hence

$$\mathbf{v}_P = \mathbf{v}_i + \dot{\alpha}\mathbf{a} + \dot{\beta}\mathbf{b} \tag{4.31}$$

We now define a vector $\mathbf{v}_{rel}$ by

$$\mathbf{v}_{rel} = \dot{\alpha}\mathbf{a} + \dot{\beta}\mathbf{b}$$

so that equation (4.31) becomes

$$\mathbf{v}_P = \mathbf{v}_i + \mathbf{v}_{rel} \tag{4.32}$$

The reader should note that (4.32) cannot be differentiated as it holds only when Q and Q_i are coincident. The vector $\mathbf{v}_{rel}$ is the velocity of the particle relative to an observer fixed in the body at the point Q_i.

Differentiating both sides of (4.30) with respect to time leads to

$$\begin{aligned}\mathbf{f}_P = \dot{\mathbf{v}}_P = \dot{\mathbf{v}}_0 &+ \ddot{\theta}\mathbf{k} \times (\alpha\mathbf{a} + \beta\mathbf{b}) + \dot{\theta}\mathbf{k} \times (\alpha\dot{\mathbf{a}} + \beta\dot{\mathbf{b}} + \dot{\alpha}\mathbf{a} + \dot{\beta}\mathbf{b}) \\ &+ \dot{\alpha}\dot{\mathbf{a}} + \dot{\beta}\dot{\mathbf{b}} + \ddot{\alpha}\mathbf{a} + \ddot{\beta}\mathbf{b}\end{aligned}$$

Using the relations in (4.5), the right-hand side of this relation can be simplified to give

$$\begin{aligned}\mathbf{f}_P = \dot{\mathbf{v}}_0 &+ \dot{\boldsymbol{\omega}} \times (\alpha\mathbf{a} + \beta\mathbf{b}) + \boldsymbol{\omega} \times (\boldsymbol{\omega} \times (\alpha\mathbf{a} + \beta\mathbf{b})) \\ &+ 2\boldsymbol{\omega} \times (\dot{\alpha}\mathbf{a} + \dot{\beta}\mathbf{b}) + \ddot{\alpha}\mathbf{a} + \ddot{\beta}\mathbf{b}\end{aligned} \tag{4.33}$$

It follows from (4.22) that

$$\mathbf{f}_i = \dot{\mathbf{v}}_0 + \dot{\boldsymbol{\omega}} \times (\alpha_i\mathbf{a} + \beta_i\mathbf{b}) + \boldsymbol{\omega} \times (\boldsymbol{\omega} \times (\alpha_i\mathbf{a} + \beta_i\mathbf{b}))$$

Hence again putting $\alpha = \alpha_i$ and $\beta = \beta_i$ we have, when Q and Q_i are coincident,

$$\mathbf{f}_P = \mathbf{f}_i + 2\boldsymbol{\omega} \times \mathbf{v}_{rel} + \mathbf{f}_{rel} \tag{4.34}$$

where

$$\mathbf{f}_{rel} = \ddot{\alpha}\mathbf{a} + \ddot{\beta}\mathbf{b} \tag{4.35}$$

Noting (4.4) we see that the quantity $\mathbf{f}_{rel}$ represents the acceleration of the particle P relative to a reference frame, defined here by the unit vectors **a** and $\mathbf{b}$, that is fixed in the body. The term $2\boldsymbol{\omega} \times \mathbf{v}_{rel}$ occurring in the right-hand side of (4.34) is called the Coriolis acceleration.

4.4 MECHANISMS

A mechanism consists of an assembly of connected elements called links. The links are assumed to behave like rigid bodies while the connections, as yet undefined, permit relative motion between the links. Examples of mechanisms are provided by aircraft undercarriages, excavators, knitting machines and robots. One link of a mechanism is fixed, this term implying that the link is fixed in space or fixed with respect to another body whose motion for the kinematic analysis in hand may be temporarily disregarded. We can, for example, determine the velocities and accelerations of the moving parts of the engine of a car relative to the car. If the car is stationary (relative to the earth) these quantities are absolute quantities, while if the car is in motion its motion must be taken into account to determine absolute velocities and accelerations.

In many mechanisms the motion of the links is two-dimensional with the links having a common (fixed) plane of motion (see Section 4.1). Such mechanisms are described as planar mechanisms and their kinematic analysis, as we shall show shortly, can be undertaken using the results derived in Sections 4.2 and 4.3. This analysis is carried through by proceeding from one link to another that is connected to it. At each connection we continue to think of two points, one in each of the links, that are coincident at the instant being considered: the nature of the connection will impose constraints on the relative motion of these points. In reality it is impossible for the links to occupy the same space and have coincident points, but hypothetically we can assume that they do.

Two types of connection, the revolute joint and the prismatic joint, are in common use in mechanisms.

Revolute joint

This joint is the most widely used in engineering practice. Its simplest form can be realized by slotting a pin through holes, centres A_1, A_2, cut in two links (see Fig. 4.8). In practice the joint is a highly sophisticated piece of engineering, but for modelling the relative motion of the two links it is sufficient to think in the simple terms just described. The pin permits the links to move in parallel planes, the axis (of symmetry) of the pin being perpendicular to these planes. We assume that there is no relative motion between the point A_1 (fixed in one link) and the point A_2 (fixed in the

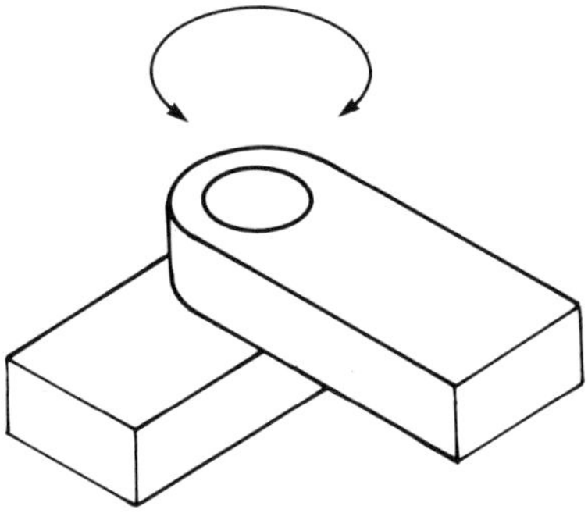

Fig. 4.8

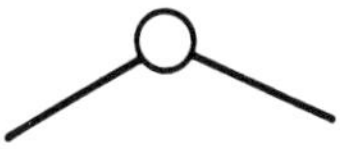

Fig. 4.9

other link) and take them to be coincident. In practical terms this means that the deflections arising from the forces acting on the joint and the relative motion within the clearance of the joint (called backlash, play or lost motion) are assumed to be negligible. A revolute joint is represented diagrammatically in Fig. 4.9, the straight lines that denote the links defining the plane of the relative motion.

Since the points A_1 and A_2 are taken to be coincident at all times, it follows that

$$\mathbf{v}_{A_1} = \mathbf{v}_{A_2}, \qquad \mathbf{f}_{A_1} = \mathbf{f}_{A_2} \tag{4.36}$$

It is common practice, as the points are coincident, to omit the subscripts and write $\mathbf{v}_A$ and $\mathbf{f}_A$ for the respective velocities and accelerations, it being understood that A may be considered to be a point of either link. We shall generally follow this practice but there will be occasions when it is necessary to depart from it.

Prismatic joint

Figure 4.10 shows two rigid links, one of which (link 1) contains a slot into which the other (link 2) just fits. The slot permits the links to slide over each other so that they move in parallel planes. This simple arrangement is

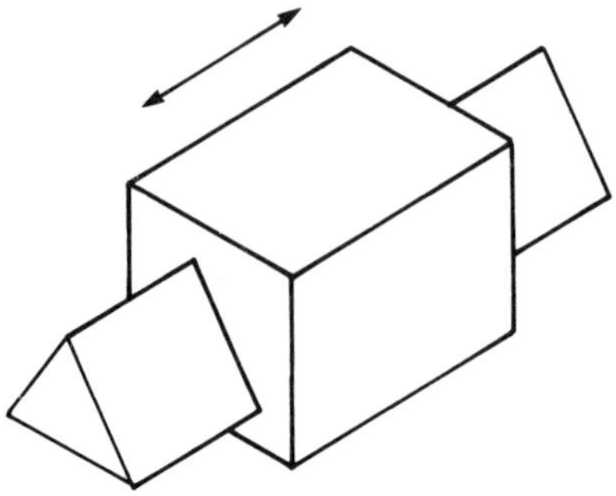

Fig. 4.10

Fig. 4.11

an example of a prismatic (or sliding) joint. A prismatic joint is represented diagrammatically by the symbol in Fig. 4.11.

When modelling the motion of two links that are connected by a prismatic joint we make two assumptions. Firstly we assume that the angle between a line fixed in one link and a line fixed in the other (both lines being parallel to the fixed plane of motion) is constant. This implies that the angular velocities of the two links are equal. Secondly we assume that

$$\mathbf{v}_{A_2} - \mathbf{v}_{A_1} = v\mathbf{n} \tag{4.37}$$

where $\mathbf{n}$ is a unit vector fixed with respect to each link (by definition being parallel to the plane of motion of the links), v is a variable and A_1 is a point of one link that is instantaneously in contact with the point A_2 of the other link.

We now show that the accelerations of the points of coincidence are connected by the relation

$$\mathbf{f}_{A_2} - \mathbf{f}_{A_1} = 2v\boldsymbol{\omega} \times \mathbf{n} + \dot{v}\mathbf{n} \tag{4.38}$$

where $\boldsymbol{\omega}$ is the common angular velocity of the links. Let O_1, A_1 be points fixed in one link and O_2, A_2 be points in the second link. For the moment we do not assume A_1 and A_2 to be coincident. With the help of (4.19) we can write

$$\mathbf{v}_{A_1} = \mathbf{v}_1 + \boldsymbol{\omega} \times \underline{O_1A_1}$$

$$\mathbf{v}_{A_2} = \mathbf{v}_2 + \boldsymbol{\omega} \times \underline{O_2A_2}$$

where $\mathbf{v}_1$, $\mathbf{v}_2$ denote the velocities of the points O_1, O_2. Hence

$$\mathbf{v}_{A_2} - \mathbf{v}_{A_1} = \mathbf{v}_2 - \mathbf{v}_1 + \boldsymbol{\omega} \times (\underline{O_2A_2} - \underline{O_1A_1})$$

$$= \mathbf{v}_2 - \mathbf{v}_1 + \boldsymbol{\omega} \times (\underline{O_2O_1} + \underline{O_1A_2} - \underline{O_1A_1}) \tag{4.39}$$

When A_1, A_2 are coincident this becomes, with the help of (4.37),

$$\mathbf{v}_2 - \mathbf{v}_1 + \boldsymbol{\omega} \times \underline{O_2O_1} = v\mathbf{n} \tag{4.40}$$

The relation (4.40) can be differentiated with respect to time t even though it is derived with the aid of (4.37) which holds for a given instant only. Replacing A_1, A_2 by B_1, B_2 in the analysis leading to (4.40) produces a relation of precisely the same form but holding for a different value of the time (see Example 4.2 for a similar situation). Differentiating both sides of (4.40) with respect to time yields

$$\dot{\mathbf{v}}_2 - \dot{\mathbf{v}}_1 + \dot{\boldsymbol{\omega}} \times \underline{O_2O_1} + \boldsymbol{\omega} \times \underline{\dot{O_2O_1}} = \dot{v}\mathbf{n} + v\dot{\mathbf{n}}$$

which becomes, on noting (4.5),

$$\mathbf{f}_2 - \mathbf{f}_1 + \dot{\boldsymbol{\omega}} \times \underline{O_2O_1} + \boldsymbol{\omega} \times (\mathbf{v}_1 - \mathbf{v}_2) = \dot{v}\mathbf{n} + v\boldsymbol{\omega} \times \mathbf{n} \tag{4.41}$$

Using (4.20) we can write

$$\mathbf{f}_{A_1} = \mathbf{f}_1 + \dot{\boldsymbol{\omega}} \times \underline{O_1A_1} + \boldsymbol{\omega} \times (\boldsymbol{\omega} \times \underline{O_1A_1})$$

$$\mathbf{f}_{A_2} = \mathbf{f}_2 + \dot{\boldsymbol{\omega}} \times \underline{O_2A_2} + \boldsymbol{\omega} \times (\boldsymbol{\omega} \times \underline{O_2A_2})$$

so that when A_1, A_2 are coincident we have

$$\mathbf{f}_{A_2} - \mathbf{f}_{A_1} = \mathbf{f}_2 - \mathbf{f}_1 + \dot{\boldsymbol{\omega}} \times \underline{O_2O_1} + \boldsymbol{\omega} \times (\boldsymbol{\omega} \times \underline{O_2O_1})$$

Eliminating $\boldsymbol{\omega} \times \underline{O_2O_1}$ from this relation by means of (4.40) we deduce that

$$\mathbf{f}_{A_2} - \mathbf{f}_{A_1} = \mathbf{f}_2 - \mathbf{f}_1 + \dot{\boldsymbol{\omega}} \times \underline{O_2O_1} + \boldsymbol{\omega} \times (v\mathbf{n} + \mathbf{v}_1 - \mathbf{v}_2)$$

which reduces in turn, noting (4.41), to

$$\mathbf{f}_{A_2} - \mathbf{f}_{A_1} = \dot{v}\mathbf{n} + 2v\boldsymbol{\omega} \times \mathbf{n}$$

as required.

If we write $\mathbf{v}_{rel} = v\mathbf{n}$ and $\mathbf{f}_{rel} = \dot{v}\mathbf{n}$ the relation (4.38) becomes

$$\mathbf{f}_{A_2} = \mathbf{f}_{A_1} + 2\boldsymbol{\omega} \times \mathbf{v}_{rel} + \mathbf{f}_{rel} \tag{4.42}$$

This relation can be compared in form with the relation in (4.34). However, a word of caution is necessary. The direction of the vector $\mathbf{f}_{rel}$ in (4.34) is not necessarily fixed relative to the rigid body.

It commonly happens that the unit vector $\mathbf{n}$ is directed along a line that represents one of the links. In this case the other (second) link is frequently described as sliding on the first. The terminology is confusing since, as we pointed out earlier, either link may be regarded as sliding over the other.

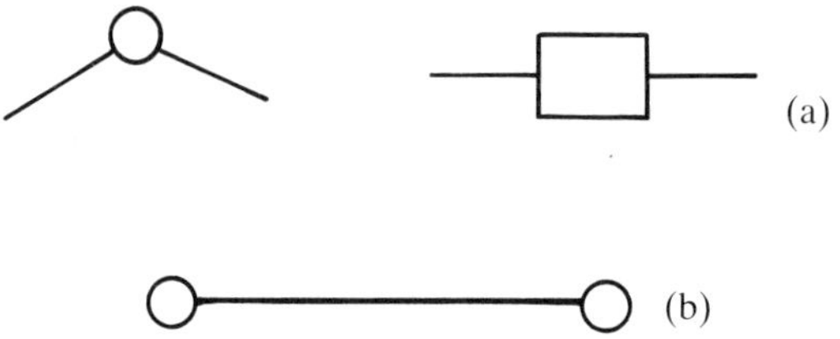

Fig. 4.12

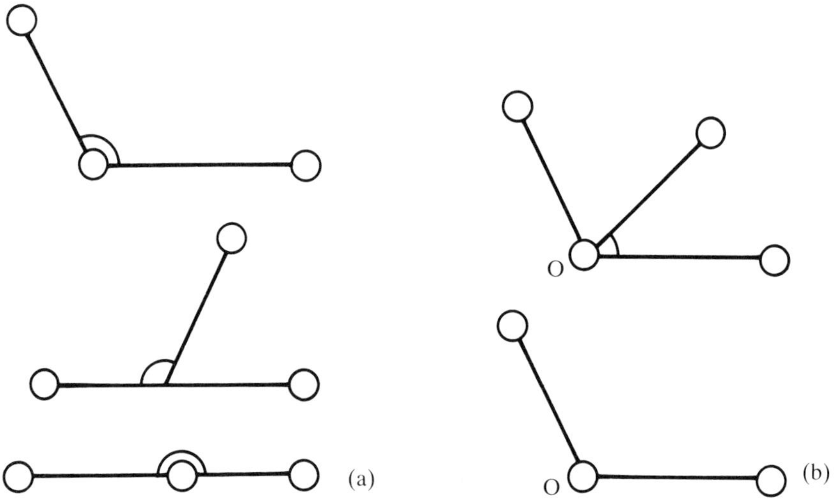

Fig. 4.13

Conventions and notation used in drawing diagrams of planar mechanisms

For the purpose of the kinematic analysis of a mechanism the detailed drawings and manufacturing information necessary for the mechanism to be made are not required. It is sufficient for the geometry of the mechanism to be defined, and this is most conveniently done by providing a diagram that shows the links and the types of connections between them, together with a list of linear and angular dimensions. The symbols for revolute (R) and prismatic (P) joints have already been introduced and are shown again in Fig. 4.12(a). A link may be connected to several other links. The symbol in Fig. 4.12(b) represents a link that can be connected to two other links by revolute joints. Each of the symbols in Fig. 4.13(a) also represents a single link (with three revolute connections), the arc between lines denoting that the angle between them is fixed. In Fig. 4.13(b) each of the symbols represents two links connected by a revolute joint at O.

Each link of a mechanism is numbered and each joint is labelled with a letter. The number 1 is reserved for the fixed link and the number 2 for the driving link (this link itself being driven by the application of an external force and/or couple). The letters O, P, Q ... are reserved for revolute joints in link 1, other joints being labelled in sequence around the mechanism with letters starting at A. Cross-hatchings are attached to the line (or lines) that represent the fixed link. The point of contact of link i with another link at a joint is denoted by attaching the suffix i to the letter that labels the joint. The suffix will be omitted if there is no ambiguity (see note subsequent to the relation (4.36)). The reader should appreciate that A_4 (for example) will only have a meaning if link 4 is connected to another link at joint A.

The unit vector located in one of the lines that represents link i is labelled $\mathbf{s}_i$ while the further unit vector $\mathbf{t}_i$ is defined by $\mathbf{s}_i \times \mathbf{t}_i = \mathbf{k}$, where $\mathbf{k}$ is perpendicular to the plane of motion of the links. The angle between the unit vector $\mathbf{s}_i$ and the axis Ox is θ_i (the angle being measured positive counter-clockwise from Ox). The axis Ox is usually along, or parallel to, one of the lines that represent the fixed link. Noting the relations in (4.18) it follows that the angular velocity $\boldsymbol{\omega}_i$ of link i is given by $\boldsymbol{\omega}_i = \omega_i\mathbf{k}$, where $\omega_i = \dot{\theta}_i$.

Each of Fig. 4.14(a), (b) and (c) shows a diagram of a mechanism using the conventions and notation just described. In Fig. 4.14(b) the length AB is fixed so that the relative motion between links at B is directed along link 4, while in Fig. 4.14(c) the length PB is fixed so that the relative motion at B is directed along link 3. A particularly important and widely used form of the mechanism shown in Fig. 4.14(b) is obtained by taking the points A and B to coincide so that link 3 moves relatively to link 4 along the line of link 4, which passes through the pin connecting link 2 to link 3. The mechanism is shown in Fig. 4.15.

Figure 4.16 shows a diagram of a further mechanism, again using the conventions and notation that have been defined. It well illustrates the complexities that can arise. The mechanism has six links with five revolute joints and two prismatic joints. The revolute joints are at O between links 1 and 2, at A between links 2 and 3, at B between links 4 and 5, at C between links 5 and 6, and at P between links 1 and 4. The prismatic joints are at A, where the relative motion between links 3 and 4 is along link 4; and at C, where the relative motion between links 1 and 6 is along link 1. If the lengths of OP, OA, PB and BC and the angles CPO and PBA are known then the position of the mechanism is defined when the value of the angle θ_2 is given. For the purposes of kinematic analysis three coincident points of contact would be considered at the point A: A_2 in link 2, A_3 in link 3 and A_4 in link 4. For the revolute joint between links 2 and 3 the points A_2 and A_3 are coincident at all times, while for the prismatic joint between links 3 and 4 the point A_4 is instantaneously coincident with A_3.

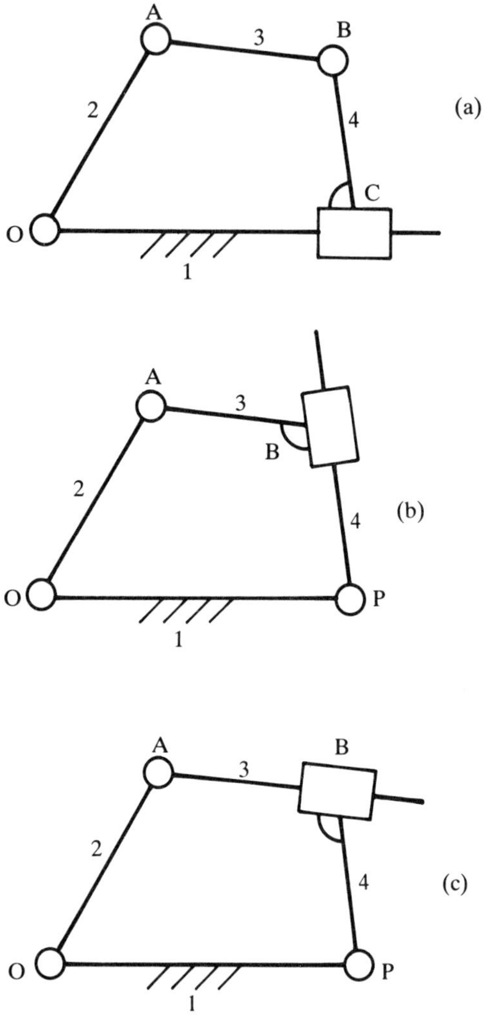

Fig. 4.14

It is convenient at this stage to obtain expressions for the scalar products $\mathbf{s}_i \cdot \mathbf{s}_j$, $\mathbf{t}_i \cdot \mathbf{t}_j$, $\mathbf{s}_i \cdot \mathbf{t}_j$ that arise frequently in analysis. Referred to the unit vectors $\mathbf{i}$, $\mathbf{j}$, $\mathbf{k}$,

$$\mathbf{s}_i = [\cos\theta_i, \sin\theta_i, 0], \qquad \mathbf{s}_j = [\cos\theta_j, \sin\theta_j, 0],$$
$$\mathbf{t}_i = [-\sin\theta_i, \cos\theta_i, 0], \qquad \mathbf{t}_j = [-\sin\theta_j, \cos\theta_j, 0]$$

so that

$$\mathbf{s}_i \cdot \mathbf{s}_j = \cos(\theta_i - \theta_j) = \mathbf{t}_i \cdot \mathbf{t}_j, \qquad \mathbf{s}_i \cdot \mathbf{t}_j = \sin(\theta_i - \theta_j) \tag{4.43}$$

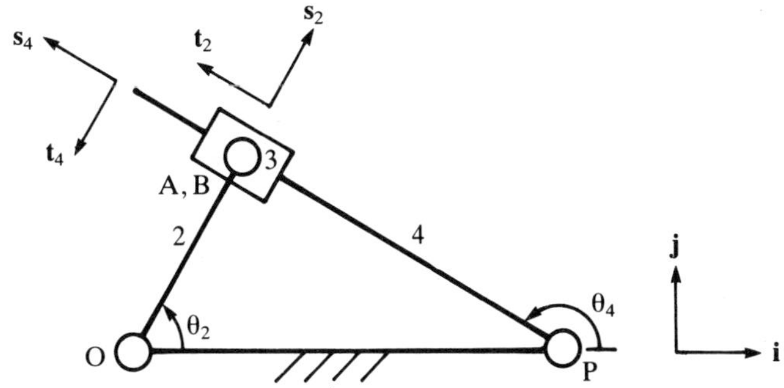

Fig. 4.15

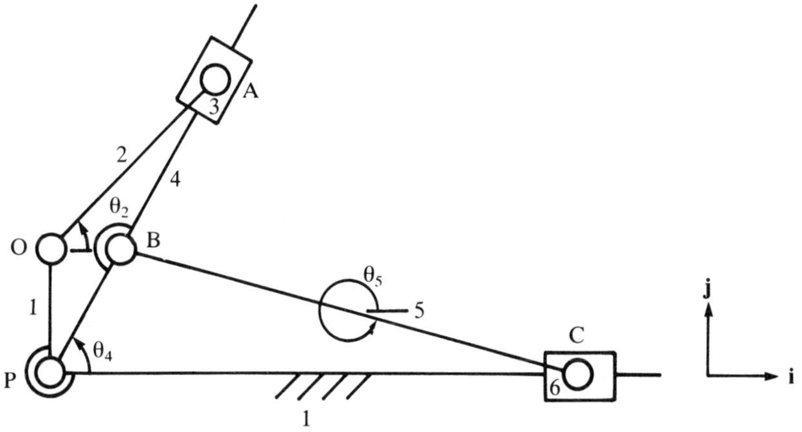

Fig. 4.16

Example 4.3

Figure 4.17 shows a diagram of a mechanism that has four links and four revolute joints. Derive expressions for the angular velocities $\boldsymbol{\omega}_3$, $\boldsymbol{\omega}_4$ and the angular accelerations $\dot{\boldsymbol{\omega}}_3$, $\dot{\boldsymbol{\omega}}_4$ of the links AB and BP in terms of the quantities θ_2, θ_3, θ_4, $\dot{\theta}_2$, $\ddot{\theta}_2$ and the lengths r_i ($i = 1, 2, 3, 4$) of the links.

Solution Since the mechanism contains only revolute joints the suffixes used in respect of coincident points are omitted.

As O is fixed we have, using (4.18),

$$\mathbf{v}_A = \boldsymbol{\omega}_2 \times \underline{OA} = \dot{\theta}_2\mathbf{k} \times r_2\mathbf{s}_2 = r_2\dot{\theta}_2\mathbf{t}_2$$

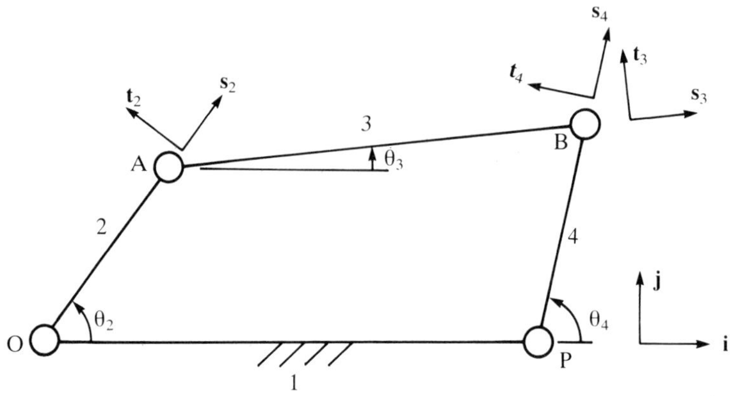

Fig. 4.17

Recalling (4.19) we obtain

$$\mathbf{v}_B = \mathbf{v}_A + \boldsymbol{\omega}_3 \times \underline{AB} = r_2\dot{\theta}_2\mathbf{t}_2 + r_3\dot{\theta}_3\mathbf{t}_3$$

Since P is fixed we also have

$$\mathbf{v}_B = \boldsymbol{\omega}_4 \times \underline{PB} = r_4\dot{\theta}_4\mathbf{t}_4$$

so that equating the two expressions for $\mathbf{v}_B$ yields

$$r_2\dot{\theta}_2\mathbf{t}_2 + r_3\dot{\theta}_3\mathbf{t}_3 = r_4\dot{\theta}_4\mathbf{t}_4$$

This is a vector equation for the two unknowns $\dot{\theta}_3$ and $\dot{\theta}_4$. To eliminate $\dot{\theta}_4$ we take the scalar product of both sides with the vector $\mathbf{s}_4$. Hence

$$r_2\dot{\theta}_2\mathbf{t}_2 \,.\, \mathbf{s}_4 + r_3\dot{\theta}_3\mathbf{t}_3 \,.\, \mathbf{s}_4 = 0$$

Noting the relations in (4.43) this leads to

$$r_3\dot{\theta}_3\sin(\theta_4 - \theta_3) = r_2\dot{\theta}_2\sin(\theta_2 - \theta_4) \tag{4.44}$$

Likewise, taking the scalar product with $\mathbf{s}_3$, we obtain

$$r_4\dot{\theta}_4\sin(\theta_4 - \theta_3) = r_2\dot{\theta}_2\sin(\theta_2 - \theta_3) \tag{4.45}$$

As O is fixed it follows, noting (4.23), that

$$\begin{aligned}\mathbf{f}_A &= \ddot{\theta}_2\mathbf{k} \times \underline{OA} - \dot{\theta}_2^2\underline{OA}\\ &= r_2\ddot{\theta}_2\mathbf{t}_2 - r_2\dot{\theta}_2^2\mathbf{s}_2\end{aligned}$$

while

$$\begin{aligned}\mathbf{f}_B &= \dot{\mathbf{v}}_A + \ddot{\theta}_3\mathbf{k} \times \underline{AB} - \dot{\theta}_3^2\underline{AB}\\ &= r_2(\ddot{\theta}_2\mathbf{t}_2 - \dot{\theta}_2^2\mathbf{s}_2) + r_3(\ddot{\theta}_3\mathbf{t}_3 - \dot{\theta}_3^2\mathbf{s}_3)\end{aligned}$$

But $\mathbf{f}_B$ is also given, as P is fixed, by

$$\mathbf{f}_B = r_4\ddot{\theta}_4 t_4 - r_4\dot{\theta}_4^2\mathbf{s}_4$$

Equating the two expressions for $\mathbf{f}_B$ yields the following vector equation for $\ddot{\theta}_3$, $\ddot{\theta}_4$:

$$r_2(\ddot{\theta}_2\mathbf{t}_2 - \dot{\theta}_2^2\mathbf{s}_2) + r_3(\ddot{\theta}_3\mathbf{t}_3 - \dot{\theta}_3^2\mathbf{s}_3) = r_4(\ddot{\theta}_4\mathbf{t}_4 - \dot{\theta}_4^2\mathbf{s}_4)$$

Taking the scalar product throughout with the vector $\mathbf{s}_4$ eliminates $\ddot{\theta}_4$. We obtain, with the help of (4.43) and some elementary manipulation and rearrangement,

$$\begin{aligned} r_3\sin(\theta_4 - \theta_3)\ddot{\theta}_3 = r_2\{\ddot{\theta}_2\sin(\theta_2 - \theta_4) + \dot{\theta}_2^2\cos(\theta_4 - \theta_2)\} \\ + r_3\dot{\theta}_3^2\cos(\theta_4 - \theta_3) - r_4\dot{\theta}_4^2 \end{aligned} \quad (4.46)$$

In a similar manner it can be deduced that

$$\begin{aligned} r_4\sin(\theta_4 - \theta_3)\ddot{\theta}_4 = r_2\{\ddot{\theta}_2\sin(\theta_2 - \theta_3) + \dot{\theta}_2^2\cos(\theta_3 - \theta_2)\} \\ - r_4\dot{\theta}_4^2\cos(\theta_4 - \theta_3) + r_3\dot{\theta}_3^2 \end{aligned} \quad (4.47)$$

Hence the angular velocities and accelerations of links 3 and 4 have effectively been found.

This example prompts a number of remarks. The particular mechanism considered is frequently described as an RRRR mechanism because its four connections are all revolute joints.

In Fig. 4.17 it is evident that once the position of OA is fixed relative to link 1 then the position of the links AB and BP are also fixed (there are in fact two geometrically possible positions but only one is applicable for a given RRRR mechanism). The mechanism is said to have one degree of freedom. When n independent variables are required to specify the position of a mechanism, the mechanism is said to have n degrees of freedom. We limit our consideration of mechanisms to those having one degree of freedom.

If we wish to implement numerically the formulae for angular velocity and acceleration (see (4.44) to (4.47)) it is necessary to know the values of θ_3 and θ_4 for a given θ_2. The problem of finding these values is known as the problem of position (or sometimes as the problem of the geometry). Most problems of position in mechanisms can be solved by elementary trigonometry without recourse, as is sometimes proposed, to the techniques of numerical analysis. The difficulties that arise do so because trigonometric equations invariably have multiple solutions. It is not the purpose of this book to look at problems of position in any detail, and we limit our discussions to a consideration of the RRRR mechanism (additional material is to be found in the problems). The details are given in Section 4.5 and take the form of sequential steps that can be used as the basis of a computer program that solves the problem of the geometry. One other aspect of the geometry of mechanisms is also considered. We show that if the driving link OA (sometimes referred to as the crank) is to make complete revolutions then the lengths of the links must satisfy certain inequalities.

Example 4.4

Find, for the mechanism shown in Fig. 4.15, expressions for the angular velocity $\boldsymbol{\omega}_4$ and the angular acceleration $\dot{\boldsymbol{\omega}}_4$ of link 4 in terms of θ_2, θ_4, $\dot{\theta}_2$, $\ddot{\theta}_2$, the lengths r_1 and r_2 of links 1 and 2, and r_4 where $r_4 = |\underline{\mathrm{PA}}|$.

▷ *Solution* Since O is fixed we have, using (4.18),

$$\mathbf{v}_{\mathrm{A}_2} = \boldsymbol{\omega}_2 \times \underline{\mathrm{OA}_2} = \omega_2 \mathbf{k} \times r_2 \mathbf{s}_2 = \omega_2 r_2 \mathbf{t}_2 \tag{4.48}$$

The relative motion of link 3 to link 4 at the joint A is along the line of link 4 and it thus follows, with the help of (4.37), that

$$\mathbf{v}_{\mathrm{A}_3} = \mathbf{v}_{\mathrm{A}_4} + v\mathbf{s}_4 \tag{4.49}$$

where v is, at present, an unknown scalar.

Now P is a fixed point and therefore, using (4.18) once more,

$$\mathbf{v}_{\mathrm{A}_4} = \boldsymbol{\omega}_4 \times \underline{\mathrm{PA}_4} = \omega_4 r_4 \mathbf{t}_4 \tag{4.50}$$

Substituting the expressions for $\mathbf{v}_{\mathrm{A}_2} = \mathbf{v}_{\mathrm{A}_3}$ and $\mathbf{v}_{\mathrm{A}_4}$ given by (4.48) and (4.50) into (4.49) yields

$$\omega_2 r_2 \mathbf{t}_2 = \omega_4 r_4 \mathbf{t}_4 + v\mathbf{s}_4$$

This is a vector equation for the unknowns ω_4 and v. Taking the scalar product of both sides of it with $\mathbf{t}_4$ leads to

$$\omega_2 r_2 \mathbf{t}_2 \cdot \mathbf{t}_4 = \omega_4 r_4$$

from which we deduce

$$r_4 \omega_4 = r_2 \omega_2 \cos(\theta_2 - \theta_4) \tag{4.51}$$

Hence the angular velocity $\boldsymbol{\omega}_4$ has effectively been determined.

Taking the scalar product of both sides of the equation for ω_4 and v with $\mathbf{s}_4$ yields

$$v = r_2 \omega_2 \sin(\theta_4 - \theta_2) \tag{4.52}$$

Recalling the formula (4.25) we have

$$\mathbf{f}_{A_3} = \mathbf{f}_{A_2} = \dot{\omega}_2 \mathbf{k} \times \underline{OA_2} - \omega_2^2 r_2 \mathbf{s}_2$$
$$= r_2(\dot{\omega}_2 \mathbf{t}_2 - \omega_2^2 \mathbf{s}_2)$$

and

$$\mathbf{f}_{A_4} = \dot{\omega}_4 \mathbf{k} \times \underline{PB} - \omega_4^2 r_4 \mathbf{s}_4$$
$$= r_4(\dot{\omega}_4 \mathbf{t}_4 - \omega_4^2 \mathbf{s}_4)$$

Applying the result in (4.38) gives

$$\mathbf{f}_{A_3} = \mathbf{f}_{A_4} + 2\boldsymbol{\omega}_4 \times v\mathbf{s}_4 + \dot{v}\mathbf{s}_4$$

Substituting the expressions for $\mathbf{f}_{A_3}$ and $\mathbf{f}_{A_4}$ into this last relation leads to

$$r_2(\dot{\omega}_2 \mathbf{t}_2 - \omega_2^2 \mathbf{s}_2) = r_4(\dot{\omega}_4 \mathbf{t}_4 - \omega_4^2 \mathbf{s}_4) + 2\omega_4 v \mathbf{t}_4 + \dot{v}\mathbf{s}_4$$

We eliminate the unknown quantity $\dot{v}$ from this vector equation by taking the scalar product of both sides with $\mathbf{t}_4$. After some simplification and rearrangement of terms it is found that

$$r_4 \dot{\omega}_4 = r_2\{\dot{\omega}_2 \cos(\theta_2 - \theta_4) - \omega_2^2 \sin(\theta_2 - \theta_4)\} - 2\omega_4 v$$

This relation can, with the help of (4.52), be written in the alternative form

$$r_4 \dot{\omega}_4 = r_2 \dot{\omega}_2 \cos(\theta_2 - \theta_4) - r_2 \sin(\theta_2 - \theta_4)(\omega_2^2 - 2\omega_2\omega_4)$$

Thus the angular acceleration of link 4 is now known.

We conclude with the observation that the solution of the problem of position is straightforward. For a given θ_2 the quantities r_4 and θ_4 are determined from the formulae

$$r_4^2 = r_1^2 + r_2^2 - 2r_1 r_2 \cos\theta_2$$
$$2r_1 r_4 \cos\theta_4 = r_2^2 - r_4^2 - r_1^2$$

We now proceed to consider a mechanism that has more than four links and four joints.

Example 4.5

Figure 4.18 shows a diagram of a mechanism using the defined conventions and notation. It consists of six links with five revolute joints and two prismatic joints. At B the relative motion between links 5 and 6 is along the line of link 6.

Derive the following relations:

$$\omega_2 r_2 \cos\theta_2 = -\omega_3 r_3 \cos\theta_3$$
$$\omega_6 r_6 = \omega_2 r_2 \cos(\theta_2 - \theta_6) + \omega_3 |\underline{AB}| \cos(\theta_3 - \theta_6)$$

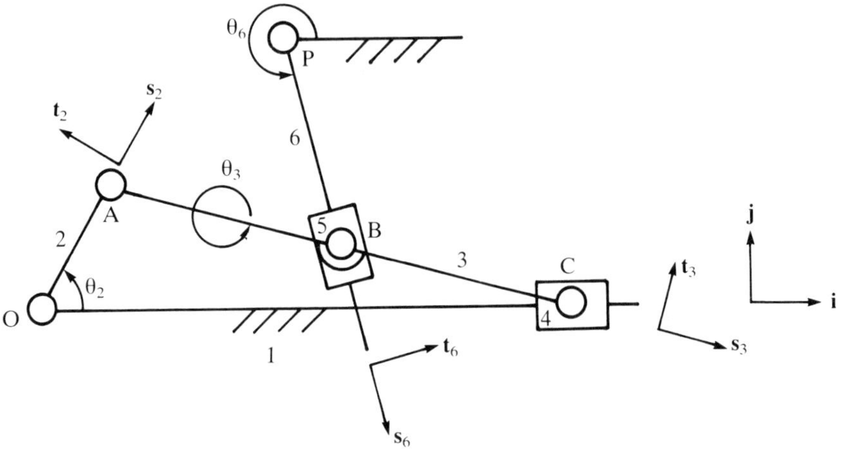

Fig. 4.18

where r_2, r_3 are the lengths of links 2 and 3, and $r_6 = |\underline{\mathrm{PB}}|$.

▷ *Solution* Since O is fixed we have, using (4.18),

$$\mathbf{v}_A = \boldsymbol{\omega}_2 \times \underline{\mathrm{OA}} = r_2\omega_2\mathbf{t}_2$$

while (4.19) gives

$$\mathbf{v}_{C_3} = \mathbf{v}_A + \boldsymbol{\omega}_3 \times \underline{\mathrm{AC}_3}$$
$$= r_2\omega_2\mathbf{t}_2 + \omega_3 r_3\mathbf{t}_3$$

Since link 1 is fixed

$$\mathbf{v}_{C_3} = v\mathbf{i}$$

where v is an unknown scalar and $\mathbf{i}$ (as shown in Fig. 4.18) is along the line of the fixed link. Equating the two expressions for $\mathbf{v}_{C_3}$ yields

$$v\mathbf{i} = r_2\omega_2\mathbf{t}_2 + \omega_3 r_3\mathbf{t}_3$$

Taking the scalar product of both sides of this equation with the vector $\mathbf{j}$ leads to

$$0 = r_2\omega_2\mathbf{t}_2 \cdot \mathbf{j} + \omega_3 r_3\mathbf{t}_3 \cdot \mathbf{j}$$

so that, recalling (4.43), we have

$$\omega_2 r_2\cos\theta_2 = -r_3\omega_3\cos\theta_3$$

as required.

Using (4.18) it follows that

$$\mathbf{v}_{B_6} = \boldsymbol{\omega}_6 \times \underline{PB_6} = r_6\omega_6\mathbf{t}_6$$

while by (4.19)

$$\mathbf{v}_{B_3} = \mathbf{v}_A + \boldsymbol{\omega}_3 \times \underline{AB}$$
$$= r_2\omega_2\mathbf{t}_2 + \omega_3|\underline{AB}|\mathbf{t}_3$$

But

$$\mathbf{v}_{B_5} = \mathbf{v}_{B_6} + v\mathbf{s}_6$$

where v is some unknown scalar. Now $\mathbf{v}_{B_5} = \mathbf{v}_{B_3}$, so that

$$\omega_2 r_2\mathbf{t}_2 + \omega_3|\underline{AB}|\mathbf{t}_3 = \omega_6 r_6\mathbf{t}_6 + v\mathbf{s}_6$$

Taking the scalar product throughout with $\mathbf{t}_6$ yields, recalling (4.43),

$$\omega_6 r_6 = \omega_2 r_2\cos(\theta_2 - \theta_6) + \omega_3|\underline{AB}|\cos(\theta_3 - \theta_6)$$

as required.

4.5 THE SOLUTION OF THE PROBLEM OF POSITION FOR THE RRRR MECHANISM

We observe that for a given angle θ_2 there are two possible configurations of the mechanism (see Fig. 4.19(a) and (b)). In Fig. 4.19(a) the vector $\underline{BA} \times \underline{BP}$ is out of the page and in Fig. 4.19(b) it is into the page. We assume for a *given* θ_2 that the mechanism is assembled in one of the two configurations. It is not difficult to see, either by drawing the mechanism in a succession of positions or by constructing a simple cardboard model with drawing pins for the revolute joints, that unless the points A, B, P of the links become collinear in the cycle the direction of the vector $\underline{BA} \times \underline{BP}$ is is unchanged throughout the cycle of motion. (For an alternative approach see Problem 4.17.) Thus the configuration is unaltered and is therefore defined by the initial assembly of the links. We distinguish between the two configurations by the introduction of a (configuration) number N whose value is $+1$ in Fig. 4.19(a) and -1 in Fig. 4.19(b).

The formulae that lead to the determination of the angles θ_3 and θ_4 for a given value of θ_2 are now stated. It may be necessary to add 2π to some of the expressions for θ_3 and θ_4 if negative values of these angles are to be avoided. The reader should note that mechanisms in which A, B, P become collinear are excluded from our consideration.

Let

$$\mu = \cos(\theta_4 - \theta_3) = \frac{r_3^2 + r_4^2 - r_1^2 - r_2^2}{2r_3r_4} + \frac{r_1r_2}{r_3r_4}\cos\theta_2$$

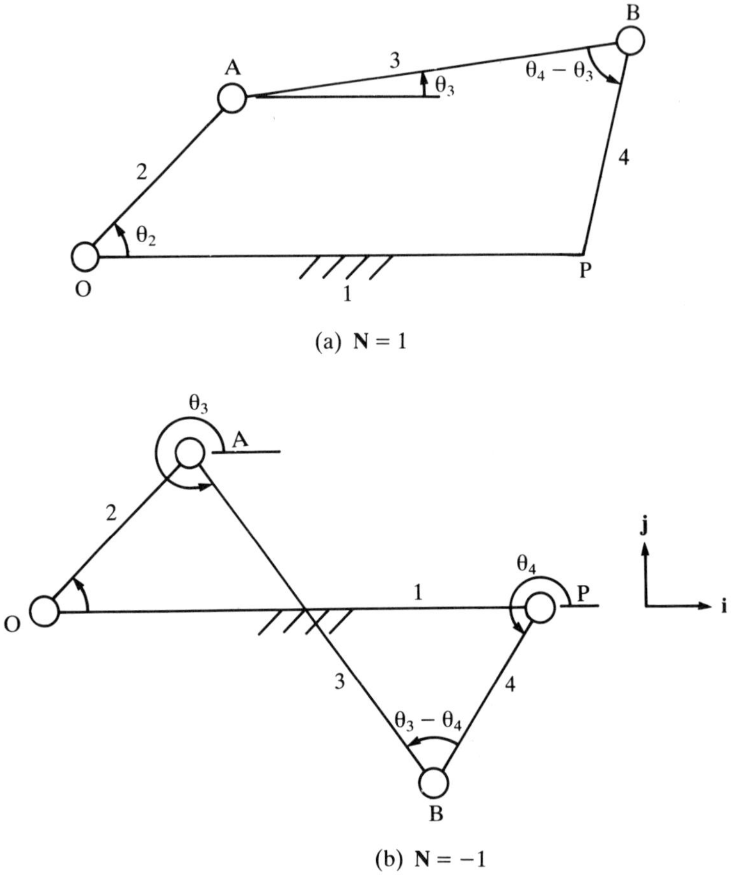

Fig. 4.19

where r_i $(i = 1, 2, 3, 4)$ are the lengths of the links,

$$\alpha = \arccos\mu, \qquad 0 < \alpha < \pi$$

$$\Gamma = \arcsin\{(r_1 - r_2\cos\theta_2)/\sqrt{(r_1^2 + r_2^2 - 2r_1r_2\cos\theta_2)}\}, \qquad -\frac{\pi}{2} < \Gamma < \frac{\pi}{2}$$

$$\delta = \arctan\{(r_4 - r_3\mu)/r_3\sqrt{(1-\mu^2)}\}, \qquad -\frac{\pi}{2} < \delta < \frac{\pi}{2}$$

then

$$\left.\begin{aligned} \theta_4 &= \Gamma + \delta, & 0 \le \theta_2 \le \pi \\ &= \pi - \Gamma + \delta, & \pi \le \theta_2 \le 2\pi \end{aligned}\right\} N = 1$$

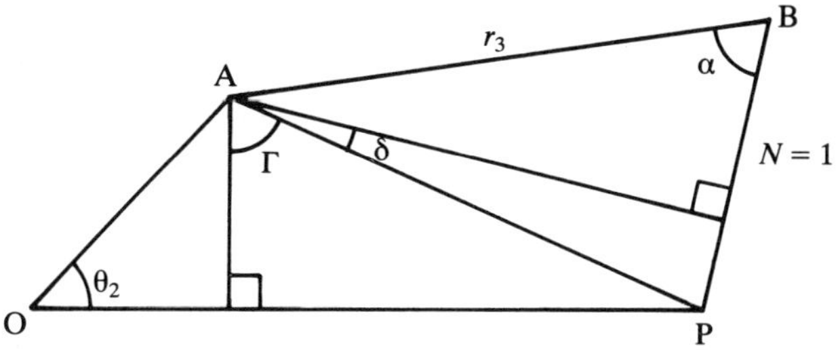

Fig. 4.20

$$\left.\begin{aligned}\theta_4 &= \pi + \Gamma - \delta, \qquad & 0 \le \theta_2 \le \pi \\ &= -\Gamma - \delta, \qquad & \pi \le \theta_2 \le 2\pi\end{aligned}\right\} N = -1$$

$$\theta_3 = \theta_4 - N\alpha$$

These formulae are in fact not difficult to derive. Their derivation, for example when $N = 1$ and $0 \le \theta_2 \le \pi$, can be carried out with the help of Fig. 4.20 (this shows the angles Γ and δ). Equating the two expressions that are obtained for $\underline{AP}^2$ by applying the cosine rule to the triangles OAP and ABP yields the formula for $\mu = \cos(\theta_4 - \theta_3)$.

An alternative way of verifying the formulae for θ_3 and θ_4 is suggested in Problem 4.16.

The reader should note that there are certain constraints on the lengths of the links if the link OA is to make complete revolutions relative to the link OP. Since we have already excluded the possibility of the points A, B, P being collinear, that is $\cos(\theta_4 - \theta_3) \ne \pm 1$ for any value of θ_2, we require

$$-1 < \mu < 1 \qquad \text{for all } \theta_2$$

or

$$-1 < \frac{r_3^2 + r_4^2 - r_1^2 - r_2^2}{2r_3 r_4} + \frac{r_1 r_2}{r_3 r_4}\cos\theta_2 < 1 \qquad \text{for all } \theta_2$$

Now μ is least when $\theta_2 = \pi$ and greatest when $\theta_2 = 0$. Since the least value must be greater than -1 and the greatest less than $+1$ we obtain, after some elementary manipulation, the inequalities

$$r_3 + r_4 > r_1 + r_2$$

$$(r_1 - r_2)^2 - (r_3 - r_4)^2 > 0$$

The second inequality can be written as

$$(r_1 - r_2 - r_3 + r_4)(r_1 - r_2 + r_3 - r_4) > 0$$

The expressions within the brackets must be either both positive or both negative. We are thus led to the following sets of inequalities:

Either	$r_3 + r_4 > r_1 + r_2$	or	$r_3 + r_4 > r_1 + r_2$
	$r_1 + r_4 > r_2 + r_3$		$r_2 + r_3 > r_1 + r_4$
	$r_1 + r_3 > r_2 + r_4$		$r_2 + r_4 > r_1 + r_3$

It is left as an exercise for the reader to show that the first set of inequalities implies that link 2 is the shortest link, while the second set implies that the fixed link is the shortest link. When the fixed link is the shortest link the mechanism is known as a drag-link, with link 4 making complete revolutions. If link 2 is the shortest link then link 4 oscillates to and fro and the mechanism is a crank-rocker. It is also left as an exercise for the reader to show that if r_1 or r_2 is least then the relevant set of inequalities is satisfied if the sum of the intermediate lengths of the links is greater than the sum of the lengths of the shortest and longest links. This form of the inequalities is due to Grashof (1885).

The following numerical details are provided for readers who wish to check computer programs that they have written to solve the problem of position. All angles are measured in degrees.

$r_1 = 0.19\,\text{m},\ r_2 = 0.04\,\text{m},\ r_3 = 0.16\,\text{m},\ r_4 = 0.08\,\text{m}$

N	θ_2	0	30	90	150	210	270	330
1	θ_3	29.69	21.91	11.79	8.79	18.96	33.57	36.58
1	θ_4	97.90	94.96	114.66	146.25	156.42	138.44	109.63
−1	θ_3	330.31	323.42	324.43	341.04	351.21	348.21	338.09
−1	θ_4	262.10	250.37	221.56	203.57	213.75	245.34	265.04

The angular velocities and accelerations (for given $\dot{\theta}_2$ and $\ddot{\theta}_2$) of links 3 and 4 can be found with the help of formulae (4.44) to (4.47). The mechanism given above is a crank-rocker, and readers can verify (for $N = 1$) that $\dot{\theta}_4 = 0$ when $\theta_2 = 23.44$ and $\theta_2 = 194.74$.

The problem of position in mechanisms can usually be solved by graphical (that is drawing) methods. It is not difficult to see from Fig. 4.19(a) and Fig. 4.19(b) how this can be achieved for the RRRR mechanism. The methods have severe limitations if the kinematic analysis of a complete cycle of a mechanism is required. The figure must be drawn for each value of the single parameter (θ_2 for the RRRR mechanism) that defines its position, while accuracy in measuring angles is limited.

4.6 CAMS

At the beginning of Section 4.4 a mechanism was defined as an assembly of connected elements called links, and so far we have considered mechanisms with two types of connections. Figure 4.21(a) to (e) illustrates further mechanisms known as cams. Contact between some of the links is maintained by the use of prismatic and revolute joints, but for the remainder contact is maintained by other means (not shown). For example, if in Fig. 4.22 link 4 is vertical and link 2 rotates sufficiently slowly about a horizontal axis, then links 2 and 3 would remain in contact due to the action of gravity.

A cam mechanism consists essentially of two parts: the cam, which is a body that rotates about a fixed axis, and the follower. The cross-sections of the cam with the axis of rotation as normal are, in two-dimensional motion, identical. The curve bounding any one of the cross-sections is called the cam profile. The follower may take any one of a number of forms, as in Fig. 4.21(a) to (e). The purpose of the mechanism is to impart to the follower a prescribed planar motion, the plane of this motion having the axis of the cam as normal. Clearly the motion of a particular follower depends on the shape of the cam profile and, as we now show, if this is suitably chosen a prescribed follower motion can be produced. Our analysis here, in contrast to earlier work on mechanisms, is directed towards finding profile shapes and not determining velocities and accelerations. The approach is systematic and the techniques used in the two examples that follow can be applied equally well elsewhere. We continue to use, where appropriate, the notation and conventions defined earlier for mechanisms but, as readers will see, links are no longer necessarily represented by straight lines. The cam is taken to be link 2 while any fixed body forms part of link 1. A point of contact is always interpreted as a joint and therefore labelled by a single letter with suffixes being used as appropriate.

As a preliminary we derive a result that is related to rotation theory (see Chapter 5) and is required for both examples and problems. Figure 4.23 shows two vectors $\underline{OL}$, $\underline{OM}$ of equal magnitudes lying in the Ox, Oy plane, the angle between them being equal to ϕ. If

$$\underline{OL} = \alpha\mathbf{i} + \beta\mathbf{j}, \qquad \underline{OM} = x\mathbf{i} + y\mathbf{j}$$

then

$$x = \alpha\cos\phi + \beta\sin\phi, \qquad y = \beta\cos\phi - \alpha\sin\phi \tag{4.53}$$

Proof Using the notation of Fig. 4.23 we have

$$\begin{aligned}\alpha &= |\underline{OL}|\cos(\theta + \phi)\\ &= |\underline{OM}|(\cos\theta\cos\phi - \sin\theta\sin\phi)\\ &= x\cos\phi - y\sin\phi\end{aligned}$$

Likewise

$$\beta = x\sin\phi + y\cos\phi$$

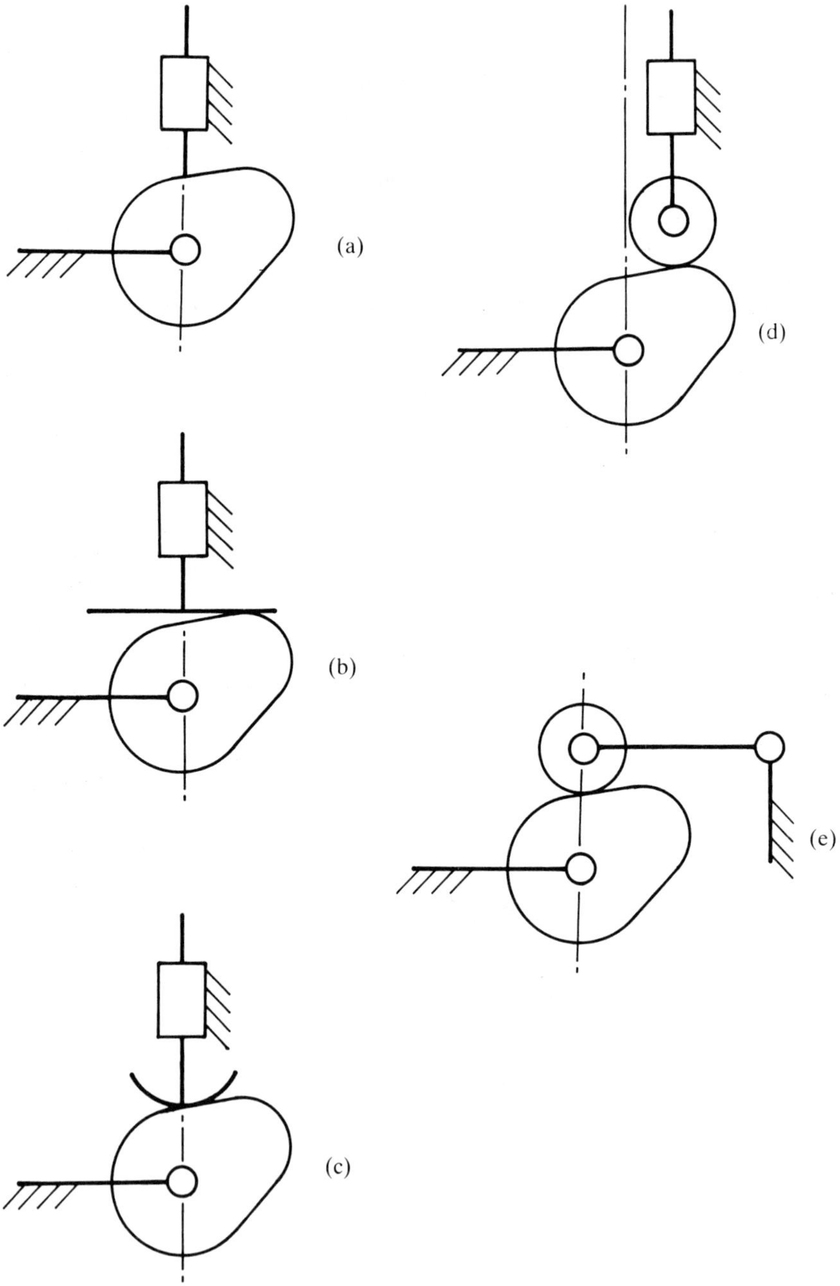

Fig. 4.21

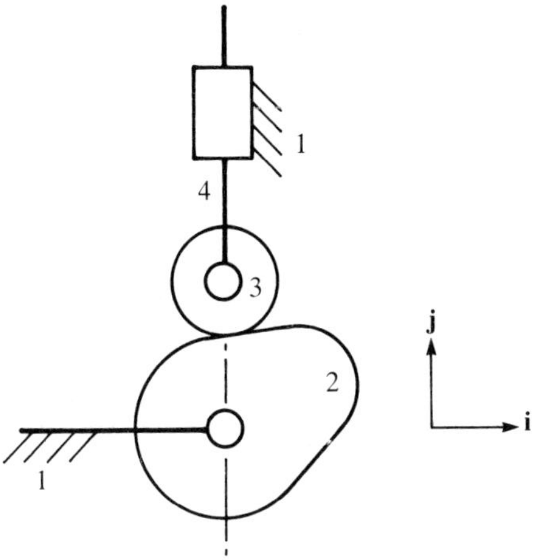

Fig. 4.22

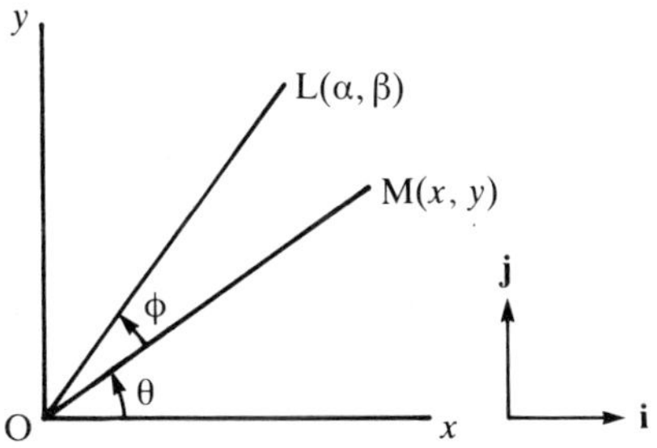

Fig. 4.23

Hence solving for x and y in terms of α and β, the relations in (4.53) are established.

Cam with mushroom follower

Figure 4.24(a) and (b) shows the profile of a cam (link 2) and a cross-section of its follower (link 3), the axis of rotation of the cam passing through the point O. In Fig. 4.24(a) the follower is in the position where the (variable) radius measured from O to the point of contact of cam and follower is least. Figure 4.24(b) shows the positions of the cam and its follower when the former has turned through an angle ϕ counter-clockwise from its position in Fig. 4.24(a). The centre of the circular arc DE is C

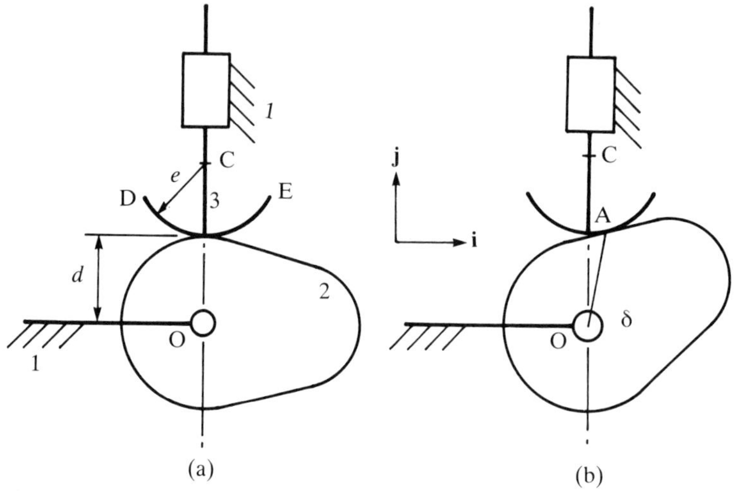

Fig. 4.24

while its radius is e. The motion of link 3 is one of rectilinear translation, the point C moving along the straight line through O in the direction of the unit vector **j**. We label the point of contact between the cam and follower as A. The displacement of the point C at time t due to the rotation of the cam is $s(t)\mathbf{j}$, where $s(t)$ is a prescribed function. The time t is measured from the instant $\phi = 0$ so that $s(0) = 0$. It follows from Fig. 4.24(a) and (b) that if d is the minimum radius of the cam then at time t

$$\underline{OC} = (s(t) + d + e)\mathbf{j} \tag{4.54}$$

Hence

$$\mathbf{v}_C = \dot{s}(t)\mathbf{j}$$

while

$$\mathbf{v}_{A_2} = \dot{\phi}\mathbf{k} \times \underline{OA_2} = \dot{\phi}\mathbf{k} \times \underline{OA}$$

Since the follower and cam are constrained so as to remain in contact and their deformations at the point of contact are assumed to be negligible there can be no component of the relative velocity $\mathbf{v}_{A_3} - \mathbf{v}_{A_2}$ in the direction of the common normal CA at A to the arc DE and the cam profile. Now $\mathbf{v}_{A_3} = \mathbf{v}_C$ as the follower's motion is one of rectilinear translation, and hence

$$(\dot{\phi}\mathbf{k} \times \underline{OA} - \dot{s}(t)\mathbf{j}) \cdot \mathbf{n} = 0$$

where **n** is a unit vector in the direction of $\underline{CA}$. Dividing throughout by $\dot{\phi}$ yields the alternative form

$$\left(\mathbf{k} \times \underline{\mathrm{OA}} - \frac{\mathrm{d}s}{\mathrm{d}\phi}\mathbf{j}\right) . \mathbf{n} = 0 \tag{4.55}$$

where s is now regarded as a (known) function of ϕ. The relations (4.54) and (4.55) together with the (closed-loop) vector relation

$$\underline{\mathrm{OC}} + \underline{\mathrm{CA}} = \underline{\mathrm{OA}} \tag{4.56}$$

are sufficient to find the vector $\underline{\mathrm{OA}}$.

Noting $\underline{\mathrm{CA}} = e\mathbf{n}$ and writing $\mathbf{n} = a\mathbf{i} + b\mathbf{j}$, where a and b are scalars to be found, leads with the help of (4.54) and (4.56) to

$$\underline{\mathrm{OA}} = ea\mathbf{i} + \{\mathrm{s} + \mathrm{d} + \mathrm{e}(1 + b)\}\mathbf{j} \tag{4.57}$$

Substituting this expression for $\underline{\mathrm{OA}}$ into (4.55) yields after some elementary vector manipulation

$$a(s + d + e) = -b\frac{\mathrm{d}s}{\mathrm{d}\phi}$$

Since $a^2 + b^2 = 1$ we deduce

$$\mathbf{n} = \pm\left\{\frac{\mathrm{d}s}{\mathrm{d}\phi}\mathbf{i} - (s + d + e)\mathbf{j}\right\} \Big/ \sqrt{\left\{\left(\frac{\mathrm{d}s}{\mathrm{d}\phi}\right)^2 + (s + d + e)^2\right\}} \tag{4.58}$$

From Fig. 4.24(b) it is evident that the $\mathbf{j}$ component of $\underline{\mathrm{CA}}$ and therefore that of $\mathbf{n}$ is negative, and hence the plus sign must be taken in the right-hand side of (4.58) to given $\mathbf{n}$ correctly. It follows now from (4.57) that

$$\underline{\mathrm{OA}} = e\left\{\frac{\mathrm{d}s}{\mathrm{d}\phi}\mathbf{i} - (s + d + e)\mathbf{j}\right\} \Big/ \sqrt{\left\{\left(\frac{\mathrm{d}s}{\mathrm{d}\phi}\right)^2 + (s + d + e)^2\right\}} + (s + d + e)\mathbf{j}$$

Using the results contained in (4.53) we see that the coordinates (x, y) of the position of the point A_2 of the cam at time $t = 0$ are given in terms of the angle ϕ by

$$\begin{aligned} x &= \frac{(s + d + e)\sin\phi + e\left\{\dfrac{\mathrm{d}s}{\mathrm{d}\phi}\cos\phi - (s + d + e)\sin\phi\right\}}{\sqrt{\left\{\left(\dfrac{\mathrm{d}s}{\mathrm{d}\phi}\right)^2 + (s + d + e)^2\right\}}} \\ y &= \frac{(s + d + e)\cos\phi - e\left\{\dfrac{\mathrm{d}s}{\mathrm{d}\phi}\sin\phi - (s + d + e)\cos\phi\right\}}{\sqrt{\left\{\left(\dfrac{\mathrm{d}s}{\mathrm{d}\phi}\right)^2 + (s + d + e)^2\right\}}} \end{aligned} \tag{4.59}$$

The angle ϕ can be regarded as a parameter, so that by taking values of it between 0 and 2π we obtain the locus of the points with coordinates (x, y); this locus is the profile of the cam.

If the follower is at rest for a range of values of ϕ then s is constant and hence $ds/d\phi$ is zero for these values. Reference to (4.59) shows that

$$x = (s + d)\sin\phi, \qquad y = (s + d)\cos\phi$$

giving

$$x^2 + y^2 = (s + d)^2$$

Thus part of the profile will (as might be anticipated) be an arc of a circle. When $ds/d\phi = 0$ the follower is said to dwell. A typical motion of a follower will consist of movement interspersed with dwells, the number (usually two) and duration of the dwells depending on the physical function of the follower.

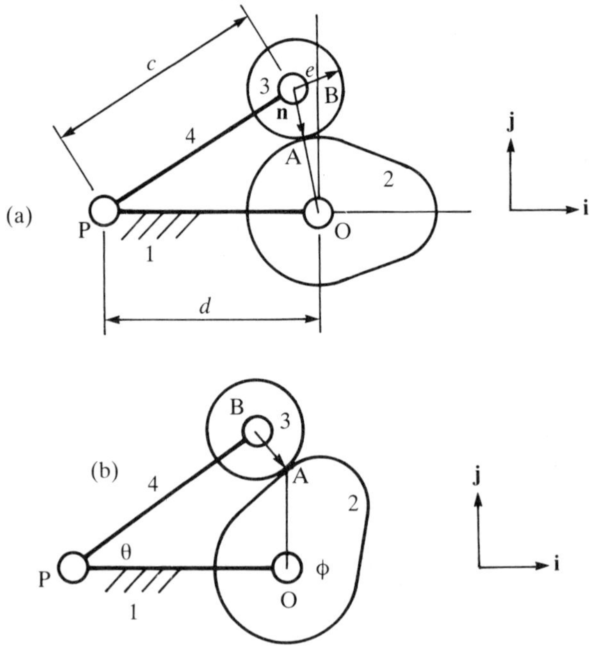

Fig. 4.25

Cam with oscillating roller follower

Each of Fig. 4.25(a) and (b) shows the profile of a cam with an oscillating roller follower. The follower consists of an arm PB (link 4) and a circular disc, the roller, of radius e. Link 4 is free to rotate about an axis that passes through the fixed point P and is parallel to the axis of the cam. The roller (link 3) is attached at its centre B to link 4 by a revolute joint so that the roller and the arm PB have a common plane of motion. Figure 4.25(a) shows the position of the cam and its follower when O, B and A (the point of contact between cam and roller) are collinear, while Fig. 4.25(b) shows a general position obtained by turning the cam through an angle ϕ (counter-clockwise) from its position in Fig. 4.25(a). The function of the cam is to impart (by means of the roller) oscillating rotation to the arm PB, and hence the angle θ (see Fig. 4.25(b)) will be a prescribed function of the time t. Now

$$\mathbf{v}_B = \dot{\theta}\mathbf{k} \times \underline{PB}, \qquad \mathbf{v}_{A_2} = \dot{\phi}\mathbf{k} \times \underline{OA}$$
$$\mathbf{v}_{A_3} = \mathbf{v}_B + \boldsymbol{\omega}_3 \times \underline{BA} = \dot{\theta}\mathbf{k} \times \underline{PB}\boldsymbol{\omega}_3 \times \underline{BA}$$

Writing $\mathbf{n}$ for the unit vector that is along the common normal to the roller and the cam at A, it follows that

$$(\mathbf{v}_{A_3} - \mathbf{v}_{A_2}) \cdot \mathbf{n} = 0$$

Hence as $\underline{BA} = e\mathbf{n}$ we obtain

$$(\dot{\theta}\mathbf{k} \times \underline{PB} + \boldsymbol{\omega}_3 \times e\mathbf{n} - \dot{\phi}\mathbf{k} \times \underline{OA}) \cdot \mathbf{n} = 0$$

This reduces to

$$\left(\frac{d\theta}{d\phi}\mathbf{k} \times \underline{PB} - \mathbf{k} \times \underline{OA}\right) \cdot \mathbf{n} = 0 \tag{4.60}$$

Putting $\mathbf{n} = a\mathbf{i} + b\mathbf{j}$ and writing (for brevity) $|\underline{PB}| = c$, $|\underline{OP}| = d$, it follows from the vector relation

$$\underline{OA} = \underline{OP} + \underline{PB} + \underline{BA}$$

that

$$\underline{OA} = (-d + c\cos\theta + ea)\mathbf{i} + (c\sin\theta + eb)\mathbf{j} \tag{4.61}$$

Substituting the expression in (4.61) for $\underline{OA}$ into (4.60) we obtain, after some vector manipulation,

$$\left(\frac{d\theta}{d\phi} - 1\right)c\sin\theta\, a = \left\{d + c\cos\theta\left(\frac{d\theta}{d\phi} - 1\right)\right\}b$$

Hence

$$\mathbf{n} = \frac{\pm\left[\left\{d + \left(\frac{d\theta}{d\phi} - 1\right)c\cos\theta\right\}\mathbf{i} + c\left(\frac{d\theta}{d\phi} - 1\right)\sin\theta\mathbf{j}\right]}{\sqrt{\left[d^2 + c^2\left(\frac{d\theta}{d\phi} - 1\right)^2 + 2cd\cos\theta\left(\frac{d\theta}{d\phi} - 1\right)\right]}} \tag{4.62}$$

Since **n** is a unit vector its two components cannot be simultaneously zero. It follows, as **n** is varying continuously with ϕ, that only one sign can apply, otherwise on a change of sign there would be a sudden jump in value in a non-zero component. Now in Fig. 4.25(a) the **j** component of **n** is negative, and thus if $(\mathrm{d}\theta/\mathrm{d}\phi) - 1 < 0$ we take the plus sign for **n** in the expression on the right-hand side of (4.62). Taking the plus sign in the expression for **n** we obtain, after some manipulation, the following parametric form of the coordinates of the cam profile:

$$x = -d\cos\phi + c\cos(\theta - \phi) + e\left\{d\cos\phi + c\cos(\theta - \phi)\left(\frac{\mathrm{d}\theta}{\mathrm{d}\phi} - 1\right)\right\}\Big/ u$$

$$y = d\sin\phi + c\sin(\theta - \phi) + e\left\{-d\sin\phi + c\sin(\theta - \phi)\left(\frac{\mathrm{d}\theta}{\mathrm{d}\phi} - 1\right)\right\}\Big/ u$$

where

$$u = \sqrt{\left\{c^2\left(\frac{\mathrm{d}\theta}{\mathrm{d}\phi} - 1\right)^2 + d^2 + 2cd\cos\theta\left(\frac{\mathrm{d}\theta}{\mathrm{d}\phi} - 1\right)\right\}}$$

4.7 KINEMATICS OF GEAR TRAINS

Gears transmit motion through gear teeth that are said to mesh. As two gears rotate about their axes, which need not be parallel, each tooth on one gear comes into contact with a tooth on the other and transmits motion until it comes out of contact. By then the next tooth is in contact and takes up the transmission, and so on for each tooth in turn to give continuous motion from one gear to another. When the teeth mesh so that the ratio of the angular speeds of the gears is constant, the gears are said to be correctly geared. The requirements of correct gearing impose a geometrical constraint on the choice of shape for the profile of the teeth. We determine the nature of the constraint for gears that rotate about parallel axes.

Condition for the correct gearing of gears that rotate about parallel axes

The notation developed earlier in this chapter for the study of mechanisms and cams is most suitable for the task in hand. Gears, like links, are referred to by numbers and points of contact distinguished by the use of suffixes. In Fig. 4.26 gear 2 rotates about a fixed axis passing through the point O_2, and at the point A it comes into contact with gear 3, rotating about a parallel axis through O_3. The unit vector in the direction of the

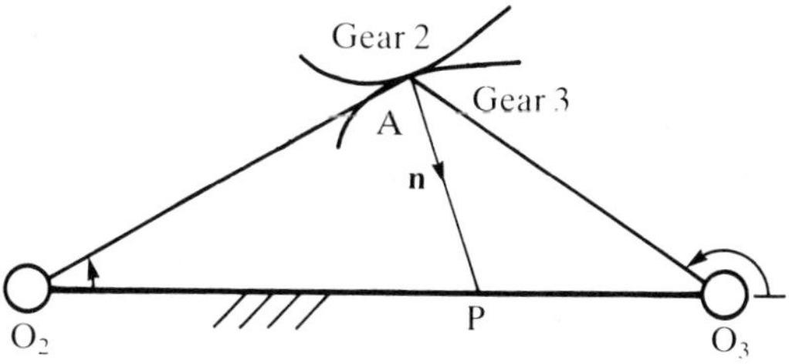

Fig. 4.26

common normal to the gear profiles at A is denoted by **n**. The common normal intersects the line O_2O_3, which is perpendicular to the axes of rotation (see Fig. 4.26), at the point P. Writing $\omega_2\mathbf{k}$ and $\omega_3\mathbf{k}$ for the angular velocities of gears 2 and 3, it follows that

$$\mathbf{v}_{A_2} = \omega_2\mathbf{k} \times \underline{O_2A_2} = \omega_2\mathbf{k} \times \underline{O_2A}$$

$$\mathbf{v}_{A_3} = \omega_3\mathbf{k} \times \underline{O_3A_3} = \omega_3\mathbf{k} \times \underline{O_3A}$$

We require $(\mathbf{v}_{A_2} - \mathbf{v}_{A_3}) \cdot \mathbf{n} = 0$, so that

$$(\omega_2\mathbf{k} \times \underline{O_2A} - \omega_3\mathbf{k} \times \underline{O_3A}) \cdot \mathbf{n} = 0$$

Using property (1.20) of triple-scalar products we deduce that

$$\mathbf{k} \cdot ((\omega_2\underline{O_2A} - \omega_3\underline{O_3A}) \times \mathbf{n}) = 0$$

Now

$$\underline{O_2A} \times \mathbf{n} = (\underline{O_2P} + \underline{PA}) \times \mathbf{n} = \underline{O_2P} \times \mathbf{n}$$

$$\underline{O_3A} \times \mathbf{n} = \underline{O_3P} \times \mathbf{n}$$

Hence we have

$$\mathbf{k} \cdot ((\omega_2\underline{O_2P} - \omega_3\underline{O_3P}) \times \mathbf{n}) = 0 \tag{4.62(a)}$$

The vector $\omega_2\underline{O_2P} - \omega_3\underline{O_3P}$ is parallel to $\underline{O_2O_3}$, and hence its vector product with **n** cannot be perpendicular to **k**. Thus the vector $(\omega_2\underline{O_2P} - \omega_3\underline{O_3P}) \times \mathbf{n}$ must be equal to the zero vector, from which we infer that **n** must be parallel to $\underline{O_2O_3}$ or

$$\omega_2\underline{O_2P} - \omega_3\underline{O_3P} = \mathbf{0} \tag{4.63}$$

If **n** is parallel to $\underline{O_2O_3}$ either $\underline{AP}$ is parallel to $\underline{O_2O_3}$, in which case P is a point at infinity in O_2O_3 (or O_3O_2) produced, or A and P both lie in O_2O_3. When the latter holds, any pair of touching circles centres O_2, O_3 may be taken for the gear profiles. Whilst these profiles ensure that the kinematic condition (4.62(a)) is satisfied, they are clearly not a practical solution because of the absence of teeth. Accordingly we examine the condition contained in the vector relation (4.63). If the ratio ω_2/ω_3 is constant the relation implies that the point P is a fixed point in the line O_2O_3. Hence for correct gearing the common normal to the profiles at the point of contact must always pass through a fixed point in O_2O_3. Curves called epicycloids and hypocycloids satisfy this condition, and for this reason they are sometimes chosen as profiles for gear teeth (see problem 4.5). The point P is referred to as the pitch point while the circles with centres O_2, O_3 and radii O_2P, O_3P are called the pitch circles. From the relation (4.63) we deduce that P is an interior point of the line O_2O_3 if ω_2, ω_3 are of opposite sign (see Fig. 4.27(a)) and

$$\omega_2 O_2 P = -\omega_3 O_3 P \tag{4.64}$$

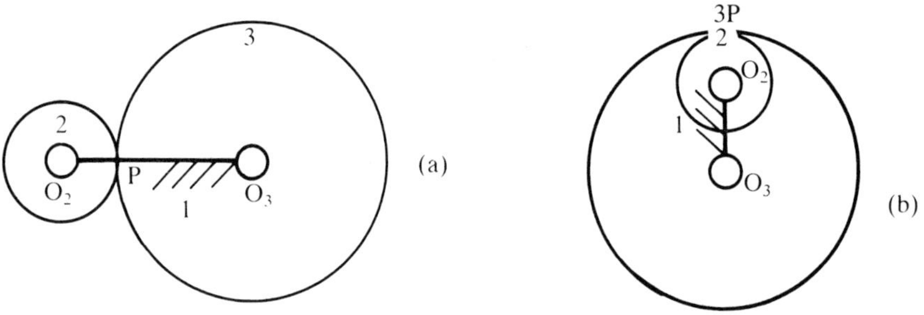

Fig. 4.27

When ω_2, ω_3 are of the same sign P is an exterior point of O_2O_3 see Fig. 4.27(b)) and gear 3 is called an annular gear. For an annular gear

$$\omega_2 O_2 P = \omega_3 O_3 P \tag{4.65}$$

When ω_2 and ω_3 are nearly equal (for annular gearing) the point P is at a great distance from the points O_2, O_3 and may be equated with the point at infinity.

The conditions in (4.64) and (4.65) can be *interpreted* as stating that there is no velocity of slip at P between two circles, radii O_2P and O_3P, rotating about axes through O_2, O_3 with (signed) angular speeds ω_2 and ω_3. This simple interpretation of gear motion forms, as we will show, the basis of the kinematical analysis of gear trains (see Fig. 4.28(a) and (b)) whose gears connect parallel shafts rotating about fixed axes. The condition (4.63) should not be dismissed completely as it serves as a useful reminder of how matters of sign are resolved.

The diameter d of a pitch circle is given by $d = mN$, where N is the number of teeth on the gear and m is the tooth module, the module being a

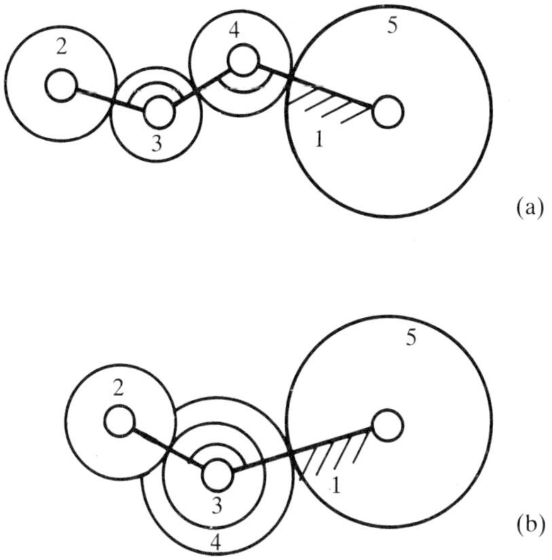

Fig. 4.28

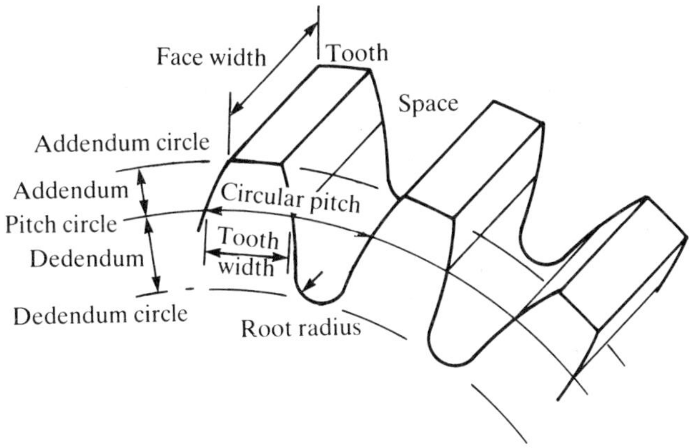

Fig. 4.29

measure of the tooth size (see Fig. 4.29). If the gear teeth are of standard form (not always so) then gears that mesh together must have the same module, and if the shaft carrying a gear rotates through more than one revolution the number of teeth on the gear must be an integer.

We use the suffix notation when considering a number of gears, so that the pitch circle diameter, module and number of teeth of the ith gear are denoted by d_i, m_i and N_i. Thus in Fig. 4.27(a) and (b) $m_2 = d_2/N_2$ and $m_3 = d_3/N_3$, giving, as the gears mesh,

$$\frac{d_2}{d_3} = \frac{N_2}{N_3} \tag{4.66}$$

Using condition (4.64) we deduce for the gears in Fig. 4.27(a)

$$\frac{\omega_3}{\omega_2} = -\frac{N_2}{N_3} \tag{4.67}$$

while for those in Fig. 4.27(b)

$$\frac{\omega_3}{\omega_2} = \frac{N_2}{N_3} \tag{4.68}$$

The gear train shown diagrammatically in Fig. 4.28(a) consists of four gears attached to shafts rotating about parallel fixed axes, each driving the next so that the motion is transmitted from gear 2 (called the input) to gear 5 (called the output). The ratio ω_5/ω_2, called the overall velocity ratio, is given by

$$\begin{aligned}\omega_5/\omega_2 &= (\omega_3/\omega_2)(\omega_4/\omega_3)(\omega_5/\omega_4)\\ &= (-N_2/N_3)(-N_3/N_4)(-N_4/N_5) \qquad \text{using (4.67)}\\ &= -N_2/N_5\end{aligned}$$

It is seen that the number of teeth on gears 3 and 4 do not influence the overall velocity ratio, and for this reason they are called idler gears.

In some gear trains two or more gears are fixed together to make what is called a *compound* gear. For example, in the gear train shown in Fig. 4.28(b) gears 3 and 4 are a compound gear (so that $\omega_3 = \omega_4$) with gears 2 and 3 meshing and gears 4 and 5 meshing. If gear 2 is the input and gear 5 the output then the overall velocity ratio is given by

$$\begin{aligned}\omega_5/\omega_2 &= (\omega_3/\omega_2)(\omega_4/\omega_3)(\omega_5/\omega_4)\\ &= (-N_2/N_3)(-N_4/N_5) \qquad \text{using (4.67)}\\ &= N_2N_4/N_3N_5\end{aligned}$$

Here it is seen that the numbers of teeth in the intermediate gears 3 and 4 do influence the overall velocity ratio.

In epicyclic gear trains (see Fig. 4.30) some of the gears turn about axes that are attached to links that themselves rotate. To carry out the kinematic analysis of such trains we have first to modify the result contained in (4.63). Returning to Fig. 4.26 we now take O_2O_3 as a link rotating about the fixed axis through O_2, coincident with the axis of the shaft carrying gear 2. As before gear 2 meshes with gear 3, the latter being attached to the link O_2O_3 by means of a revolute joint at O_3. We have

$$\mathbf{v}_{A_2} = \omega_2\mathbf{k} \times \underline{O_2A}$$

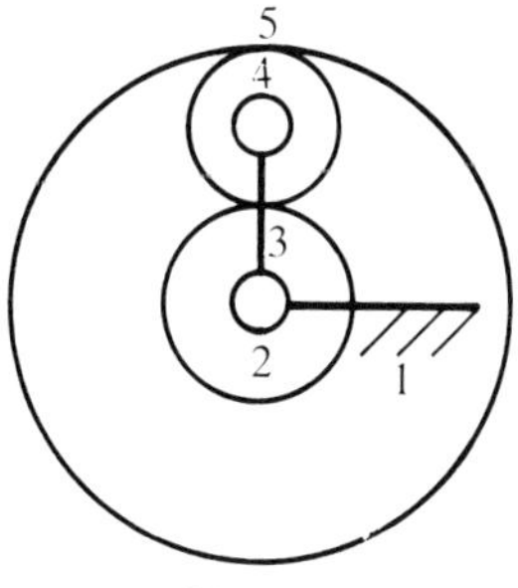

Fig. 4.30

while

$$\mathbf{v}_{A_3} = \omega_3 \mathbf{k} \times \underline{O_3A} + \omega \mathbf{k} \times \underline{O_2O_3}$$

where $\omega\mathbf{k}$ is the angular velocity of the link O_2O_3. The condition $(\mathbf{v}_{A_2} - \mathbf{v}_{A_3}) \cdot \mathbf{n} = 0$ yields

$$\mathbf{k} \times (\omega_2 \underline{O_2A} - \omega_3 \underline{O_3A} - \omega \times \underline{O_2O_3}) \cdot \mathbf{n} = 0$$

Noting once more the triple-scalar property (1.20) and the relations $\underline{O_2A} = \underline{O_2P} + \underline{PA}$, $\underline{O_3A} = \underline{O_3P} + \underline{PA}$, we deduce

$$\omega_2 \underline{O_2P} - \omega_3 \underline{O_3P} - \omega \underline{O_2O_3} = 0$$

so that

$$(\omega_2 - \omega)\underline{O_2P} = (\omega_3 - \omega)\underline{O_3P} \tag{4.69}$$

If the ratios ω_2/ω_3, ω/ω_3 are constant then the normal to the profiles at the point of contact must pass through a point fixed in the link O_2O_3. As before, P is called the pitch point and the circles centres O_2, O_3 touching at P are called pitch circles. From (4.69) we deduce

$$(\omega_2 - \omega)O_2P = (\omega_3 - \omega)O_3P, \qquad \text{P an exterior point of } O_2O_3 \tag{4.70}$$

$$(\omega_2 - \omega)O_2P = -(\omega_3 - \omega)O_3P, \qquad \text{P an interior point of } O_2O_3 \tag{4.71}$$

It is left to the reader to show, by the methods of kinematic analysis developed earlier in the chapter, that the conditions (4.70) and (4.71) can be interpreted as stating that there is no velocity of slip at P between two revolving circles centres O_2, O_3 and radii O_2P, O_3P whose angular speeds are ω_2, ω_3, the link O_2O_3 having an angular speed ω.

In Figs 4.30 and 4.31 gear 4 is attached to link 3 by a revolute joint and meshes with gear 2 and gear 5. Gear 2, link 3 and annular gear 5 rotate about the same fixed axis. It should be noted that Fig. 4.30, following the conventions used in previous figures of this section, shows the pitch circles

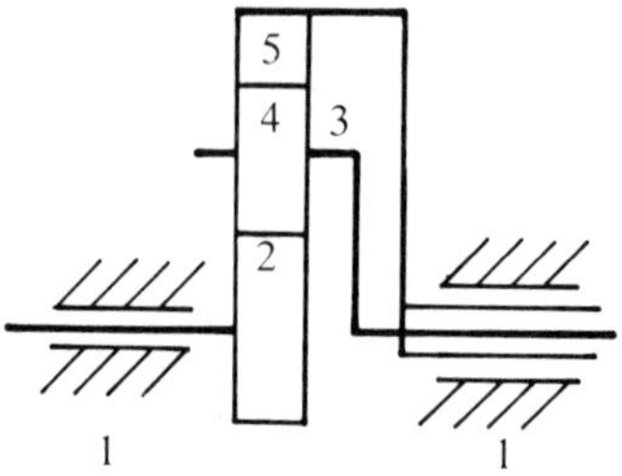

Fig. 4.31

of the gears, whereas Fig. 4.31 is drawn in a plane parallel to the axes of rotation, the pitch circles of the gears being shown as cylinders for clarity. Usually a single diagram using the conventions of Fig. 4.31 is sufficient to define the arrangement of an epicyclic gear train. Gear 2 is commonly called the sun, gear 4 the planet, and link 3 the planet carrier. Applying the result in (4.71) to gears 2 and 4, so that ω_3 is replaced by ω_4, with ω put equal to ω_3, it follows that $-(\omega_2 - \omega_3)/(\omega_4 - \omega_3)$ is equal to the ratio of the radius of the pitch circle for gear 4 to that of gear 3. Recalling (4.66) yields

$$(\omega_2 - \omega_3)/(\omega_4 - \omega_3) = -N_4/N_2 \tag{4.72}$$

Similarly applying (4.70) to gears 5 and 6 leads to

$$(\omega_4 - \omega_3)/(\omega_5 - \omega_3) = -N_5/N_4 \tag{4.73}$$

When the conditions under which the gear train is being used are given, the relations (4.72) and (4.73) enable us to determine the overall velocity ratio. For example, if the annular gear 5 is fixed and the input to the train is to the sun (gear 2) and the output is from the planet carrier (link 3) then we require the velocity ratio ω_3/ω_2. Putting $\omega_5 = 0$ in (4.73) and eliminating ω_4 from the outcome and (4.72) gives the velocity ratio

$$\omega_3/\omega_2 = N_2/(N_2 + N_5)$$

Example 4.6

Figure 4.32 shows diagrammatically an epicyclic gear train with a compound planet (gears 4 and 5). The numbers of the gear teeth are $N_2 = 24$, $N_4 = 20$, $N_5 = 30$, $N_6 = 74$. There are two inputs to the train, one to the sun (gear 2) and the other to the annular gear 6, and the output is from the planet carrier (link 3). Show that the modules of gears 2 and 5 are equal and derive an expression for the overall velocity ratio ω_3/ω_2, given that $\omega_6/\omega_2 = \alpha$, where α is a constant. Determine the velocity ratio ω_3/ω_2 when the annular gear is fixed.

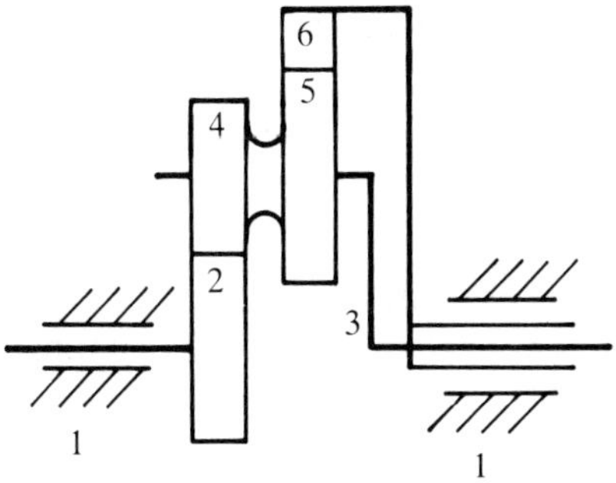

Fig. 4.32

> *Solution* Since gears 2 and 4 mesh they must have the same module. Likewise gears 5 and 6 must have the same module but this can be different from that of gears 2 and 4. Whether these two modules are equal or not the geometry of the pitch circles requires that their radii r_i be related to r_3, the length of link 3, by $r_5 + r_3 = r_6$ and $r_2 + r_4 = r_3$, so that

$$d_6 = d_2 + d_4 + d_5$$

Putting $m_2 = m_4$, $m_5 = m_6$ we deduce, since $d_i = m_i N_i$,

$$m_5(N_6 - N_5) = m_2(N_2 + N_4)$$

Inserting the given numerical values it is found that $m_2 = m_5$ as required.
Using (4.71) for gears 2 and 4 and recalling (4.66) yields

$$(\omega_2 - \omega_3)/(\omega_4 - \omega_3) = -N_4/N_2$$

Likewise using (4.70) for gears 5 and 6 leads to

$$(\omega_5 - \omega_3)/(\omega_6 - \omega_3) = N_6/N_5$$

Putting $\omega_5 = \omega_4$ and multiplying corresponding sides of the two relations just derived gives

$$(\omega_2 - \omega_3)/(\omega_6 - \omega_3) = -N_4 N_6/N_2 N_5$$

After some elementary manipulation it is found that

$$N_2 N_5 + N_4 N_6 \omega_6/\omega_2 = (\omega_3/\omega_2)(N_4 N_6 + N_2 N_5)$$

which becomes, on inserting numerical values,

$$\omega_3/\omega_2 = (74\alpha + 36)/110$$

When gear 6, the annular gear, is fixed, $\omega_6 = 0$ and therefore

$$\omega_3/\omega_2 = 0.3273$$

It is left as an exercise for the reader to verify that the velocity ratio for the planet is given by

$$\omega_4/\omega_2 = (74\alpha - 24)/50$$

We conclude this section on gears by obtaining an expression for the velocity of slip at the point of contact of two gears rotating about parallel fixed axes.

Velocity of slip at the point of contact of gears rotating about fixed parallel axes

The unit vector $\mathbf{t}$ in the direction of the common tangent to the profile at the point of contact A is chosen so that $\mathbf{t} = \mathbf{n} \times \mathbf{k}$. Now

$$\begin{aligned}\mathbf{v}_{A_2} - \mathbf{v}_{A_3} &= \mathbf{k} \times (\omega_2 \underline{O_2A} - \omega_3 \underline{O_3A}) \\ &= \mathbf{k} \times (\omega_2(\underline{O_2P} + \underline{PA}) - \omega_3(\underline{O_3P} + \underline{PA}) \\ &= \mathbf{k} \times (\omega_2 - \omega_3)\underline{PA}\end{aligned}$$

as $\omega_2 \underline{O_2P} - \omega_3 \underline{O_3P} = \mathbf{0}$ from condition (4.63). But $\underline{PA} = -|\underline{PA}|\mathbf{n}$ and hence

$$\begin{aligned}\mathbf{v}_{A_2} - \mathbf{v}_{A_3} &= -|\underline{PA}|(\omega_2 - \omega_3)\mathbf{k} \times \mathbf{n} \\ &= |\underline{PA}|(\omega_2 - \omega_3)\mathbf{t}\end{aligned}$$

The velocity of slip is equal to the zero vector only when the point of contact of the profiles coincides with the pitch point for the two gears.

PROBLEMS

Section 4.3

4.1 Three particles are initially situated at points whose coordinates referred to the axes Ox, Oy, Oz are $[-1, -1, -2]$, $[4, 9, -1]$, $[6, 5, -1]$. Subsequently the three particles are situated at points whose coordinates are $[0, 1, 1]$, $[-6, 4, 10]$, $[2, 2, 10]$. Show that the particles do not form part of a rigid body.

4.2 Four particles P_1, P_2, P_3, P_4 are initially situated at the points A_1, A_2, A_3, A_4. Subsequently the particles move to the points B_1, B_2, B_3, B_4. Referred to the axes Ox, Oy, Oz the coordinates of A_1, A_2, A_3, A_4 are $[1, 1, -2]$, $[3, 4, -1]$, $[1, -1, -1]$, $[-2, 0, 6]$ while those of B_1, B_2,

B_3, B_4 are [3, 4, −3], [1, 1, −4], [5, 4, −2], [4, 1, 5]. Show that the four particles do not form part of a rigid body.
Hint: find the body coordinates of P_4 with P_1, P_2, P_3 as reference particles in both positions.

4.3 Show that

$$\dot{\mathbf{v}}_{\rm rel} = \mathbf{f}_{\rm rel} + \boldsymbol{\omega} \times \mathbf{v}_{\rm rel}$$

where $\mathbf{v}_{\rm rel}$ and $\mathbf{f}_{\rm rel}$ are defined by the formulae (4.31), (4.32) and (4.35).

Hence show that differentiating both sides of (4.32) with respect to time gives the *erroneous* answer (see (4.34))

$$\dot{\mathbf{v}}_{\rm P} = \dot{\mathbf{v}}_i + \boldsymbol{\omega} \times \mathbf{v}_{\rm rel} + \mathbf{f}_{\rm rel}$$

4.4 The velocity of the particle P of a rigid body moving two-dimensionally with angular velocity $\boldsymbol{\omega}$ is equal to $\mathbf{v}$. Show that all particles of the rigid body that lie on the line passing through the point I, where $\omega^2 \underline{\rm PI} = \boldsymbol{\omega} \times \mathbf{v}$, parallel to the vector $\boldsymbol{\omega}$ are instantaneously at rest.
Note: this line is called the instantaneous axis of rotation, while I is called the instantaneous centre of rotation. The instantaneous centre is a particularly useful concept in the graphical approach to the kinematic analysis of mechanisms.

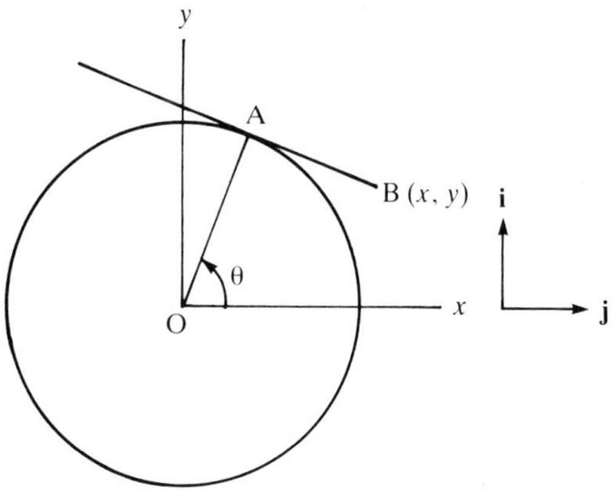

Fig. 4.33

4.5 Figure 4.33 shows a rod rolling on the outside of a fixed circle centre O and radius a. The angle between OA, where A is the point of contact of the circle and the rod, and Ox is θ, while x, y are the coordinates of the end B of the rod at time t.

Explain why the angular velocity of the rod is equal to $\dot{\theta}\mathbf{k}$. Show that

(a) $\mathbf{v}_B = AB\dot{\theta}\mathbf{a}$, where $\mathbf{a}$ is a unit vector in the direction of $\underline{OA}$ and AB denotes $|\underline{AB}|$.
(b) $x = a\cos\theta + AB\sin\theta$, $y = a\sin\theta - AB\cos\theta$.
(c) $\dot{AB} = a\dot{\theta}$. *Hint:* differentiate the expressions for x and y in (b) and use result (a).
(d) $AB = a\theta$, given $x = a$, $y = 0$ when $t = 0$.
(e) $dy/dx = \tan\theta$, and hence deduce that AB is always in the direction of the normal to the locus of B. The locus of B, which is an example of an involute curve, is used as a profile for involute gears.

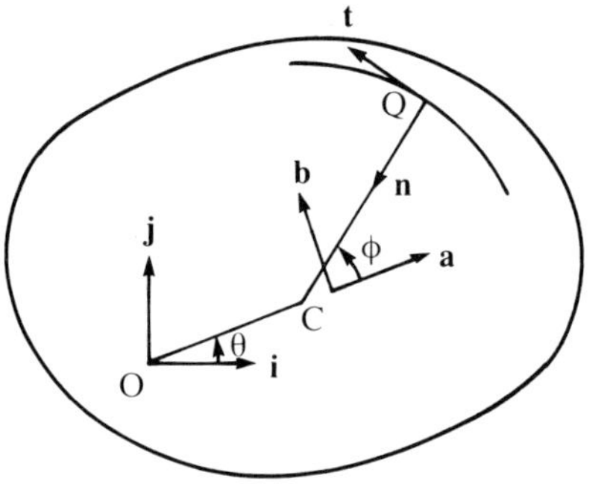

Fig. 4.34

4.6 A turntable, centre O, rotates with variable angular speed ω about an axis that is perpendicular to it. The turntable contains a shallow circular groove centre C, radius a, where $OC = c$ (see Figure 4.34). A particle moves along the groove with variable speed v relative to the groove. Determine when the particle is at the point Q the components of its absolute velocity and acceleration with respect to the unit vectors $\mathbf{i}$, $\mathbf{j}$ in terms of θ, ϕ, ω, v, $\dot{\omega}$ and $\dot{v}$, where θ and ϕ are as shown in Fig. 4.34.

Section 4.4

4.7 Show that the quantity $\dot{v}$ of Example 4.4 is given by

$$\dot{v} = \dot{\omega}_2 r_2 \sin(\theta_4 - \theta_2) - \omega_2^2 r_2 \cos(\theta_2 - \theta_4) + r_4\omega_4^2$$

4.8 The lengths of the links of the RRRR mechanism shown in Fig. 4.17 are $OA = 0.04\,\text{m}$, $AB = 0.16\,\text{m}$, $BP = 0.08\,\text{m}$ and $OP = 0.19\,\text{m}$. Determine the values of $\dot{\theta}_3$, $\dot{\theta}_4$, $\ddot{\theta}_3$, $\ddot{\theta}_4$ when $\theta_2 = 30°$, given that $\theta_3 = 21.91°$, $\theta_4 = 94.96°$, $\dot{\theta}_2 = 25\,\text{rad}\,\text{s}^{-1}$ and $\ddot{\theta}_2 = 0$.

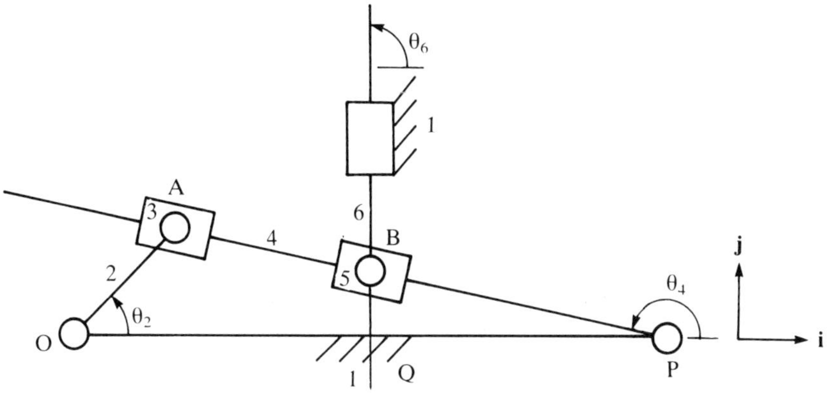

Fig. 4.35

Show that $\mathbf{f}_G$, the acceleration of G, the midpoint of AB, is given by $\mathbf{f}_G = -27.6\,\text{m s}^{-2}\mathbf{s}_3 + 2.7\,\text{m s}^{-2}\mathbf{t}_3$. Express $\mathbf{f}_G$ in terms of components respectively parallel and perpendicular to the fixed link OP.

In Problems 4.9 to 4.15 the conventions and notation defined in Section 4.4 are used.

4.9 Given that $\theta_2 = 120°$, $\omega_2 = 20\,\text{rad s}^{-1}$, $\dot{\omega}_2 = 0$, determine for the mechanism shown in Fig. 4.35 the quantities θ_4, ω_4, $\mathbf{v}_{A_4} - \mathbf{v}_{A_3}$, $\mathbf{v}_{B_5} - \mathbf{v}_{B_4}$, $\dot{\omega}_4$, $\mathbf{f}_{A_4} - \mathbf{f}_{A_3}$, $\mathbf{f}_{B_5} - \mathbf{f}_{B_4}$, $\mathbf{f}_{B_6}$, where the dimensions are: OP = 0.2 m, OQ = 0.1 m, OA = 0.04 m and $\theta_6 = 90°$ (constant).

4.10 Given that $\theta_2 = 25°$, $\omega_2 = 65\,\text{rad s}^{-1}$, $\dot{\omega}_2 = 0$, determine for the mechanism shown in Fig. 4.36 the quantities θ_4, θ_5, ω_4, ω_5, $\mathbf{v}_{C_6}$, $\mathbf{v}_{A_4} - \mathbf{v}_{A_3}$, $\dot{\omega}_4$, $\dot{\omega}_5$, $\mathbf{f}_B$, $\mathbf{f}_{A_4} - \mathbf{f}_{A_3}$, $\mathbf{f}_{C_6}$, where the dimensions are: OP = 0.068 m, OA = 0.138 m, PB = 0.080 m, BC = 0.276 m and ∠OPC = 90°.

4.11 Given that $\theta_2 = 45°$, $\omega_2 = 100\,\text{rad s}^{-1}$, $\dot{\omega}_2 = 0$, determine for the mechanism shown in Fig. 4.37 the quantities θ_4, θ_5, ω_4, ω_5, $\mathbf{v}_{C_6}$, $\mathbf{v}_{A_4} - \mathbf{v}_{A_3}$, $\dot{\omega}_4$, $\dot{\omega}_5$, $\mathbf{f}_{C_6}$, $\mathbf{f}_{A_4} - \mathbf{f}_{A_3}$.
The dimensions of the mechanism are: OP = 0.15 m, OA = 0.075 m, PB = 0.15 m, BC = 0.3 m and ∠OPC = 90°.

4.12 In the mechanism shown in Fig. 4.38, link 6 is connected to link 5 by a revolute joint and to link 1 by a prismatic joint at B. Link 5 is connected to link 4 by a prismatic joint, the relative motion at the joint B being along the line of link 4. Given that $\theta_2 = 110°$, $\omega_2 = 10\,\text{rad s}^{-1}$, $\dot{\omega}_2 = 2\,\text{rad s}^{-2}$, determine the quantities θ_4, ω_4, $\mathbf{v}_{B_6}$, $\mathbf{v}_{A_4} - \mathbf{v}_{A_3}$, $\mathbf{v}_{B_5} - \mathbf{v}_{B_4}$, $\dot{\omega}_4$, $\mathbf{f}_{B_5} - \mathbf{f}_{B_4}$, $\mathbf{f}_{B_6}$.

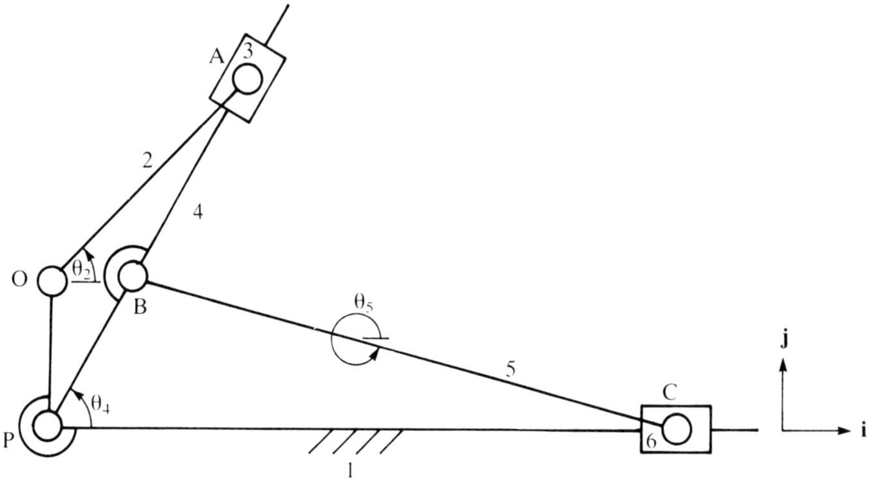

Fig. 4.36

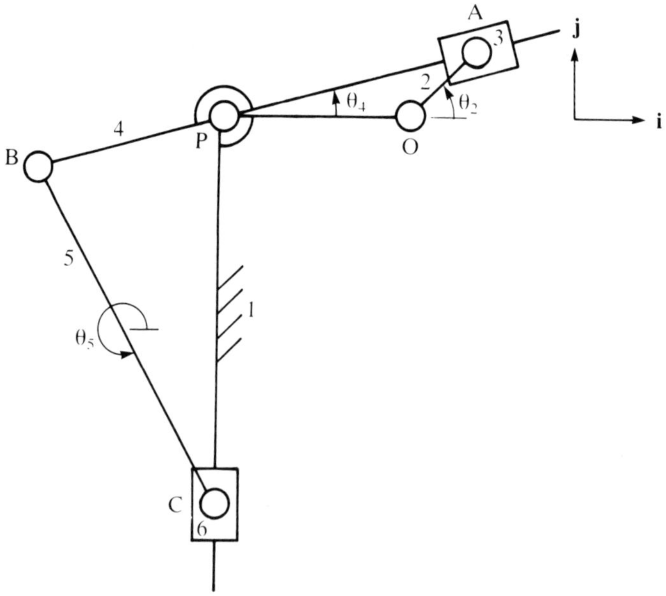

Fig. 4.37

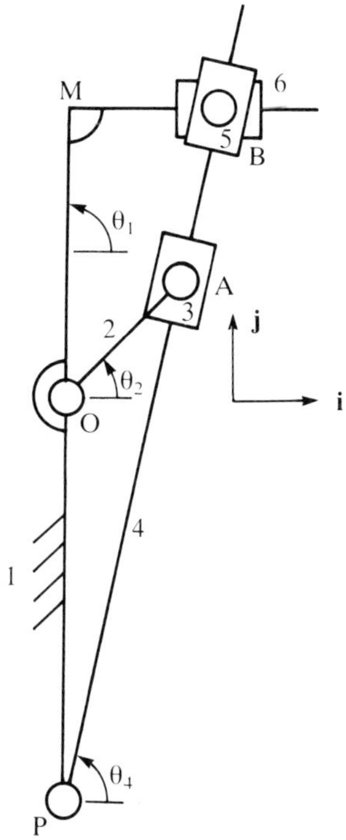

Fig. 4.38

Where the dimensions are: OP = 0.254 m, OM = 0.165 m, OA = 0.102 m, $\angle$OMB = 90° and $\theta_1 = 90°$.

4.13 In the mechanism shown in Fig. 4.39 link 4 is connected to link 3 at A by a prismatic joint and to link 5 at P also by a prismatic joint, the direction of relative motion at both points being along the line of link 4. Show that

(a) $\omega_4 AP = \omega_2 OA\cos(\theta_2 - \theta_4)$
(d) $\mathbf{v}_{C_6} = \omega_4 PC\sec\theta_4 \mathbf{j}$
(c) $\dot{\omega}_4 AP = (2\omega_2\omega_4 - \omega_2^2) OA\sin(\theta_2 - \theta_4)$
(d) $\mathbf{f}_{C_6} = PC\sec^2\theta_4(\dot{\omega}_4\cos\theta_4 + 2\omega_4^2\sin\theta_4)\mathbf{j}$.
Assume $\dot{\omega}_2 = 0$.

4.14 Derive the following results for the mechanism shown in Fig. 4.40:

(a) $\omega_5\sin(\theta_6 - \theta_5)CD = \mathbf{v}_C \cdot \mathbf{s}_6$
(b) $\omega_6\sin(\theta_5 - \theta_6)QD = \mathbf{v}_C \cdot \mathbf{s}_5$
(c) $\dot{\omega}_6\sin(\theta_5 - \theta_6)QD = \dot{\mathbf{v}}_C \cdot \mathbf{s}_5 + \omega_6^2 QD\cos(\theta_5 - \theta_6) + \omega_5^2 CD$.
Given the following data (the lengths are in metres), find the velocity

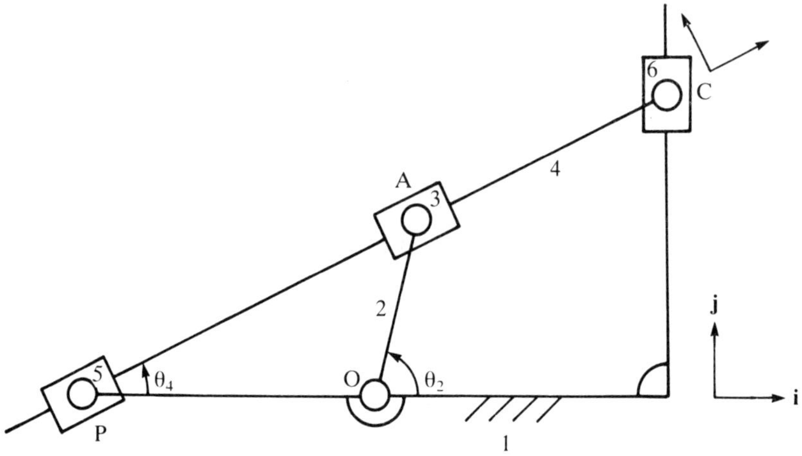

Fig. 4.39

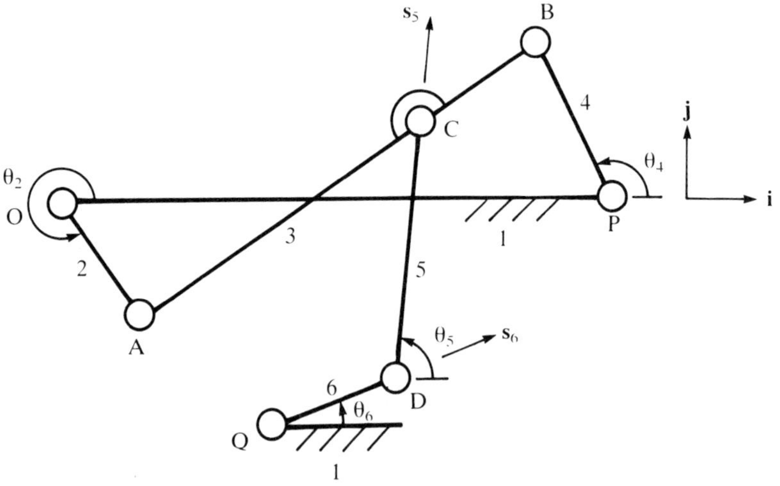

Fig. 4.40

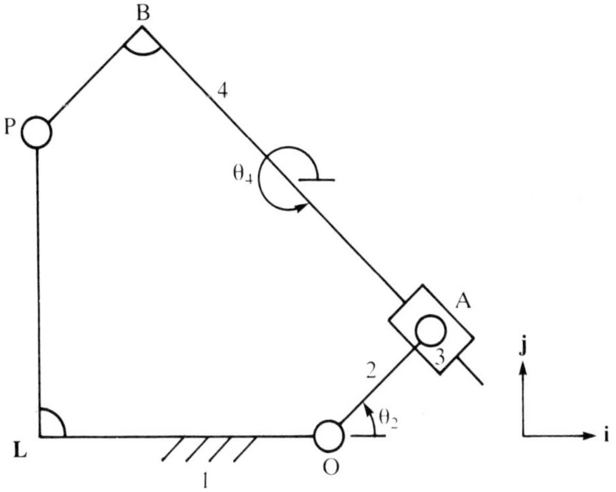

Fig. 4.41

$\mathbf{v}_C$ of C and angular speed ω_6: $\theta_2 = 315°$, $\theta_3 = 18.24°$, $\theta_4 = 73.05°$, $\theta_5 = 54.68°$, $\theta_6 = 0.75°$, $\omega_2 = 100\,\mathrm{rad\,s^{-1}}$, OA = 0.075, AB = 0.475, AC = 0.325, PB = 0.1, QD = 0.1, DC = 0.15.

4.15 Given that $\theta_2 = 180°$, $\omega_2 = 2\,\mathrm{rad\,s^{-1}}$, $\dot{\omega}_2 = 0$, determine for the mechanism shown in Fig. 4.41 the values of $\omega_4, \mathbf{V}_{A_4} - \mathbf{V}_{A_3}, \mathbf{f}_{A_4} - \mathbf{f}_{A_3}, \dot{\omega}_4$, where OA = PB = 0.1 m and LP = LO = 0.2 m.

Section 4.5

4.16 Using the notation contained in Section 4.5 on the solution of the problem of position for the RRRR mechanism, show (a)–(c) as follows:

(a) $\sin N\alpha = N\sin\alpha = N\sqrt{(1-\mu^2)}$; $\cos N\alpha = \mu$

(b) $\cos\Gamma = |r_2\sin\theta_2|/\sqrt{(r_1^2 + r_2^2 - 2r_1r_2\cos\theta_2)}$

(c) θ_3, θ_4 satisfy the equations

$$r_2\cos\theta_2 + r_3\cos\theta_3 - r_4\cos\theta_4 = r_1$$
$$r_2\sin\theta_2 + r_3\sin\theta_3 - r_4\sin\theta_4 = 0$$

(d) On putting $\theta_3 = \theta_4 - N\alpha$ in the left-hand side of the first equation in (c) and expanding $\cos(\theta_4 - N\alpha)$, show that the resulting expression is equal to

$$r_2\cos\theta_2 + N\{\sin\theta_4 r_3\sqrt{(1-\mu^2)} - N(r_4 - \mu r_3)\cos\theta_4\}$$

and that this, in turn, is equal to

$$r_2\cos\theta_2 + N\sin(\theta_4 - N\delta)\surd\{(\mu r_3 - r_4)^2 + r_3^2(1-\mu^2)\}$$

Deduce that the first equation in (c) is satisfied when θ_4 takes the value given to it, for specified θ_2 and N, in the solution of the problem of position of the RRRR mechanism.

Show, by a similar method, that the second equation in (c) is also satisfied when θ_3, θ_4 take their given values.

4.17 Using the notation of Fig. 4.17 for the RRRR mechanism, show that

(a) $\underline{\text{AB}} = |\underline{\text{AB}}|\,[\cos\theta_3,\ \sin\theta_3,\ 0]$, $\underline{\text{PB}} = |\underline{\text{PB}}|\,[\cos\theta_4,\ \sin\theta_4,\ 0]$, referred to the unit vectors **i**, **j**, **k**.

(b) $\underline{\text{BA}} \times \underline{\text{BP}} = |\underline{\text{AB}}|\,|\underline{\text{BP}}|\sin(\theta_4 - \theta_3)\mathbf{k}$.

Show further, if $\cos(\theta_4 - \theta_3) \neq \pm 1$, that $\sin(\theta_4 - \theta_3)$ must be either positive or negative in a complete cycle of the mechanism and hence that the direction of the vector $\underline{\text{BA}} \times \underline{\text{BP}}$ is fixed.

Hint: note that $\sin(\theta_4 - \theta_3)$ varies continuously with θ_2.

4.18 Derive the following result for the mechanism shown in Fig. 4.14(c):

$$r_3^2 - 2\mu r_3 r_4 + \beta = 0$$

where

$$\beta = r_4^2 - r_1^2 - r_2^2 + 2r_1 r_2\cos\theta_2$$

r_1, r_2, r_4 are the lengths of links 1, 2, 4, $r_3 = |\underline{\text{AB}}|$, θ_2 = angle POA and $\mu = \cos(\text{angle ABP})$ is a constant.

Hint: use the expression for μ contained in the solution of the geometry of the RRRR mechanism. Once r_3 has been found, the angles θ_3, θ_4 can also be found with the help of this solution. It follows from Problem 4.16 that the value of N is determined by the sign of $\sin(\theta_4 - \theta_3)$. If the quadratic for r_3 has two positive (unequal) roots for all θ_2, either the smaller or the larger value must be consistently chosen through a cycle of the mechanism as r_3 is a continuous function. The initial assembly of the mechanism determines the particular choice.

Show if the link OA makes complete revolutions that $\mu^2 r_4^2 \geq \beta$ for all θ_2 and hence deduce that

$$(r_1 - r_2)^2 \geq (1 - \mu^2) r_4^2.$$

Section 4.6

4.19 Figure 4.42 shows the profile of a cam (link 2) with a roller follower. The follower consists of a roller, centre B and radius e, and an arm BC (link 4) that is attached to the roller (link 3) by a revolute joint at B. Link 4 is constrained by a prismatic joint with link 1 to move in a

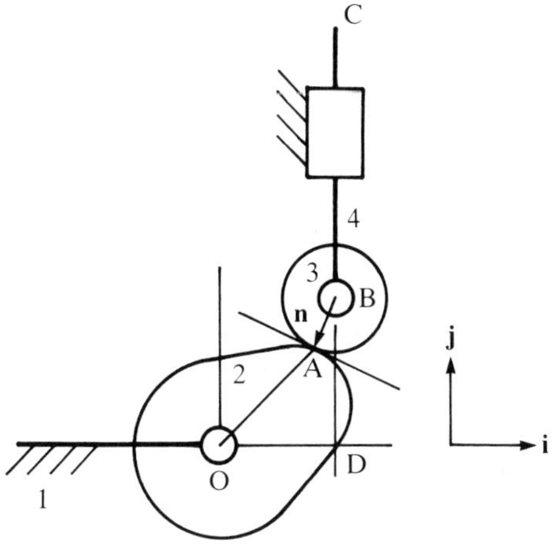

Fig. 4.42

straight line that passes through the fixed point D and is perpendicular to OD. The axis of the cam is perpendicular to the (common) plane of motion of links 3 and 4.

Given that $\underline{DB} = s(\phi)\mathbf{j}$ is a known function of the angle ϕ turned through by the cam and $\underline{OD} = d\mathbf{i}$, show, using the notation of Fig. 4.42, that

(a) $\underline{OA} = d\mathbf{i} + s\mathbf{j} + e\mathbf{n}$

(b) $\left(\dfrac{ds}{d\phi}\mathbf{j} - \mathbf{k} \times \underline{OA}\right) . \mathbf{n} = 0$

(c) $sa = \left(d - \dfrac{ds}{d\phi}\right)b \qquad$ where $\mathbf{n} = a\mathbf{i} + b\mathbf{j}$

(d) $\mathbf{n} = -\left\{\left(d - \dfrac{ds}{d\phi}\right)\mathbf{i} + s\mathbf{j}\right\}\Big/ u \qquad$ where $u^2 = \left(d - \dfrac{ds}{d\phi}\right)^2 + s^2$

(e) The parametric forms of the coordinates of a point on the cam profile are

$$x = d\cos\phi + s\sin\phi - e\left\{\left(d - \frac{ds}{d\phi}\right)\cos\phi + s\sin\phi\right\}\Big/ u$$

$$y = -d\sin\phi + s\cos\phi - e\left\{s\cos\phi - \left(d - \frac{ds}{d\phi}\right)\sin\phi\right\}\Big/ u$$

4.20 Figure 4.43 shows the profile of a cam (link 2) with a flat-faced follower (link 3). The follower is constrained by a prismatic joint with link 1 so that its motion is one of rectilinear translation parallel to the unit

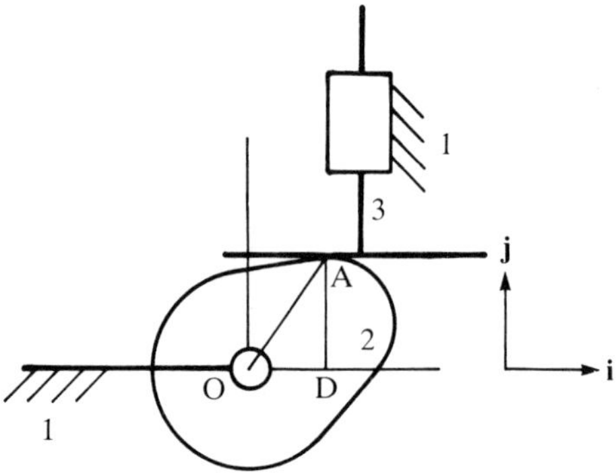

Fig. 4.43

vector **j**, which is perpendicular to the flat face. Using the notation of Fig. 4.43 and given that $\underline{\mathrm{DA}} = s(\phi)\mathbf{j}$ is a known function of the angle ϕ turned through by the cam, show that

(a) $(\mathbf{k} \times \underline{\mathrm{OA}}) \cdot \mathbf{j} = \mathrm{d}s/\mathrm{d}\phi$

(b) $\underline{\mathrm{OA}} = s\mathbf{j} + a\mathbf{i}$, where a is some (unknown) scalar

(c) $a = \mathrm{d}s/\mathrm{d}\phi$

(d) the parametric forms of the coordinates of a point on the cam profile are

$$x = \frac{\mathrm{d}s}{\mathrm{d}\phi}\cos\phi + s\sin\phi, \qquad y = s\cos\phi - \frac{\mathrm{d}s}{\mathrm{d}\phi}\sin\phi$$

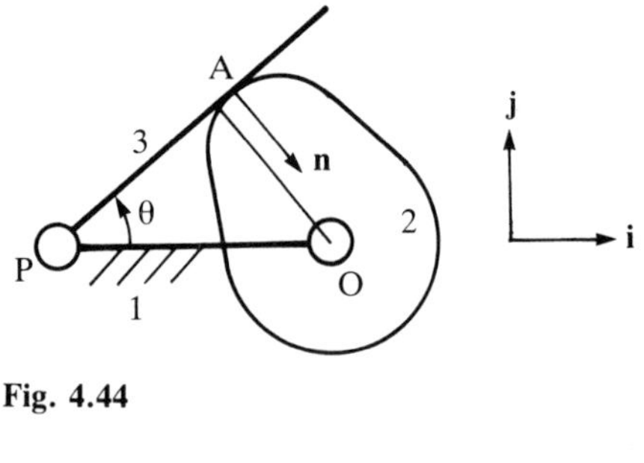

Fig. 4.44

4.21 Figure 4.44 shows the profile of a cam with a flat-faced follower which rotates about an axis that passes through the fixed point P and is parallel to the axis of the cam. The vector **n** is a unit vector, and ϕ is the angle turned through by the cam. Show, using the notation of the figure, that

(a) $\mathbf{n} = \sin\theta\mathbf{i} - \cos\theta\mathbf{j}$

(b) $\alpha\sin\theta - \beta\cos\theta = -|\underline{PO}|\sin\theta$, where $\underline{OA} = \alpha\mathbf{i} + \beta\mathbf{j}$
(*Hint:* note $\underline{PA} \,.\, \mathbf{n} = 0$

(c) $\dfrac{d\theta}{d\phi}(\mathbf{k} \times \underline{PA}) \,.\, \mathbf{n} = (\mathbf{k} \times \underline{OA}) \,.\, \mathbf{n}$

(d) $\dfrac{d\theta}{d\phi}(\mathbf{k} \times \underline{PO}) \,.\, \mathbf{n} = (\mathbf{k} \times \underline{OA}) \,.\, \mathbf{n}\left(1 - \dfrac{d\theta}{d\phi}\right)$

(e) $(\alpha\cos\theta + \beta\sin\theta)\left(1 - \dfrac{d\theta}{d\phi}\right) = PO\cos\theta\dfrac{d\theta}{d\phi}$

(f) The parametric forms of the coordinates of a point of the cam profile are

$$x = -PO\cos\phi + PO\cos\theta\cos(\theta - \phi)\Big/\left(1 - \frac{d\theta}{d\phi}\right)$$

$$y = PO\sin\phi + PO\cos\theta\sin(\theta - \phi)\Big/\left(1 - \frac{d\theta}{d\phi}\right)$$

What is the radius of the cam at a dwell?

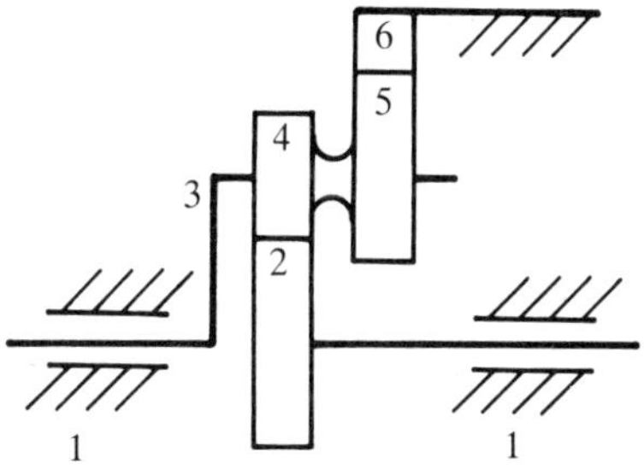

Fig. 4.45

Section 4.7

4.22 An epicyclic gear train is shown diagrammatically in Fig. 4.45. The input is to gear wheel 2 and the output is from the carrier, link 3, the annular gear 6 being fixed. Gears 4 and 5, as the diagram indicates, form a compound gear. The following are the number of teeth on each of the gear wheels:

Gear number	2	4	5	6
Number of teeth	52	15	25	92

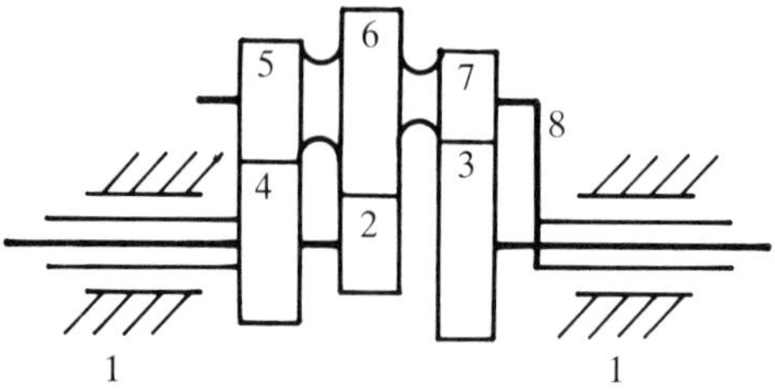

Fig. 4.46

If the input shaft has a rotational speed of 1450 rev min^{-1}, what is the speed of the output shaft?

4.23 Figure 4.46 shows diagrammatically a two-speed and reverse epicyclic gear drive. The input is to gear wheel 2 and the output is from gear wheel 3, while gears 5, 6, 7 form a compound gear carried by link 8 whose axis of rotation coincides with that of gear 2. The following are the number of teeth on each of the gear wheels:

Gear number	2	3	4	5	6	7
Number of teeth	24	46	42	28	46	24

In reverse the gear wheel 4 is fixed; in one forward speed the planet carrier, link 8, is fixed; and in the other forward speed, gear wheel 4 is connected to gear wheel 2. Determine the velocity ratio ω_2/ω_3, that is the ratio of the input rotational speed to the output rotational speed for each gear of the gear drive. (These values are the gear ratios for each gear of the epicyclic gear drive.)

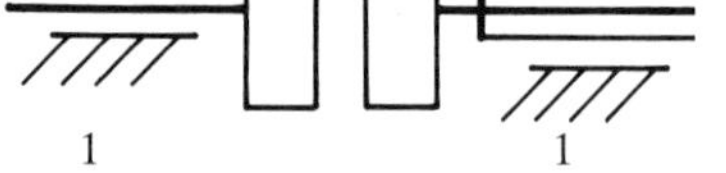

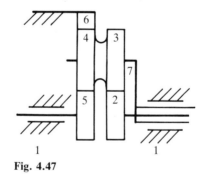

Fig. 4.47

4.24 An epicyclic gear train is shown diagrammatically in Fig. 4.47. The input is to sun gear 2, the output is from sun gear 5 and the annular gear 6 is fixed. Link 7 rotates about an axis whose line is coincident with that of the input shaft and carries the compound gear comprising gears 3 and 4. Sun gear 2 has 18 teeth and planet gear 3 has 30 teeth.

Given that the modules of all gears are equal, show that N_4, N_5, N_6, the numbers of teeth on gears 4, 5, 6, satisfy the equations

$$N_4 + N_5 = 48, \qquad N_5 + N_6 = 96$$

Obtain the values of N_4, N_5, N_6 if the velocity ratio of output to input is equal to $+0.45$.

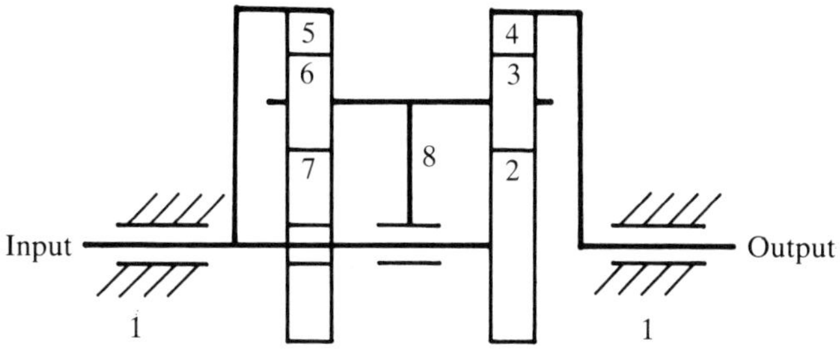

Fig. 4.48

4.25 An epicyclic gear transmission is shown diagrammatically in Fig. 4.48. Sun wheel 2 and internal wheel 5 are fixed to the input shaft and internal wheel 4 is fixed to the output shaft. Sun wheel 7 is fixed. The carrier, link 8, carries gear wheels 3 and 6 and is free to rotate about an axis whose line is coincident with that of the input and output shafts.

Show that the ratio of the output shaft speed to that of the input shaft is

$$(N_5 N_4 - N_7 N_2)/N_4(N_5 + N_7)$$

Computer exercises

4.26 The mechanism shown in Fig. 4.49 is composed of two identical RRRR mechanisms defined by the loops OABP and PDEO. As the figure shows, the angle α between the links PD and PB is fixed but the angle between OA and OE is not. Using the information provided for the solution of the geometry of the RRRR mechanism, together with the expressions for the angular speeds ω_3, ω_4 contained in (4.44) and (4.45), write a computer program that will give the ratio q of the angular speeds of links OE and OA.

Input should consist of the lengths OA, AB, BC, PO and the angle α. The angle θ_2 will constitute a variable in the program. By turning the figure through 180° it is not difficult to see that the angle OPD

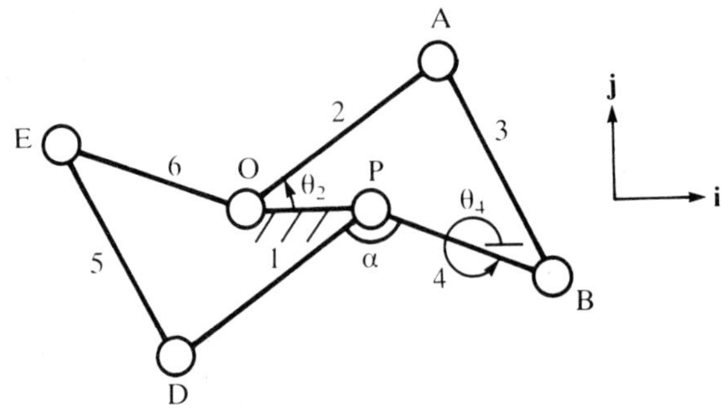

Fig. 4.49

$(\theta_4 - \pi - \alpha)$ plays the same role for the loop PDEO as θ_2 does for the loop OABP. Values of the ratio q for given θ_2 (in degrees) are provided as a check:

θ_2	0	30	90	150	210	270	330
q	2.37	2.21	1.19	0.68	0.34	0.35	1.08

The lengths of the links are given by OA = PD = 0.20 m, AB = DE = 0.15 m, BP = EO = 0.17 m, OP = 0.075 m, while $\alpha = 135°$.

4.27 Write a program that determines the profile of the cam in Fig. 4.44 (see Problem 4.21), assuming:

$$\theta = \begin{cases} a + b\{\phi - (1/4)\sin 4\phi\} & 0 \le \phi \le \pi/2 \\ \text{constant (follower dwells)} & \pi/2 \le \phi \le 3\pi/2 \\ a + b\{2\pi - \phi + (1/4)\sin 4\phi\} & 3\pi/2 \le \phi \le 2\pi \end{cases}$$

where a and b are constants.

What is the physical significance of the constants a and b? Use the following information as a check on your program; it corresponds to OP = 7.0, $a = \arcsin 5/7$ (a acute), $b = 0.1$:

ϕ	0	30°	60°	90°
x	−3.571	−0.736	1.437	3.306
y	3.499	5.164	5.441	4.649

4.28 Figure 4.50 shows a mechanism that forms the basis of a power press, the loop OABP being an RRRR mechanism. Write a program that determines the locus of C for a complete revolution of OA. This locus is an example of a coupler curve, that is a curve traced out by any point that is fixed relative to AB. Also obtain a plot of the speed of D against its displacement along QD.

The following details are provided as a check.

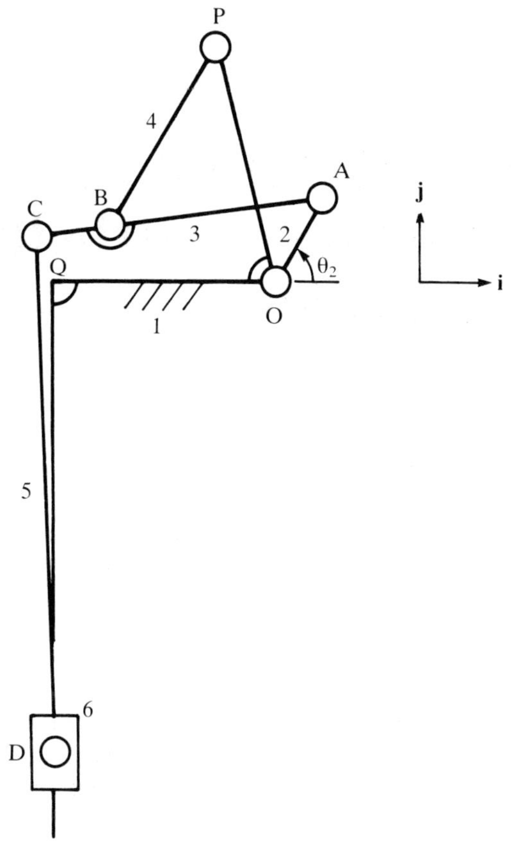

Fig. 4.50

θ_2/degree	0	90	150	210	300
y/cm	−96.50	−84.49	−57.95	−69.99	−89.04
$\dot{y}$/cm s^{-1}	−6.15	34.71	1.54	−17.04	−7.59

where $\underline{QD} = y\mathbf{j}$, $\dot{\theta}_2 = 1$ rad s^{-1} and OA = 20 cm, AB = 45 cm, AC = 61 cm, BP = 43 cm, CD = 106 cm, OP = 49.477 cm, OQ = 47 cm, $\angle$POQ = 75.96°.

Readers should note that θ_2 is not equal to the angle between the crank OA and the fixed link OP. (See Section 4.5).

5 Three-dimensional kinematics of particles and rigid bodies

5.1 INTRODUCTION

This chapter provides the theory of three-dimensional kinematics of rigid bodies and particles necessary for the three-dimensional applications of dynamics contained in Chapter 6 and the work on robots in Chapter 7. The properties of finite and infinitesimal rotations are examined in detail, and the important theorem that the displacement of a rigid body is equivalent to a translation together with a rotation is derived. There follows an extension of the one-dimensional angular velocity vector of Chapter 4 (see (4.18)–(4.20) and Problem 5.37) to a three-dimensional vector. Euler's angles, essential in many three-dimensional dynamics problems, are defined and related to the angular velocity vector of a rigid body. The worked examples are mainly of an analytical nature, while the physical applications apart from that relating to Hooke's joint are integrated with the work of Chapters 6 and 7.

We begin with a formal definition of the rotation of a particle about an axis.

5.2 ROTATION THEORY

Rotation of a particle about an axis

The axis of rotation (see Fig. 5.1) is represented by the directed line ON while the particle is denoted by P, its position prior to the rotation being the point A in space. The plane through A with ON as normal is labelled Π: it meets the axis ON in the point C. In order to define the rotation we draw the circle, centre C, radius CA that lies in the plane Π. The particle P is said to rotate about the axis ON when it moves along the circumference of this

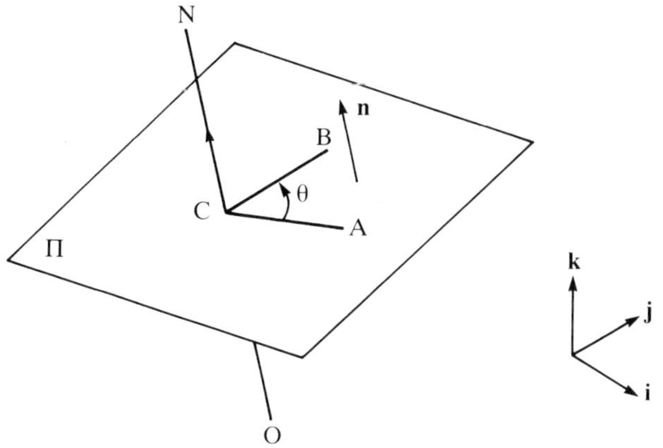

Fig. 5.1

circle. If the motion of P terminates at the point B on the circumference without making a complete revolution then the angle ACB is referred to as the angle of rotation. We denote the angle of rotation by θ and take it to be positive or negative according to the sense of the rotation. Viewed from N the motion of P from A to B as shown in Fig. 5.1 is counter-clockwise and the angle is accordingly taken to be positive. Had the motion been in the clockwise sense then the angle would have been negative. If P describes the circle once or more, the angle of rotation θ is given by $\theta = \angle ACB \pm 2k\pi$, where k is the number of complete revolutions made; the plus sign is chosen for counter-clockwise motion and the minus sign for clockwise motion. It is evident that a rotation through an angle $-\theta$ about the directed line NO is identical with a rotation of amount θ about the directed line ON.

We now state and subsequently derive a relation between $\underline{OA}$, the position vector with respect to O of the particle before the rotation, and $\underline{OB}$, its position vector after the rotation (Fig. 5.2). Letting $\underline{OA} = \mathbf{r}$ and $\underline{OB} = \mathbf{R}$, then

$$\mathbf{R} - \mathbf{r} = \tan\frac{\theta}{2}\mathbf{n} \times (\mathbf{R} + \mathbf{r}) \tag{5.1}$$

where $\mathbf{n}$ is a unit vector in the direction of the vector $\underline{ON}$. Before deriving this relation some remarks about it are necessary. The relation cannot be sensibly interpreted when $\theta = \pi$ as $\tan \theta/2$ is then infinite, and this value of θ is therefore temporarily excluded from present considerations. An alternative form of (5.1) is

$$\mathbf{d} = \mathbf{a} \times \mathbf{m} \tag{5.2}$$

where

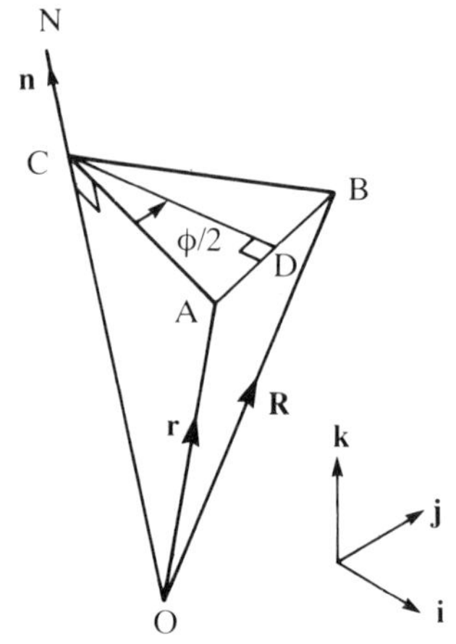

Fig. 5.2

$$\mathbf{d} = \mathbf{R} - \mathbf{r}, \qquad \mathbf{a} = 2\tan\frac{\theta}{2}\mathbf{n}, \qquad \mathbf{m} = \frac{1}{2}(\mathbf{R} + \mathbf{r}) \tag{5.3}$$

The vectors **d** and **m** are readily interpreted physically. The vector **d** is the displacement of the particle P due to the rotation, while **m** is the (arithmetic) mean of the position vectors of P before and after the rotation. The vector **a** can be regarded as defining the rotation. If **a** is known, the angle of rotation θ (apart from the addition or subtraction of multiples of 2π) can be determined from the relation $|\mathbf{a}| = 2\tan\theta/2$ if the sense of **n** is taken to be the same as that of **a**, or from $-|\mathbf{a}| = 2\tan\theta/2$ if the senses of **n** and **a** oppose each other. We can, by choosing the sense of **n** to be the same as that of **a**, take θ to be in the range 0 to π. Other rotations obtained by adding or subtracting multiples of 2π from this value (of θ) produce the same effect on the particle and may therefore be referred to as equivalent rotations. We conclude these remarks with the observation that to specify the rotation completely it is essential to know, in addition to **a**, the position of a point on the axis of rotation. A proof of (5.1) is now given.

Let D be the midpoint of AB (see Fig. 5.2). It follows from elementary geometry that $\underline{AB}$ is perpendicular to CD. Now as AB lies in the plane Π it is also perpendicular to **n** and hence we have, using the definition of the vector product, that $\underline{AB} = \alpha\mathbf{n} \times \underline{CD}$, where α is some scalar to be determined. As the vectors $\underline{AB}$ and $\mathbf{n} \times \underline{CD}$ have the same sense in Fig. 5.2, the quantity α is positive. Comparing magnitudes yields the following equation for α:

$$2|\underline{AD}| = |\underline{AB}| = \alpha|\mathbf{n}|\,|\underline{CD}|\sin\frac{\pi}{2}$$

Alternatively

$$\alpha = 2|\underline{AD}|/|\underline{CD}| = 2\tan\frac{\phi}{2}$$

where $\phi = \angle ACB$. Thus

$$\begin{aligned}\underline{AB} &= 2\tan\frac{\phi}{2}\mathbf{n} \times \underline{CD} \\ &= 2\tan\frac{\phi}{2}\mathbf{n} \times (\underline{OD} - \underline{OC}) \\ &= 2\tan\frac{\phi}{2}\mathbf{n} \times \underline{OD}\end{aligned}$$

as $\mathbf{n}$ and $\underline{OC}$ are parallel. Now

$$\underline{OD} = \underline{OA} + \underline{AD} = \underline{OA} + \frac{1}{2}\underline{AB} = \mathbf{r} + \frac{1}{2}(\mathbf{R} - \mathbf{r}) = \frac{1}{2}(\mathbf{R} + \mathbf{r})$$

while

$$\tan\frac{\theta}{2} = \tan\left(\frac{\phi}{2} \pm k\pi\right) = \tan\frac{\phi}{2}$$

so that

$$\mathbf{R} - \mathbf{r} = \tan\frac{\theta}{2}\mathbf{n} \times (\mathbf{R} + \mathbf{r})$$

as required.

The relation (5.1) can be regarded as an equation for the vector $\mathbf{R}$ when the quantities $\mathbf{n}$, $\mathbf{r}$ and θ have been specified. It has the solution

$$\mathbf{R} = \cos\theta\mathbf{r} + (1 - \cos\theta)(\mathbf{n} \cdot \mathbf{r})\mathbf{n} + \sin\theta\mathbf{n} \times \mathbf{r} \tag{5.4}$$

which is now derived. For simplicity we put $\mathbf{b} = (\tan\theta/2)\mathbf{n}$ so that the relation (5.1) becomes

$$\mathbf{R} - \mathbf{r} = \mathbf{b} \times (\mathbf{R} + \mathbf{r}) \tag{5.5}$$

If A is on the axis of rotation the position of the particle P is unaffected by the rotation. Substituting $\mathbf{r} = \mu\mathbf{n}$, where μ is some scalar, in the right-hand side of (5.4) yields the correct expression for $\mathbf{R}$, namely $\mu\mathbf{n}$. If A is not on the axis of rotation the vectors $\mathbf{r}$ and $\mathbf{b}$ are not parallel and hence $\mathbf{R}$ can be expressed in the form

$$\mathbf{R} = \alpha\mathbf{r} + \beta\mathbf{b} + \mu\mathbf{b} \times \mathbf{r}$$

where α, β, μ are scalars to be determined.

To find α, β, μ the expression for $\mathbf{R}$ is substituted into equation (5.5). Hence

$$(\alpha - 1)\mathbf{r} + \beta\mathbf{b} + \mu\mathbf{b} \times \mathbf{r} = \mathbf{b} \times \{(\alpha + 1)\mathbf{r} + \beta\mathbf{b} + \mu(\mathbf{b} \times \mathbf{r})\}$$
$$= (\alpha + 1)\mathbf{b} \times \mathbf{r} + \mu(\mathbf{b} \, . \, \mathbf{r})\mathbf{b} - \mu\mathbf{b}^2\mathbf{r}$$

Equating coefficients of corresponding vectors yields the equations

$$\alpha - 1 = -\mathbf{b}^2\mu, \qquad \beta = (\mathbf{b} \, . \, \mathbf{r})\mu, \qquad \alpha + 1 = \mu$$

Their solution is given by

$$(1 + t^2)\mu = 2, \qquad (1 + t^2)\alpha = 1 - t^2, \qquad (1 + t^2)\beta = 2t(\mathbf{n} \, . \, \mathbf{r})$$

where $t = \tan\theta/2$. Substituting for α, β, μ and $\mathbf{b}$ in the expression for $\mathbf{R}$ we obtain

$$(1 + t^2)\mathbf{R} = (1 - t^2)\mathbf{r} + 2t^2(\mathbf{n} \, . \, \mathbf{r})\mathbf{n} + 2t\mathbf{n} \times \mathbf{r}$$

which reduces to

$$\mathbf{R} = \cos\theta\mathbf{r} + (1 - \cos\theta)(\mathbf{n} \, . \, \mathbf{r})\mathbf{n} + \sin\theta\mathbf{n} \times \mathbf{r}$$

by virtue of the standard formulae $\sin\theta = 2t/(1 + t^2)$ and $\cos\theta = (1 - t^2)/(1 + t^2)$.

The result contained in (5.4) is important. It enables us to consider infinitesimal displacements arising from infinitesimal rotations and to show that the order in which such rotations are performed (about intersecting axes) is immaterial. It also enables us, when cast in matrix form, to deal with successive finite rotations about intersecting and non-intersecting axes. Before examining these and other issues in detail we show that the result contained in (5.4) holds when $\theta = \pi$.

Consideration of Fig. 5.2 shows that AB is a diameter of the circle when $\theta = \pi$ and hence D, the midpoint of AB, coincides with C. Thus $\underline{\mathrm{CD}}$ is the null vector and the method of analysis that establishes the relation (5.1) breaks down and an alternative must be sought. Since C is the midpoint of AB it follows that $\underline{\mathrm{OC}}$ is $(\mathbf{R} + \mathbf{r})/2$. Further, $\underline{\mathrm{OC}} = |\underline{\mathrm{OC}}|\mathbf{n} = (\mathbf{n} \, . \, \mathbf{r})\mathbf{n}$. Equating the two expressions for $\underline{\mathrm{OC}}$ leads to the relation

$$\mathbf{R} = -\mathbf{r} + 2(\mathbf{n} \, . \, \mathbf{r})\mathbf{n}$$

which agrees with that given by (5.4) when $\theta = \pi$.

Properties of rotations

Two particles P_1 and P_2 are rotated about the axis ON through the same angle θ. As a conseqence of the rotation P_1 moves from the point A_1 to the point B_1 and P_2 from A_2 to B_2. If $\mathbf{r}_1 = \underline{\mathrm{OA}_1}$, $\mathbf{r}_2 = \underline{\mathrm{OA}_2}$, $\mathbf{R}_1 = \underline{\mathrm{OB}_1}$, $\mathbf{R}_2 = \underline{\mathrm{OB}_2}$ and $\mathbf{n}$ is a unit vector in the direction of the vector $\underline{\mathrm{ON}}$, then

(a) $\mathbf{R}_1 \,.\, \mathbf{R}_2 = \mathbf{r}_1 \,.\, \mathbf{r}_2$
(b) $|\mathbf{R}_1| = |\mathbf{r}_1|, \qquad |\mathbf{R}_2| = |\mathbf{r}_2|$
(c) $\mathbf{R}_1 \times \mathbf{R}_2 = \cos\theta \mathbf{r}_1 \times \mathbf{r}_2 + (1 - \cos\theta)(\mathbf{n} \,.\, \mathbf{r}_1 \times \mathbf{r}_2)\mathbf{n} + \sin\theta \mathbf{n} \times (\mathbf{r}_1 \times \mathbf{r}_2)$
(d) The further particle P whose position vector before the rotation is $\alpha\mathbf{r}_1 + \beta\mathbf{r}_2 + \mu\mathbf{r}_1 \times \mathbf{r}^2$ moves as a result of the rotation to the point whose position vector is $\alpha\mathbf{R}_1 + \beta\mathbf{R}_2 + \mu\mathbf{R}_1 \times \mathbf{R}_2$.

Proofs

(a) In order to facilitate this and other proofs we write c for $\cos\theta$ and s for $\sin\theta$ in the relation (5.4). Thus

$$\mathbf{R} = c\mathbf{r} + (1 - c)(\mathbf{n} \,.\, \mathbf{r})\mathbf{n} + s\mathbf{n} \times \mathbf{r} \tag{5.6}$$

Hence

$$\begin{aligned}\mathbf{R}_1 \,.\, \mathbf{R}_2 &= \{c\mathbf{r}_1 + (1 - c)(\mathbf{n} \,.\, \mathbf{r}_1)\mathbf{n} + s\mathbf{n} \times \mathbf{r}_1\} \\ &\quad .\, \{c\mathbf{r}_2 + (1 - c)(\mathbf{n} \,.\, \mathbf{r}_2)\mathbf{n} + s\mathbf{n} \times \mathbf{r}_2\} \\ &= c^2\mathbf{r}_1 \,.\, \mathbf{r}_2 + \{2(1 - c)c + (1 - c)^2\}(\mathbf{n} \,.\, \mathbf{r}_1)(\mathbf{n} \,.\, \mathbf{r}_2) \\ &\quad + sc\{\mathbf{r}_1 \,.\, (\mathbf{n} \times \mathbf{r}_2) + (\mathbf{n} \times \mathbf{r}_1) \,.\, \mathbf{r}_2\} + s^2(\mathbf{n} \times \mathbf{r}_1) \,.\, (\mathbf{n} \times \mathbf{r}_2)\end{aligned}$$

Using property (1.20) of triple-scalar products the coefficient of sc in the expression for $\mathbf{R}_1 \,.\, \mathbf{R}_2$ is readily seen to be zero. The coefficient of s^2 is, by virtue of Example 1.8, equal to $\mathbf{r}_1 \,.\, \mathbf{r}_2 - (\mathbf{n} \,.\, \mathbf{r}_1)(\mathbf{n} \,.\, \mathbf{r}_2)$, and thus we deduce after some elementary manipulation that $\mathbf{R}_1 \,.\, \mathbf{R}_2 = \mathbf{r}_1 \,.\, \mathbf{r}_2$.
(b) It follows if $\mathbf{r}_2$, $\mathbf{R}_2$ are respectively replaced by $\mathbf{r}_1$, $\mathbf{R}_1$ throughout the proof of (a) that $\mathbf{R}_1^2 = \mathbf{r}_1^2$ and therefore $|\mathbf{R}_1| = |\mathbf{r}_1|$. Likewise $|\mathbf{R}_2| = |\mathbf{r}_2|$.
(c) We have

$$\begin{aligned}\mathbf{R}_1 \times \mathbf{R}_2 &= \{c\mathbf{r}_1 + (1 - c)(\mathbf{n} \,.\, \mathbf{r}_1)\mathbf{n} + s\mathbf{n} \times \mathbf{r}_1\} \\ &\quad \times \{c\mathbf{r}_2 + (1 - c)(\mathbf{n} \,.\, \mathbf{r}_2)\mathbf{n} + s\mathbf{n} \times \mathbf{r}_2\} \\ &= c^2\mathbf{r}_1 \times \mathbf{r}_2 + c(1 - c)\{(\mathbf{n} \,.\, \mathbf{r}_2)\mathbf{r}_1 - (\mathbf{n} \,.\, \mathbf{r}_1)\mathbf{r}_2\} \times \mathbf{n} \\ &\quad + cs\{\mathbf{r}_1 \times (\mathbf{n} \times \mathbf{r}_2) + (\mathbf{n} \times \mathbf{r}_1) \times \mathbf{r}_2\} \\ &\quad + (1 - c)s\{(\mathbf{n} \,.\, \mathbf{r}_1)(\mathbf{n} \times (\mathbf{n} \times \mathbf{r}_2)) + (\mathbf{n} \,.\, \mathbf{r}_2)((\mathbf{n} \times \mathbf{r}_1) \times \mathbf{n})\} \\ &\quad + s^2(\mathbf{n} \times \mathbf{r}_1) \times (\mathbf{n} \times \mathbf{r}_2)\end{aligned}$$

The terms in the expression for $\mathbf{R}_1 \times \mathbf{R}_2$ can be simplified by using the rule for expanding triple-vector products. The coefficient of $c(1 - c)$ is equal to $\{\mathbf{n} \times (\mathbf{r}_1 \times \mathbf{r}_2)\} \times \mathbf{n}$, which on expansion becomes $\mathbf{r}_1 \times \mathbf{r}_2 - (\mathbf{n} \,.\, \mathbf{r}_1 \times \mathbf{r}_2)\mathbf{n}$. The coefficients of cs and $(1 - c)s$ both reduce to $\mathbf{n} \times (\mathbf{r}_1 \times \mathbf{r}_2)$, whilst that of s^2 reduces to $\mathbf{n} \,.\, (\mathbf{r}_1 \times \mathbf{r}_2)\mathbf{n}$. Hence we obtain

$$\begin{aligned}\mathbf{R}_1 \times \mathbf{R}_2 &= c^2\mathbf{r}_1 \times \mathbf{r}_2 + c(1 - c)\{\mathbf{r}_1 \times \mathbf{r}_2 - (\mathbf{n} \,.\, \mathbf{r}_1 \times \mathbf{r}_2)\mathbf{n}\} \\ &\quad + s\mathbf{n} \times (\mathbf{r}_1 \times \mathbf{r}_2) + s^2(\mathbf{n} \,.\, \mathbf{r}_1 \times \mathbf{r}_2)\mathbf{n} \\ &= c\mathbf{r}_1 \times \mathbf{r}_2 + (1 - c)(\mathbf{n} \,.\, \mathbf{r}_1 \times \mathbf{r}_2)\mathbf{n} + s\mathbf{n} \times (\mathbf{r}_1 \times \mathbf{r}_2)\end{aligned}$$

as required.

(d) Using (5.6) with $\mathbf{r}$ replaced by $\alpha\mathbf{r}_1 + \beta\mathbf{r}_2 + \mu\mathbf{r}_1 \times \mathbf{r}_2$ we see, with the help of the result contained in (c), that the corresponding $\mathbf{R}$ is given by $\mathbf{R} = \alpha\mathbf{R}_1 + \beta\mathbf{R}_2 + \mu\mathbf{R}_1 \times \mathbf{R}_2$.

Rotation of a body about an axis

The body (not assumed to be rigid) is composed of $N + 1$ particles P_0, P_1, ..., P_N. Each particle of the body is rotated an amount θ about an axis that passes through O and is in the direction of the unit vector $\mathbf{n}$. The particles are situated at the points A_0, A_1, ..., A_N before the rotation and at the points B_0, B_1, ..., B_N after it.

Letting $\underline{OA_i} = \mathbf{r}_i$ and $\underline{OB_i} = \mathbf{R}_i$ ($i = 0, 1, \ldots, N$) we have, using the result (5.4),

$$\mathbf{R}_i = \cos\theta\mathbf{r}_i + (1 - \cos\theta)(\mathbf{n} \cdot \mathbf{r}_i)\mathbf{n} + \sin\theta\mathbf{n} \times \mathbf{r}_i$$
$$\mathbf{R}_j = \cos\theta\mathbf{r}_j + (1 - \cos\theta)(\mathbf{n} \cdot \mathbf{r}_j)\mathbf{n} + \sin\theta\mathbf{n} \times \mathbf{r}_j$$

Subtracting corresponding sides of these two relations leads to

$$\underline{B_iB_j} = \cos\theta\underline{A_iA_j} + (1 - \cos\theta)(\mathbf{n} \cdot \underline{A_iA_j})\mathbf{n} + \sin\theta\mathbf{n} \times \underline{A_iA_j} \tag{5.7}$$

The relation (5.7) is of considerable importance, being central to our analysis of the rotation of a body about an axis. Firstly we observe that the relation between the pair of vectors $\underline{B_iB_j}$, $\underline{A_iA_j}$ is of the same form as the relation (5.6) between the pair $\mathbf{R}$, $\mathbf{r}$. Hence the rotation properties (a) to (d) that hold for the latter pair hold *equally* well for the former pair. For example, using property (b) it follows that $|\underline{B_iB_j}| = |\underline{A_iA_j}|$ and thus distances between particles are unaltered by the rotation. In fact, as we now show, the relative orientation of particles as well as distances are preserved during the rotation. In short (see Section 4.3) the body behaves as if it were a rigid body during the rotation.

Let P_0, P_1, P_2 be three particles whose positions A_0, A_1, A_2 before the rotation are non-collinear. Now the position vector $\underline{A_0A_j}$ of the particle P_j with respect to P_0 (before the rotation) can be expressed as

$$\underline{A_0A_j} = \alpha\underline{A_0A_1} + \beta\underline{A_0A_2} + \mu\underline{A_0A_1} \times \underline{A_0A_2} \tag{5.8}$$

To find the vector $\underline{B_0B_j}$ we use the relation (5.7) with $i = 0$. Substituting the expression for $\underline{A_0A_j}$ given by the right-hand side of (5.8) into the relation yields, noting property (c),

$$\underline{B_0B_j} = \alpha\underline{B_0B_1} + \beta\underline{B_0B_2} + \mu\underline{B_0B_1} \times \underline{B_0B_2}$$

Hence the relative orientation of particles is preserved.

The right-hand side of (5.7) depends on the initial positions A_i, A_j of the particles P_i, P_j, the angle θ and the unit vector $\mathbf{n}$, but not on O. If the body

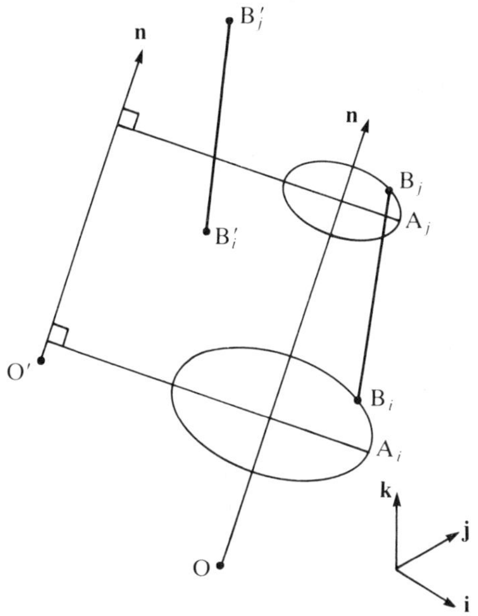

Fig. 5.3

is rotated about an axis through the point O′ with θ and **n** unchanged, the particles moving from the points A_i to the points B'_i ($i = 0, 1, 2, \ldots, N$), it follows that $\underline{B'_iB'_j} = \underline{B_iB_j}$ (see Fig. 5.3). Thus the position of the body after the rotation about the axis through O′ can be described as being parallel to its position after the rotation about the axis through O. We show that the particles of the body could be moved from one position to the other by applying a (common) translation to them. Applying (5.4) with O as origin we have

$$\underline{OB_i} = \cos\theta\underline{OA_i} + (1 - \cos\theta)(\mathbf{n} \cdot \underline{OA_i})\mathbf{n} + \sin\theta\mathbf{n} \times \underline{OA_i}$$

while

$$\underline{O'B'_i} = \cos\theta\underline{O'A_i} + (1 - \cos\theta)(\mathbf{n} \cdot \underline{O'A_i})\mathbf{n} + \sin\theta\mathbf{n} \times \underline{O'A_i}$$

Since $\underline{O'B'_i} = \underline{O'O} + \underline{OB'_i}$ we obtain, by subtracting corresponding sides of these two results,

$$\underline{OO'} + \underline{B'_iB_i} = \cos\theta\underline{OO'} + (1 - \cos\theta)(\mathbf{n} \cdot \underline{OO'})\mathbf{n} + \sin\theta\mathbf{n} \times \underline{OO'}$$

or

$$\underline{B'_iB_i} = (\cos\theta - 1)\underline{OO'} + (1 - \cos\theta)(\mathbf{n} \cdot \underline{OO'})\mathbf{n} + \sin\theta\mathbf{n} \times \underline{OO'}$$

The right-hand side of this expression is the same for all i and therefore the body can be moved from one position to the other by a translation.

Since systems of mutually perpendicular right-handed axes can be regarded as particular examples of a rigid body, the results derived here have implications for rotating frames of reference. We observe that if a set of axes is rotated about an axis, the direction of any one of the associated unit vectors after the rotation depends on its direction before the rotation, the angle θ and the unit vector $\mathbf{n}$, but not on the actual location of the axis of rotation in space. We can therefore, if it is convenient, choose the axis of rotation to pass through the origin of the axes when determining directions. This result is useful in robot theory.

Determination of the angle and axis of rotation when the displacement of two particles is known

We assume that the positions of the particles P_1, P_2 are known before and after the rotation. It is also assumed that the position of O, a point on the axis of rotation, is known. Hence the four vectors $\mathbf{r}_1$, $\mathbf{r}_2$, $\mathbf{R}_1$, $\mathbf{R}_2$ (defined in the properties of rotations) are also known. Noting (5.3) we define the vectors $\mathbf{d}_i$, $\mathbf{m}_i$ ($i = 1, 2$) by

$$\mathbf{d}_i = \mathbf{R}_i - \mathbf{r}_i, \qquad 2\mathbf{m}_i = \mathbf{R}_i + \mathbf{r}_i \tag{5.9}$$

We now show that these vectors satisfy the three relations

$$\mathbf{d}_1 \cdot \mathbf{m}_1 = \mathbf{d}_2 \cdot \mathbf{m}_2 = \mathbf{d}_1 \cdot \mathbf{m}_2 + \mathbf{d}_2 \cdot \mathbf{m}_1 = 0 \tag{5.10}$$

Recalling property (b) $\mathbf{R}_1^2 = \mathbf{r}_1^2$ for rotations, it follows that $(\mathbf{R}_1 - \mathbf{r}_1) \cdot (\mathbf{R}_1 + \mathbf{r}_1) = 0$ and therefore by (5.9) $\mathbf{d}_1 \cdot \mathbf{m}_1 = 0$. Likewise $\mathbf{d}_2 \cdot \mathbf{m}_2 = 0$. Recalling property (a) gives

$$0 = 2(\mathbf{R}_1 \cdot \mathbf{R}_2 - \mathbf{r}_1 \cdot \mathbf{r}_2) = (\mathbf{R}_1 - \mathbf{r}_1) \cdot (\mathbf{R}_2 + \mathbf{r}_2) + (\mathbf{R}_2 - \mathbf{r}_2)(\mathbf{R}_1 + \mathbf{r}_1)$$

so that $\mathbf{d}_1 \cdot \mathbf{m}_2 + \mathbf{d}_2 \cdot \mathbf{m}_1 = 0$ as required.

In order to find the quantities θ and $\mathbf{n}$ that define the rotation we have, noting (5.2), either to solve the pair of equations $\mathbf{a} \times \mathbf{m}_i = \mathbf{d}_i$ ($i = 1, 2$) for the vector $\mathbf{a}$ or to show that the rotation is of amount π about an axis to be identified. The equations $\mathbf{a} \times \mathbf{m}_1 = \mathbf{d}_1$ and $\mathbf{a} \times \mathbf{m}_2 = \mathbf{d}_2$ can be compared directly with the equations $\mathbf{X} \times \mathbf{A} = \mathbf{B}$ and $\mathbf{X} \times \mathbf{C} = \mathbf{D}$ considered in Example 1.14. If $\mathbf{A}$ and $\mathbf{C}$ are not parallel, a unique solution exists for $\mathbf{X}$ only if $\mathbf{A} \cdot \mathbf{B} = \mathbf{C} \cdot \mathbf{D} = \mathbf{A} \cdot \mathbf{D} + \mathbf{B} \cdot \mathbf{C} = 0$. Thus if $\mathbf{m}_1$ and $\mathbf{m}_2$ are not parallel, it follows by virtue of (5.10) that a unique solution for $\mathbf{a}$ exists and is given (see Example 1.14) by

$$|\mathbf{m}_1 \times \mathbf{m}_2|^2\mathbf{a} = (\mathbf{d}_2 \cdot \mathbf{m}_1 \times \mathbf{m}_2)\mathbf{m}_1 + (\mathbf{d}_1 \cdot \mathbf{m}_2 \times \mathbf{m}_1)\mathbf{m}_2 + (\mathbf{d}_1 \cdot \mathbf{m}_2)(\mathbf{m}_1 \times \mathbf{m}_2) \tag{5.11}$$

If the vectors $\mathbf{m}_1$ and $\mathbf{m}_2$ are parallel, that is $\mathbf{m}_1 = c\mathbf{m}_2$ where c is some scalar, the equations $\mathbf{d}_i = \mathbf{a} \times \mathbf{m}_i$ have a solution only if $\mathbf{d}_1 = c\mathbf{d}_2$. Moreover, this solution is not unique. If $\mathbf{m}_1 = c\mathbf{m}_2$ and $\mathbf{d}_1 = c\mathbf{d}_2$ then $\mathbf{r}_1 = c\mathbf{r}_2$ and $\mathbf{R}_1 = c\mathbf{R}_2$, so that the particles P_1, P_2 and the origin O are collinear before and after the rotation.

If $\mathbf{m}_1$ and $\mathbf{m}_2$ are parallel with $\mathbf{m}_1 = c\mathbf{m}_2$ then, provided $\mathbf{d}_1 \neq c\mathbf{d}_2$, the rotation is unique, being of amount π with the axis of rotation parallel to the directions of $\mathbf{m}_1$ and $\mathbf{m}_2$.

Since the vectors $\mathbf{m}_1$ and $\mathbf{m}_2$ are parallel we can express them in the form $\mathbf{m}_1 = \alpha_1\mathbf{N}$ and $\mathbf{m}_2 = \alpha_2\mathbf{N}$, where $\mathbf{N}$ is a unit vector and α_1, α_2 are scalars which may be positive or negative. It therefore follows, recalling (5.9), that

$$\mathbf{R}_1 = -\mathbf{r}_1 + 2\alpha_1\mathbf{N}, \qquad \mathbf{R}_2 = -\mathbf{r}_2 + 2\alpha_2\mathbf{N} \tag{5.12}$$

Substituting $\mathbf{m}_1 = \alpha_1\mathbf{N}$ in the first of the relations in (5.10) we deduce that $\mathbf{N}\,.\,\mathbf{r}_1 = \mathbf{N}\,.\,\mathbf{R}_1$. Likewise $\mathbf{N}\,.\,\mathbf{R}_2 = \mathbf{N}\,.\,\mathbf{r}_2$. Taking the scalar product of each side of the relations in (5.12) with $\mathbf{N}$ leads to the results $\alpha_1 = \mathbf{N}\,.\,\mathbf{r}_1$ and $\alpha_2 = \mathbf{N}\,.\,\mathbf{r}_2$. Hence

$$\mathbf{R}_1 = -\mathbf{r}_1 + 2(\mathbf{N}\,.\,\mathbf{r}_1)\mathbf{N}, \qquad \mathbf{R}_2 = -\mathbf{r}_2 + 2(\mathbf{N}\,.\,\mathbf{r}_2)\mathbf{N}$$

Recalling (5.4) and putting $\theta = \pi$ we can infer that the rotation is of amount π about an axis in the direction of the unit vector $\mathbf{N}$. Uniqueness follows from the fact that as $\mathbf{d}_1 \neq c\mathbf{d}_2$ the equations $\mathbf{d}_i = \mathbf{a} \times \mathbf{m}_i$ have no solution. The above analysis is illustrated in the numerical example that follows.

Example 5.1(a), (b), (c)

Two particles P_1, P_2 are rotated through the same angle about the axis ON. Before the rotation the particles are situated at the points A_1, A_2 and after it at the points B_1, B_2. Find the angle of rotation and the direction of the axis of rotation when, referred to the unit vectors **i**, **j**, **k**,

(a) $\underline{OA_1} = [1, -1, 3]$, $\quad \underline{OA_2} = [-2, -2, 4]$
$\quad\;\;\underline{OB_1} = [3, -1, 1]$, $\quad \underline{OB_2} = [2, -4, 2]$
(b) $\underline{OA_1} = [2, 0, 1]$, $\quad \underline{OA_2} = [4, 0, 2]$
$\quad\;\;\underline{OB_1} = [1, 0, 2]$, $\quad \underline{OB_2} = [2, 0, 4]$
(c) $\underline{OA_1} = [3, 3, -3]$, $\quad \underline{OA_2} = [4, 1, -1]$
$\quad\;\;\underline{OB_1} = [1, 1, 5]$, $\quad \underline{OB_2} = [0, 3, 3]$

Solution

(a) The following are readily found:

$$\mathbf{m}_1 = [2, -1, 2], \qquad \mathbf{d}_1 = [2, 0, -2], \qquad \mathbf{m}_2 = [0, -3, 3],$$
$$\mathbf{d}_2 = [4, -2, -2]$$

from which we verify that $\mathbf{m}_1\,.\,\mathbf{d}_1 = \mathbf{m}_2\,.\,\mathbf{d}_2 = \mathbf{m}_1\,.\,\mathbf{d}_2 + \mathbf{m}_2\,.\,\mathbf{d}_1 = 0$. The vectors $\mathbf{m}_1$ and $\mathbf{m}_2$ are not parallel and therefore the vector $\mathbf{a}$ can be determined by using the result in (5.11). Straightforward calculation yields

$$\mathbf{m}_1 \times \mathbf{m}_2 = [3, -6, -6], \qquad \mathbf{d}_2 \,.\, \mathbf{m}_1 \times \mathbf{m}_2 = 36, \qquad \mathbf{d}_1 \,.\, \mathbf{m}_2 \times \mathbf{m}_1 = -18,$$
$$\mathbf{d}_1 \,.\, \mathbf{m}_2 = -6$$

Hence

$$81\mathbf{a} = 36[2, -1, 2] - 18[0, -3, 3] - 6[3, -6, -6] = 54[1, 1, 1]$$

so that

$$2\tan\frac{\theta}{2}\mathbf{n} = \mathbf{a} = \frac{2}{3}[1, 1, 1]$$

from which we deduce that $\theta = \pi/3$ and $(\sqrt{3})\mathbf{n} = [1, 1, 1]$.

(b) Since $\underline{OA_2} = 2\underline{OA_1}$ and $\underline{OB_2} = 2\underline{OB_1}$, the particles P_1, P_2 and the point O are collinear before and after the rotation so that there is more than one possible rotation. We have

$$2\mathbf{m}_1 = [3, 0, 3], \qquad \mathbf{d}_1 = [-1, 0, 1], \qquad \mathbf{m}_2 = [3, 0, 3], \qquad \mathbf{d}_2 = [-2, 0, 2]$$

giving $\mathbf{m}_1 \,.\, \mathbf{d}_1 = \mathbf{m}_2 \,.\, \mathbf{d}_2 = \mathbf{m}_1 \,.\, \mathbf{d}_2 + \mathbf{m}_2 \,.\, \mathbf{d}_1 = 0$, $\mathbf{m}_2 = 2\mathbf{m}_1$ and $\mathbf{d}_2 = 2\mathbf{d}_1$. The equations $\mathbf{a} \times \mathbf{m}_1 = \mathbf{d}_1$ and $\mathbf{a} \times \mathbf{m}_2 = \mathbf{d}_2$ are equivalent so that we need only consider one of them to find $\mathbf{a}$. It is readily verified by direct substitution (or see Example 1.13) that $\mathbf{a}$, given by $\mathbf{m}_1{}^2\mathbf{a} = \mathbf{m}_1 \times \mathbf{d}_1$, is a solution of $\mathbf{a} \times \mathbf{m}_1 = \mathbf{d}_1$. To this solution we can add any multiple of $\mathbf{m}_1$. We have

$$\mathbf{a} = (\mathbf{m}_1 \times \mathbf{d}_1)/\mathbf{m}_1^2 = \frac{2}{3}[0, -1, 0]$$

so that one possible rotation is of amount 2 arc tan1/3 about an axis in the direction of the vector $-\mathbf{j}$.

As we have just indicated, other solutions for $\mathbf{a}$ may be obtained by adding multiples of $\mathbf{m}_1$ to the one already found. Taking the multiple (for arithmetical convenience) to be 4/9, we see that a second rotation is given by $(3\tan\theta/2)\mathbf{n} = [1, -1, 1]$. Hence $\theta = \pi/3$ and $(\sqrt{3})\mathbf{n} = [1, -1, 1]$. The reader can verify the correctness of the answer by applying result (5.4). A further solution is given by $\theta = \pi$ with $\mathbf{n}$ in the direction of $\mathbf{m}_1$ so that $(\sqrt{2})\mathbf{n} = [1, 0, 1]$.

(c) Using the given data yields

$$\mathbf{m}_1 = [2, 2, 1], \qquad \mathbf{d}_1 = [-2, -2, 8], \qquad \mathbf{m}_2 = [2, 2, 1], \qquad \mathbf{d}_2 = [-4, 2, 4]$$

so that the conditions $\mathbf{m}_1 \,.\, \mathbf{d}_1 = \mathbf{m}_2 \,.\, \mathbf{d}_2 = \mathbf{m}_1 \,.\, \mathbf{d}_2 + \mathbf{m}_2 \,.\, \mathbf{d}_1 = 0$ are satisfied. The vectors $\mathbf{m}_1$ and $\mathbf{m}_2$ are equal (that is $c = 1$) and therefore parallel while $\mathbf{d}_1 \neq c\mathbf{d}_2$. Thus the rotation is unique of amount π, and $\mathbf{n}$, which is parallel to $\mathbf{m}_1$, is given by $3\mathbf{n} = [2, 2, 1]$.

Example 5.1(d)

Determine for part (a) of Example 5.1 the position of the further particle P_3 after the rotation, given that its position before the rotation is the point A_3, where $\underline{OA_3} = [5, -9, -5]$. In order to illustrate the theory we give two solutions to the problem.

Solution 1 Since $\theta = \pi/3$ and $(\sqrt{3})\mathbf{n} = [1, 1, 1]$ we obtain, on applying (5.4),

$$\mathbf{R} = \frac{1}{2}[5, -9, -5] - \frac{3}{2}[1, 1, 1] + \frac{1}{2}[4, 10, -14] = [3, -1, -11]$$

Solution 2 We begin by finding scalars of α, β, μ such that

$$\underline{OA_3} = \alpha\underline{OA_1} + \beta\underline{OA_2} + \mu\underline{OA_1} \times \underline{OA_2}$$

Now $\underline{OA_1} \times \underline{OA_2} = [2, -10, -4]$, so that α, β, μ are given by

$$[5, -9, -5] = \alpha[1, -1, 3] + \beta[-2, -2, 4] + \mu[2, -10, -4]$$

Taking the scalar product of both sides of the vector equation with the vector $[2, -10, -4]$ gives $\mu = 1$. Equating components yields the three equations

$$\alpha - 2\beta = 3, \qquad -\alpha - 2\beta = 1, \qquad 3\alpha + 4\beta = -1$$

from which we deduce $\alpha = 1$ and $\beta = -1$.

Since the relative orientation of the particles P_1, P_2, P_3 and the point O is preserved in the rotation (property (d) of rotations) we have that the particle P_3 moves to the point B_3 as a result of the rotation, where

$$\begin{aligned}\underline{OB_3} &= \underline{OB_1} - \underline{OB_2} + \underline{OB_1} \times \underline{OB_2}\\ &= [3, -1, 1] - [2, -4, 2] + [2, -4, -10] = [3, -1, -11]\end{aligned}$$

The result in solution 1 is thus confirmed.

The relation (5.4) which gives $\mathbf{R}$ when θ, $\mathbf{n}$ and $\mathbf{r}$ are known is cumbersome to use when successive rotations about intersecting axes are being considered. Matrix methods are, as we now show, preferable.

The rotation matrix

If $\mathbf{R} = [X, Y, Z]$, $\mathbf{r} = [x, y, z]$ and $\mathbf{n} = [l_1, l_2, l_3]$ referred to the unit vectors $\mathbf{i}$, $\mathbf{j}$, $\mathbf{k}$, then the relation (5.4) can be put in the form

$$[X, Y, Z] = \cos\theta[x, y, z] + (1 - \cos\theta)(l_1x + l_2y + l_3z)[l_1, l_2, l_3]$$
$$+ \sin\theta \begin{vmatrix} \mathbf{i} & \mathbf{j} & \mathbf{k} \\ l_1 & l_2 & l_3 \\ x & y & z \end{vmatrix}$$

Equating corresponding components leads to

$$X = \{\cos\theta + (1 - \cos\theta)l_1^2\}x$$
$$+ \{(1 - \cos\theta)l_1l_2 - l_3\sin\theta\}y + \{(1 - \cos\theta)l_1l_3 + l_2\sin\theta\}z$$
$$Y = \{(1 - \cos\theta)l_2l_1 + l_3\sin\theta\}x$$
$$+ \{\cos\theta + (1 - \cos\theta)l_2^2\}y + \{(1 - \cos\theta)l_2l_3 - l_1\sin\theta\}z$$
$$Z = \{(1 - \cos\theta)l_3l_1 - l_2\sin\theta\}x$$
$$+ \{(1 - \cos\theta)l_2l_3 + l_1\sin\theta\}y + \{\cos\theta + (1 - \cos\theta)l_3^2\}z$$

These relations can be put into a much more compact form by the introduction of the three matrices **A**, **B**, **C** defined by

$$\mathbf{A} = [a_{ij}] = \cos\theta\mathbf{I} + (1 - \cos\theta)\mathbf{B} + \sin\theta\mathbf{C}$$

$$\mathbf{B} = [b_{ij}] = \begin{bmatrix} l_1^2 & l_1l_2 & l_1l_3 \\ l_2l_1 & l_2^2 & l_2l_3 \\ l_3l_1 & l_3l_2 & l_3^2 \end{bmatrix} \tag{5.13}$$

$$\mathbf{C} = [c_{ij}] = \begin{bmatrix} 0 & -l_3 & l_2 \\ l_3 & 0 & -l_1 \\ -l_2 & l_1 & 0 \end{bmatrix}$$

With the help of these definitions we can write

$$\begin{bmatrix} X \\ Y \\ Z \end{bmatrix} = \mathbf{A} \begin{bmatrix} x \\ y \\ z \end{bmatrix} \tag{5.14}$$

The matrix **A** is an example of a rotation matrix. Its elements $a_{ij}(i, j = 1, 2, 3)$ depend on θ and the components l_1, l_2, l_3 of **n** *referred* to the unit vectors **i**, **j**, **k**. If the set **i**, **j**, **k** is replaced by a second set of vectors **I**, **J**, **K** then the elements of **A** will be changed. In order to avoid possible confusion we adopt the following notation. Rotation matrices will be denoted by such expressions as $\mathbf{A}(\theta, \mathbf{n})$, $\mathbf{A}(\phi, \mathbf{l})$, $\mathbf{A}(\theta_1, \mathbf{n}_1)$ when we do not wish to imply any particular choice of (reference) unit vectors. The angles θ, ϕ, θ_1 are of course the respective angles of rotation, while $\mathbf{n}$, $\mathbf{l}$, $\mathbf{n}_1$ are unit vectors in the directions of the respective axes of rotation. The symbol **A** will be reserved for a rotation matrix when its elements have been determined with **i**, **j**, **k** as the reference set of unit vectors. If we are considering several rotation matrices then we use suffixes as in $\mathbf{A}_1, \mathbf{A}_2, \ldots, \mathbf{A}_s$. The elements of $\mathbf{A}(\theta, n)$ when **I**, **J**, **K** are taken as the reference set may be obtained from the elements of **A** by an elementary matrix transformation. This transformation is defined in property P8 of the properties (to be given shortly) of the rotation matrix.

The relation (5.14) can be written in the form

$$\mathbf{R}^{\mathrm{T}} = \mathbf{A}(\theta, \mathbf{n})\mathbf{r}^{\mathrm{T}}$$

where $\mathbf{r}^{\mathrm{T}}$, $\mathbf{R}^{\mathrm{T}}$ are the transposes of the vectors **r**, **R**. This form has the particular merit that it does not imply the use of any particular coordinate system. We find it convenient to drop the transpose symbol and write the relation as

$$\mathbf{R} = \mathbf{A}(\theta, \mathbf{n})\mathbf{r} \tag{5.15}$$

where it is understood that the vectors **r**, **R** are to be interpreted as column vectors when any particular choice of axes is made.

Noting our concluding remarks in the discussion of the rotation of a body about an axis, it follows that if the unit vector **p** is rotated by an amount θ about any axis in the direction of the unit vector **n** the outcome is a second vector **q** given by

$$\mathbf{q} = \mathbf{A}(\theta, \mathbf{n})\mathbf{p} \tag{5.16}$$

As in (5.15) the vectors **p**, **q** are to be interpreted as column vectors. The results in (5.15) and (5.16) form the cornerstones of our study of the kinematics of robots. The reader should appreciate that the result (5.15) is a statement about localized vectors, the position vectors **r**, **R** being tied to the origin O, while (5.16) is a statement about free vectors.

The matrices **A**, **B**, **C** formally defined in (5.13) were considered in Examples 1.23 and 1.24 and a number of their properties derived. A number of additional properties are obtained here for the rotation matrix $\mathbf{A}(\theta, \mathbf{n})$. Those previously derived are restated for completeness. The properties are:

P1 trace $\mathbf{A}(\theta, \mathbf{n}) = a_{11} + a_{22} + a_{33} = 2\cos\theta + 1$

P2 $\mathbf{A}^{\mathrm{T}}(\theta, \mathbf{n}) = \mathbf{A}(-\theta, \mathbf{n})$

P3 $\mathbf{A}(\theta_1, \mathbf{n})\mathbf{A}(\theta_2, \mathbf{n}) = \mathbf{A}(\theta_1 + \theta_2, \mathbf{n})$

P4 $\mathbf{A}(\theta, \mathbf{n})\mathbf{A}^{\mathrm{T}}(\theta, \mathbf{n}) = \mathbf{A}(0, \mathbf{n}) = \mathbf{I}$

P5 $\det \mathbf{A}(\theta, \mathbf{n}) = 1$

P6 $\det(\mathbf{A}(\theta, \mathbf{n}) - \mathbf{I}) = 0$

P7 $\mathbf{A} - \mathbf{T}^{\mathrm{T}} = 2\sin\theta \begin{bmatrix} 0 & -l_3 & l_2 \\ l_3 & 0 & -l_1 \\ -l_2 & l_1 & 0 \end{bmatrix}$

P8 If the elements of $\mathbf{A}(\theta, \mathbf{n})$ are given by the elements of **A** when **i**, **j**, **k** are taken as reference vectors then its elements are given by $\mathbf{LAL}^{\mathrm{T}}$ when **I**, **J**, **K** are taken as reference vectors, the matrix **L** being defined by the relation (1.24).

P9 The cofactor of any element of $\mathbf{A}(\theta, \mathbf{n})$ is equal to that element (see Problem 1.41 for the definition of a cofactor).

Properties P4, P5, P6 have already been derived in Example 1.24. The derivations of P2, P3, P7, P9 are straightforward and are left as exercises for the reader. In deriving P9 it is sufficient to show that the result holds for a diagonal element and a non-diagonal element of $\mathbf{A}(\theta, \mathbf{n})$.

To establish the property P8 we let $[X^1, Y^1, Z^1]$ and $[x^1, y^1, z^1]$ be the respective components of the vectors $\mathbf{R}$ and $\mathbf{r}$ referred to the unit vectors $\mathbf{I}$, $\mathbf{J}$, $\mathbf{K}$. Using the transformation law for the components of a vector given in (1.34) and noting that $\mathbf{L}^{\mathrm{T}} = \mathbf{M}$, we have

$$\begin{bmatrix} X \\ Y \\ Z \end{bmatrix} = \mathbf{L}^{\mathrm{T}} \begin{bmatrix} X^1 \\ Y^1 \\ Z^1 \end{bmatrix}, \qquad \begin{bmatrix} x \\ y \\ z \end{bmatrix} = \mathbf{L}^{\mathrm{T}} \begin{bmatrix} x^1 \\ y^1 \\ z^1 \end{bmatrix}$$

Thus we obtain, with the help of (5.14),

$$\mathbf{L}^{\mathrm{T}} \begin{bmatrix} X^1 \\ Y^1 \\ Z^1 \end{bmatrix} = \mathbf{A}\mathbf{L}^{\mathrm{T}} \begin{bmatrix} x^1 \\ y^1 \\ z^1 \end{bmatrix}$$

or, since $\mathbf{L}\mathbf{L}^{\mathrm{T}} = \mathbf{I}$ (see (1.26)),

$$\begin{bmatrix} X^1 \\ Y^1 \\ Z^1 \end{bmatrix} = \mathbf{L}\mathbf{A}\mathbf{L}^{\mathrm{T}} \begin{bmatrix} x^1 \\ y^1 \\ z^1 \end{bmatrix}$$

Hence the elements of $\mathbf{A}(\theta, \mathbf{n})$ are given by the elements of $\mathbf{L}\mathbf{A}\mathbf{L}^{\mathrm{T}}$ when $\mathbf{I}$, $\mathbf{J}$, $\mathbf{K}$ form the reference set of unit vectors.

We note in passing that, given a rotation matrix, property P1 enables us to determine the angle of rotation θ; while property P7 enables us, when $\theta \neq \pi$, to determine a unit vector in the direction of the axis of rotation. When $\theta = \pi$ it follows from (5.13) that $\mathbf{A} + \mathbf{I} = 2\mathbf{B}$ and therefore the ratios $l_1{:}l_2{:}l_3$ are given by the ratios of the elements of any row (or column) of $\mathbf{A} + \mathbf{I}$.

Example 5.2

Obtain the rotation matrices (referred to $\mathbf{i}, \mathbf{j}, \mathbf{k}$) for rotations of amounts θ, ϕ, ψ about Ox, Oy, Oz respectively. Write down expressions for these matrices when each of the angles is equal to $\pi/2$.

▷ *Solution* When $\mathbf{n} = \mathbf{i}$ we have, taking $\mathbf{i}$, $\mathbf{j}$, $\mathbf{k}$ as reference vectors and using the notation of (5.13), that $l_1 = 1$ and $l_2 = 0 = l_3$. Hence

$$\mathbf{B} = \begin{bmatrix} 1 & 0 & 0 \\ 0 & 0 & 0 \\ 0 & 0 & 0 \end{bmatrix}, \qquad \mathbf{C} = \begin{bmatrix} 0 & 0 & 0 \\ 0 & 0 & -1 \\ 0 & 1 & 0 \end{bmatrix}$$

so that **A** is given by

$$\mathbf{A} = \cos\begin{bmatrix} 1 & 0 & 0 \\ 0 & 1 & 0 \\ 0 & 0 & 1 \end{bmatrix} + (1 - \cos\theta)\begin{bmatrix} 1 & 0 & 0 \\ 0 & 0 & 0 \\ 0 & 0 & 0 \end{bmatrix} + \sin\theta\begin{bmatrix} 0 & 0 & 0 \\ 0 & 0 & -1 \\ 0 & 1 & 0 \end{bmatrix}$$

$$= \begin{bmatrix} 1 & 0 & 0 \\ 0 & \cos\theta & -\sin\theta \\ 0 & \sin\theta & \cos\theta \end{bmatrix}$$

Likewise the rotation matrix for a rotation of amount ϕ about Oy is

$$\begin{bmatrix} \cos\phi & 0 & \sin\phi \\ 0 & 1 & 0 \\ -\sin\phi & 0 & \cos\phi \end{bmatrix}$$

and that for a rotation of amount ψ about Oz is

$$\begin{bmatrix} \cos\psi & -\sin\psi & 0 \\ \sin\psi & \cos\psi & 0 \\ 0 & 0 & 1 \end{bmatrix}$$

Putting each of θ, ϕ, ψ equal to $\pi/2$ yields the matrices

$$\begin{bmatrix} 1 & 0 & 0 \\ 0 & 0 & -1 \\ 0 & 1 & 0 \end{bmatrix}, \qquad \begin{bmatrix} 0 & 0 & 1 \\ 0 & 1 & 0 \\ -1 & 0 & 0 \end{bmatrix}, \qquad \begin{bmatrix} 0 & -1 & 0 \\ 1 & 0 & 0 \\ 0 & 0 & 1 \end{bmatrix}$$

for rotations of amount $\pi/2$ about the coordinate axes Ox, Oy, Oz. The reader will find it instructive to verify for some of these matrices the properties P1 to P7 and property P9 of rotation matrices.

Example 5.3

Find the rotation matrix **A** for the rotation of amount $\pi/2$ about an axis ON in the direction of the unit vector **n**, where $3\mathbf{n} = [2, 2, 1]$ referred to the unit vectors **i**, **j**, **k**. The vectors **I**, **J**, **K** are defined by $\mathbf{I} = \mathbf{n}$, $3\mathbf{J} = [1, -2, 2]$, $3\mathbf{K} = [2, -1, -2]$ referred to **i**, **j**, **k**. Verify, for the rotation about the axis ON, the transformation law stated in property P8.

> *Solution* Substituting $l_1 = l_2 = 2l_3 = 2/3$ in (5.13) we obtain

$$9\mathbf{B} = \begin{bmatrix} 4 & 4 & 2 \\ 4 & 4 & 2 \\ 2 & 2 & 1 \end{bmatrix}, \qquad 3\mathbf{C} = \begin{bmatrix} 0 & -1 & 2 \\ 1 & 0 & -2 \\ -2 & 2 & 0 \end{bmatrix}$$

which leads to

$$\mathbf{A} = \mathbf{B} + \mathbf{C} = \frac{1}{9}\begin{bmatrix} 4 & 1 & 8 \\ 7 & 4 & -4 \\ -4 & 8 & 1 \end{bmatrix}$$

The relation between the unit vectors **i**, **j**, **k** and the unit vectors **I**, **J**, **K** can be written as

$$\begin{bmatrix} \mathbf{I} \\ \mathbf{J} \\ \mathbf{K} \end{bmatrix} = \frac{1}{3}\begin{bmatrix} 2 & 2 & 1 \\ 1 & -2 & 2 \\ 2 & -1 & -2 \end{bmatrix}\begin{bmatrix} \mathbf{i} \\ \mathbf{j} \\ \mathbf{k} \end{bmatrix}$$

and hence noting (1.24) we see that **L** is defined by

$$3\mathbf{L} = \begin{bmatrix} 2 & 2 & 1 \\ 1 & -2 & 2 \\ 2 & -1 & -2 \end{bmatrix}$$

Using property P8 it is found that the elements of $\mathbf{A}(\pi/2, \mathbf{n})$ referred to the vectors **I**, **J**, **K** are given by the elements of $\mathbf{LAL}^{\mathrm{T}}$, where

$$81\mathbf{LAL}^{\mathrm{T}} = \begin{bmatrix} 2 & 2 & 1 \\ 1 & -2 & 2 \\ 2 & -1 & -2 \end{bmatrix}\begin{bmatrix} 4 & 1 & 8 \\ 7 & 4 & -4 \\ -4 & 8 & 1 \end{bmatrix}\begin{bmatrix} 2 & 1 & 2 \\ 2 & -2 & -1 \\ 1 & 2 & -2 \end{bmatrix}$$

$$= 81\begin{bmatrix} 1 & 0 & 0 \\ 0 & 0 & -1 \\ 0 & 1 & 0 \end{bmatrix}$$

Now by virtue of the work contained in Example 5.2 the elements of $\mathbf{A}(\pi/2, \mathbf{i})$ referred to **i**, **j**, **k** are given by the matrix

$$\begin{bmatrix} 1 & 0 & 0 \\ 0 & 0 & -1 \\ 0 & 1 & 0 \end{bmatrix}$$

and therefore the elements of $\mathbf{A}(\pi/2, \mathbf{n}) = \mathbf{A}(\pi/2, \mathbf{I})$ referred to **I**, **J**, **K** are also given by the matrix

$$\begin{bmatrix} 1 & 0 & 0 \\ 0 & 0 & -1 \\ 0 & 1 & 0 \end{bmatrix}$$

Thus the transformation law is verified.

Successive rotations about intersecting axes

We begin with a statement of a theorem that is central to the work on successive rotations about intersecting axes:

If $\mathbf{A}$ is a 3×3 matrix with the two properties (a) $\mathbf{A}\mathbf{A}^{\mathrm{T}} = \mathbf{I}$, (b) $\det \mathbf{A} = 1$, then it is possible to find quantities l_1, l_2, l_3, θ such that

$$\mathbf{A} = \cos\theta\mathbf{I} + (1 - \cos\theta)\mathbf{B} + \sin\theta\mathbf{C}$$

where

$$\mathbf{B} = \begin{bmatrix} l_1^2 & l_1l_2 & l_1l_3 \\ l_1l_2 & l_2^2 & l_2l_3 \\ l_3l_1 & l_3l_2 & l_3^2 \end{bmatrix}, \qquad \mathbf{C} = \begin{bmatrix} 0 & -l_3 & l_2 \\ l_3 & 0 & -l_1 \\ -l_2 & l_1 & 0 \end{bmatrix}$$

and $l_1^2 + l_2^2 + l_3^2 = 1$

Alternatively the theorem may be stated as follows:

Any 3×3 matrix $\mathbf{A}$ with the properties $\mathbf{A}\mathbf{A}^{\mathrm{T}} = \mathbf{I}$ and $\det \mathbf{A} = 1$ defines a rotation.

A proof of the theorem is not given here as it is not essential for our purposes. Readers who wish to derive a proof of the theorem can do so by successively working Problems 5.23 to 5.27.

Let the matrices that define the successive rotations about the intersecting axes ON_i $(i = 1, 2, \ldots, p)$ be denoted by $\mathbf{A}_i$. Using (5.14) we have, for a particle that moves from the point $[x, y, z]$ to the point $[X_1, Y_1, Z_1]$ as a result of the first rotation,

$$\begin{bmatrix} X_1 \\ Y_1 \\ Z_1 \end{bmatrix} = \mathbf{A}_1 \begin{bmatrix} x \\ y \\ z \end{bmatrix}$$

If as a result of the second rotation (defined by $\mathbf{A}_2$) the particle moves to the point with coordinates $[X_2, Y_2, Z_2]$ then

$$\begin{bmatrix} X_2 \\ Y_2 \\ Z_2 \end{bmatrix} = \mathbf{A}_2 \begin{bmatrix} X_1 \\ Y_1 \\ Z_1 \end{bmatrix} = \mathbf{A}_2\mathbf{A}_1 \begin{bmatrix} x \\ y \\ z \end{bmatrix}$$

It is evident that after p rotations the particle will arrive at the point with coordinates $[X_p, Y_p, Z_p]$, where

$$\begin{bmatrix} X_p \\ Y_p \\ Z_p \end{bmatrix} = \mathbf{A}_p\mathbf{A}_{p-1} \ldots \mathbf{A}_2\mathbf{A}_1 \begin{bmatrix} x \\ y \\ z \end{bmatrix}$$

In general $\mathbf{A}_i\mathbf{A}_j \neq \mathbf{A}_j\mathbf{A}_i$ $(i \neq j;\ i, j = 1, 2, \ldots, p)$ and therefore the order in which the rotations are performed is important. If the order is changed the outcome will, for the most part, be different (see Example 5.4 for an exception).

We now show that the p rotations are *equivalent* to a single rotation about an axis through O. Let $\mathbf{A} = \mathbf{A}_p\mathbf{A}_{p-1} \ldots \mathbf{A}_2\mathbf{A}_1$. It is sufficient to show that $\mathbf{A}$ has the two properties $\mathbf{A}\mathbf{A}^{\mathrm{T}} = \mathbf{I}$ and $\det\mathbf{A} = 1$ to establish the result. We have

$$\mathbf{A}\mathbf{A}^{\mathrm{T}} = (\mathbf{A}_p\mathbf{A}_{p-1}\ldots\mathbf{A}_2\mathbf{A}_1)(\mathbf{A}_p\mathbf{A}_{p-1}\ldots\mathbf{A}_2\mathbf{A}_1)^{\mathrm{T}}$$
$$= \mathbf{A}_p\mathbf{A}_{p-1}\ldots\mathbf{A}_2\mathbf{A}_1\mathbf{A}_1^{\mathrm{T}}\mathbf{A}_2^{\mathrm{T}}\ldots\mathbf{A}_p^{\mathrm{T}}$$

using the rule for the transpose of the product of several matrices. But $\mathbf{A}_i\mathbf{A}_i^{\mathrm{T}} = \mathbf{I}$ for $i = 1, 2, \ldots, p$, and hence $\mathbf{A}\mathbf{A}^{\mathrm{T}} = \mathbf{I}$. Next,

$$\det\mathbf{A} = \det(\mathbf{A}_p\mathbf{A}_{p-1}\ldots\mathbf{A}_2\mathbf{A}_1)$$
$$= \det\mathbf{A}_p\det\mathbf{A}_{p-1}\ldots\det\mathbf{A}_1 = 1$$

Thus the matrix $\mathbf{A}$ represents a rotation.

Example 5.4

Show that a particle returns to its original position after successive rotations each of amount π about the axes Ox, Oy, Oz.

▷ *Solution* Putting $\theta = \pi$ and $\mathbf{n}$ successively equal to $\mathbf{i}$, $\mathbf{j}$, $\mathbf{k}$ in (5.13) we find that the matrices for the three rotations are given by

$$\begin{bmatrix} 1 & 0 & 0 \\ 0 & -1 & 0 \\ 0 & 0 & -1 \end{bmatrix}, \quad \begin{bmatrix} -1 & 0 & 0 \\ 0 & 1 & 0 \\ 0 & 0 & -1 \end{bmatrix}, \quad \begin{bmatrix} -1 & 0 & 0 \\ 0 & -1 & 0 \\ 0 & 0 & 1 \end{bmatrix}$$

The combined effect of the rotations is given by the matrix product

$$\begin{bmatrix} -1 & 0 & 0 \\ 0 & -1 & 0 \\ 0 & 0 & 1 \end{bmatrix}\begin{bmatrix} -1 & 0 & 0 \\ 0 & 1 & 0 \\ 0 & 0 & -1 \end{bmatrix}\begin{bmatrix} 1 & 0 & 0 \\ 0 & -1 & 0 \\ 0 & 0 & -1 \end{bmatrix}$$

This reduces to the unit matrix $\mathbf{I}$, and hence the result is established.

The reader should note that the matrices here are diagonal matrices (see Problem 1.43 for a formal definition) and therefore varying the order in which the rotations are made does not affect the outcome.

Example 5.5

Show that a rotation of amount $\pi/2$ about Ox followed by a rotation of amount $\pi/2$ about Oy is equivalent to a rotation of amount $2\pi/3$ about an axis through O in the direction of the vector $\mathbf{i} + \mathbf{j} - \mathbf{k}$.

▷ *Solution* The rotation matrices $\mathbf{A}(\pi/2, \mathbf{i})$, $\mathbf{A}(\pi/2, \mathbf{j})$ are given (see Example 5.2) by

$$\begin{bmatrix} 1 & 0 & 0 \\ 0 & 0 & -1 \\ 0 & 1 & 0 \end{bmatrix}, \quad \begin{bmatrix} 0 & 0 & 1 \\ 0 & 1 & 0 \\ -1 & 0 & 0 \end{bmatrix}$$

with $\mathbf{i}$, $\mathbf{j}$, $\mathbf{k}$ as reference vectors. The matrix $\mathbf{A}$ defined by

$$\mathbf{A} = \begin{bmatrix} 0 & 0 & 1 \\ 0 & 1 & 0 \\ -1 & 0 & 0 \end{bmatrix} \begin{bmatrix} 1 & 0 & 0 \\ 0 & 0 & -1 \\ 0 & 1 & 0 \end{bmatrix} = \begin{bmatrix} 0 & 1 & 0 \\ 0 & 0 & -1 \\ -1 & 0 & 0 \end{bmatrix}$$

determines the effect of the combined rotations. Using property P1 for rotation matrices yields

$$0 = \text{trace}\,\mathbf{A} = 2\cos\theta + 1$$

so that $\cos\theta = -1/2$ or $\theta = 2\pi/3$. Using property P7 gives

$$\mathbf{A} - \mathbf{A}^{\mathrm{T}} = \begin{bmatrix} 0 & 1 & 1 \\ -1 & 0 & -1 \\ -1 & 1 & 0 \end{bmatrix} = 2\sin\theta \begin{bmatrix} 0 & -l_3 & l_2 \\ l_3 & 0 & -l_1 \\ -l_2 & l_1 & 0 \end{bmatrix}$$

Thus $l_1 = l_2 = -l_3 = 1/\sqrt{3}$ and therefore the axis of rotation is in the direction of the vector $\mathbf{i} + \mathbf{j} - \mathbf{k}$.

Successive rotations about non-intersecting axes

In order to illustrate the method of analysis we consider successive rotations about three non-intersecting axes. The axes of rotation are the directed lines O_1N_1, O_2N_2, O_3N_3, the matrices that define the respective rotations about these axes being $\mathbf{A}(\theta_1, \mathbf{n}_1)$, $\mathbf{A}(\theta_2, \mathbf{n}_2)$, $\mathbf{A}(\theta_3, \mathbf{n}_3)$. The coordinates of the points O_1, O_2, O_3 referred to the axes Ox, Oy, Oz are respectively equal to $[a_1, b_1, c_1]$, $[a_2, b_2, c_2]$, $[a_3, b_3, c_3]$. If the particle P, situated at the point A with coordinates $[x, y, z]$, is rotated successively about the axes O_1N_1, O_2N_2, O_3N_3 so that it moves to the point with coordinates $[X, Y, Z]$, then

$$\begin{bmatrix} X \\ Y \\ Z \end{bmatrix} = \mathbf{A}_3\mathbf{A}_2\mathbf{A}_1 \begin{bmatrix} x \\ y \\ z \end{bmatrix} + \mathbf{A}_3\mathbf{A}_2(\mathbf{I} - \mathbf{A}_1) \begin{bmatrix} a_1 \\ b_1 \\ c_1 \end{bmatrix} + \mathbf{A}_3(\mathbf{I} - \mathbf{A}_2) \begin{bmatrix} a_2 \\ b_2 \\ c_2 \end{bmatrix}$$

$$+ (\mathbf{I} - \mathbf{A}_3) \begin{bmatrix} a_3 \\ b_3 \\ c_3 \end{bmatrix} \tag{5.17}$$

where $\mathbf{A}_1$, $\mathbf{A}_2$, $\mathbf{A}_3$ are, in accordance with our notation, equal to the rotation matrices $\mathbf{A}(\theta_1, \mathbf{n}_1)$, $\mathbf{A}(\theta_2, \mathbf{n}_2)$, $\mathbf{A}(\theta_3, \mathbf{n}_3)$ when $\mathbf{i}$, $\mathbf{j}$, $\mathbf{k}$ are taken as reference vectors. Using (5.15) we have

$$\underline{O_1B} = \mathbf{A}(\theta_1, \mathbf{n}_1)\, \underline{O_1A}$$
$$\underline{O_2C} = \mathbf{A}(\theta_2, \mathbf{n}_2)\, \underline{O_2B}$$
$$\underline{O_3D} = \mathbf{A}(\theta_3, \mathbf{n}_3)\, \underline{O_3C}$$

where B, C, D are the respective positions of the particle after the first, second and third rotations. From the first of these three relations we have

$$\underline{O_1O} + \underline{OB} = \mathbf{A}(\theta_1, \mathbf{n}_1)(\underline{O_1O} + \underline{OA})$$

which yields

$$\underline{OB} = \mathbf{A}(\theta_1, \mathbf{n}_1)\underline{OA} + (\mathbf{I} - \mathbf{A}(\theta_1, \mathbf{n}_1))\underline{OO_1} \tag{5.18}$$

Likewise

$$\underline{OC} = \mathbf{A}(\theta_2, \mathbf{n}_2)\underline{OB} + (\mathbf{I} - \mathbf{A}(\theta_2, \mathbf{n}_2))\underline{OO_2}$$
$$\underline{OD} = \mathbf{A}(\theta_3, \mathbf{n}_3)\underline{OC} + (\mathbf{I} - \mathbf{A}(\theta_3, \mathbf{n}_3))\underline{OO_3}$$

Substituting the expressing for $\underline{OB}$ into that for $\underline{OC}$ gives

$$\begin{aligned}\underline{OC} = {} & \mathbf{A}(\theta_2, \mathbf{n}_2)\mathbf{A}(\theta_1, \mathbf{n}_1)\underline{OA} \\ & + \mathbf{A}(\theta_2, \mathbf{n}_2)(\mathbf{I} - \mathbf{A}(\theta_1, \mathbf{n}_1))\underline{OO_1} + (\mathbf{I} - \mathbf{A}(\theta_2, \mathbf{n}_2))\underline{OO_2}\end{aligned}$$

Finally, substituting this expression for $\underline{OC}$ into that for $\underline{OD}$ leads to

$$\begin{aligned}\underline{OD} = {} & \mathbf{A}(\theta_3, \mathbf{n}_3)\mathbf{A}(\theta_2, \mathbf{n}_2)\mathbf{A}(\theta_1, \mathbf{n}_1)\underline{OA} \\ & + \mathbf{A}(\theta_3, \mathbf{n}_3)\mathbf{A}(\theta_2, \mathbf{n}_2)(\mathbf{I} - \mathbf{A}(\theta_1, \mathbf{n}_1))\underline{OO_1} \\ & + \mathbf{A}(\theta_3, \mathbf{n}_3)(\mathbf{I} - \mathbf{A}(\theta_2, \mathbf{n}_2))\underline{OO_2} + (\mathbf{I} - \mathbf{A}(\theta_3, \mathbf{n}_3))\underline{OO_3}\end{aligned} \tag{5.19}$$

from which (5.17) immediately follows.

In the greater part of our work on rotation theory we have taken the origin for position vectors to be a point on the axis of rotation. The relation (5.18) shows what modification has to be made to (5.15) when this is not the case.

We now proceed to consider the three-dimensional kinematics of a rigid body.

5.3 THREE-DIMENSIONAL KINEMATICS OF A RIGID BODY

We begin by stating and proving two theorems that form the basis of the work on three-dimensional rigid body kinematics. The first theorem is known as Euler's theorem.

Euler's theorem

This theorem states that:

If a rigid body is displaced in any manner such that a specified particle P_0 of it remains fixed during the displacement then the displacement is *equivalent* to a rotation about an axis through P_0.

The angle of rotation and the direction of the axis of rotation can be determined from a knowledge of the positions of any two particles of the body before and after the displacement provided that these two particles and the particle P_0 are non-collinear.

☐ *Proof* Let the positions of the two particles, labelled P_1, P_2, be A_1, A_2 before the displacement and B_1, B_2 after it. The fixed position of the particle P_0 is taken as origin O. Since distances between pairs of particles remain fixed in the displacement it follows that

$$\mathbf{R}_1^2 = \mathbf{r}_1^2, \qquad \mathbf{R}_2^2 = \mathbf{r}_2^2, \qquad (\mathbf{R}_1 - \mathbf{R}_2)^2 = (\mathbf{r}_1 - \mathbf{r}_2)^2 \tag{5.20}$$

where $\underline{OA_1} = \mathbf{r}_1$, $\underline{OA_2} = \mathbf{r}_2$, $\underline{OB_1} = \mathbf{R}_1$, $\underline{OB_2} = \mathbf{R}_2$. The third condition in (5.20) reduces with the help of the first two conditions to $\mathbf{R}_1 \,.\, \mathbf{R}_2 = \mathbf{r}_1 \,.\, \mathbf{r}_2$. Thus the displacements $\mathbf{d}_i$ and the mean positions $\mathbf{m}_i$ $(i = 1, 2)$ defined in (5.9) satisfy the three conditions stated in (5.10).

Provided $\mathbf{m}_1$, $\mathbf{m}_2$ are not parallel, these conditions ensure the existence of a unique vector $\mathbf{a}$ such that $\mathbf{a} \times \mathbf{m}_1 = \mathbf{d}_1$ and $\mathbf{a} \times \mathbf{m}_2 = \mathbf{d}_2$. This implies that there exists a rotation about an axis through O that can take P_1 from A_1 to B_1 and P_2 from A_2 to B_2. The vector $\mathbf{a}$ is given by the relation (5.11).

If $\mathbf{m}_1$, $\mathbf{m}_2$ are parallel, that is $\mathbf{m}_1 = c\mathbf{m}_2$, where c is some constant, then $\mathbf{d}_1 \neq c\mathbf{d}_2$ otherwise the particles P_0, P_1, P_2 would be collinear. Noting the analysis subsequent to (5.11) we can assert that a rotation of amount π about an axis through O in the direction of $\mathbf{m}_1$ will take P_1 from A_1 to B_1 and P_2 from A_2 to B_2. Furthermore this rotation is unique.

Since distances between particles and relative orientations are preserved in a rotation, the rotation that takes the particles P_1, P_2 from their (initial) positions A_1, A_2 to their displaced positions B_1, B_2 will also take all other particles of the rigid body from their initial positions to their displaced positions. The proof of the theorem is therefore complete.

Chasles's theorem

This theorem states that:

If a rigid body is displaced in any manner whatsoever then the displacement is equivalent to a translation followed by a rotation.

☐ *Proof* In general no particle of the body will be fixed during the displacement. Accordingly the notation adopted in the proof of Euler's theorem is modified slightly. We take the fixed point O as origin and let the particles $P_0, P_1, \ldots, P_N$ be situated at the points $A_0, A_1, \ldots, A_N$ before the displacement and at the points $B_0, B_1, \ldots, B_N$ after it. Let $\underline{OA_i} = \mathbf{r}_i$, $\underline{OB_i} = \mathbf{R}_i$, $\mathbf{d}_i = \mathbf{R}_i - \mathbf{r}_i$, $2\mathbf{m}_i = \mathbf{R}_i + \mathbf{r}_i$ $(i = 0, 1, 2, \ldots, N)$. Thus $\mathbf{d}_i$ and $\mathbf{m}_i$

represent displacement and mean position as before. The first step in the proof is to let all particles undergo a translation $\mathbf{d}_0$ from their initial positions. Clearly this translation takes the particle P_0 from its initial position A_0 to its final position B_0. All the other particles P_i will move to points whose position vectors with respect to O are equal to $\underline{OA_i} + \mathbf{d}_0$ ($i = 1, 2, \ldots, N$). These points, designated C_i so that $\underline{OC_i} = \mathbf{r}_i + \mathbf{d}_0$, constitute an intermediate position of the rigid body. It remains now to displace the body from this intermediate position, keeping P_0 fixed to its final position. By virtue of Euler's theorem this can be achieved with a unique rotation about an axis through B_0. The direction of the axis of rotation and the angle of rotation can, as we have seen, be determined by consideration of the displacements of any two particles that are not collinear with P_0. As before these particles are taken to be P_1 and P_2. In order to make use of the results contained in the proof of Euler's theorem it is necessary for B_0 to be the origin for position vectors. We have

$$\begin{aligned}\underline{B_0C_i} &= \underline{OC_i} - \underline{OB_0} = \underline{OA_i} + \mathbf{d}_0 - \mathbf{R}_0 = \mathbf{r}_i - \mathbf{r}_0 \\ \underline{B_0B_i} &= \mathbf{R}_i - \mathbf{R}_0\end{aligned} \tag{5.21}$$

The mean position vector of the particle P_i with respect to B_0 during the rotation is given by $(\underline{B_0C_i} + \underline{B_0B_i})/2$, which reduces with the help of (5.21) to $\mathbf{m}_i - \mathbf{m}_0$. The displacement of the particle P_i during the rotation is $\underline{C_iB_i} = \underline{B_0B_i} - \underline{B_0C_i} = \mathbf{d}_i - \mathbf{d}_0$. Thus the quantities $\mathbf{m}_1 - \mathbf{m}_0$, $\mathbf{m}_2 - \mathbf{m}_0$, $\mathbf{d}_1 - \mathbf{d}_0$, $\mathbf{d}_2 - \mathbf{d}_0$ assume the roles that the quantities $\mathbf{m}_1$, $\mathbf{m}_2$, $\mathbf{d}_1$, $\mathbf{d}_2$ played in the proof of Euler's theorem.

Provided $\mathbf{m}_1 - \mathbf{m}_0$, $\mathbf{m}_2 - \mathbf{m}_0$ are not parallel, the pair of vector equations

$$\mathbf{a} \times (\mathbf{m}_1 - \mathbf{m}_0) = \mathbf{d}_1 - \mathbf{d}_0, \qquad \mathbf{a} \times (\mathbf{m}_2 - \mathbf{m}_0) = \mathbf{d}_2 - \mathbf{d}_0$$

has a unique solution for the vector $\mathbf{a}$. This solution can be obtained by appropriate amendment of the relation (5.11). If $\mathbf{m}_1 - \mathbf{m}_0$, $\mathbf{m}_2 - \mathbf{m}_0$ are parallel, the rotation is of amount π about an axis parallel to the vectors $\mathbf{m}_1 - \mathbf{m}_0$, $\mathbf{m}_2 - \mathbf{m}_0$.

Clearly the translation $\mathbf{d}_0$ followed by the (unique) rotation about the axis through B_0 could be replaced by a translation $\mathbf{d}_s$ (this takes the particle P_s from A_s to B_s) followed by a rotation that is also unique about an axis through B_s. Thus there are many possible translations. However, as we now prove, all the rotations associated with these translations are of the same amount while the axes of rotation are in the same direction.

Let θ be the angle of rotation and $\mathbf{n}$ the unit vector in the direction of the axis of rotation for the rotation corresponding to the translation $\mathbf{d}_0$. We have, using (5.18) with A, B, O_1 respectively replaced by C_i, B_i, B_0 ($i = 1, 2, \ldots, N$),

$$\mathbf{R}_i = \underline{OB_i} = \mathbf{A}(\theta, \mathbf{n})\underline{OC_i} + (\mathbf{I} - \mathbf{A}(\theta, \mathbf{n}))\underline{OB_0}$$
$$= \mathbf{A}(\theta, \mathbf{n})(\mathbf{r}_i + \mathbf{d}_0) + (\mathbf{I} - \mathbf{A}(\theta, \mathbf{n}))\underline{OB_0} \tag{5.22}$$

In particular

$$\mathbf{R}_s = \underline{OB_s} = \mathbf{A}(\theta, \mathbf{n})(\mathbf{r}_s + \mathbf{d}_0) + (\mathbf{I} - \mathbf{A}(\theta, \mathbf{n}))\underline{OB_0}$$

so that subtracting we obtain

$$\mathbf{R}_s - \mathbf{R}_i = \underline{B_iB_s} = \mathbf{A}(\theta, \mathbf{n})\underline{A_iA_s} \tag{5.23}$$

To complete the proof it is sufficient to show that a translation $\mathbf{d}_s$ followed by a rotation of amount θ about an axis through B_s in the direction of the unit vector $\mathbf{n}$ takes the particle P_i from A_i to B_i. Let the particle go from A_i to B_i^1 as a result of the translation and rotation. We have, noting (5.22),

$$\underline{OB_i^1} = \mathbf{A}(\theta, \mathbf{n})(\mathbf{r}_i + \mathbf{d}_s) + (\mathbf{I} - \mathbf{A}(\theta, \mathbf{n}))\underline{OB_s} \tag{5.24}$$

Subtracting corresponding sides of the relations (5.22) and (5.24) yields

$$\underline{B_iB_i^1} = \mathbf{A}(\theta, \mathbf{n})(\mathbf{d}_s - \mathbf{d}_0) + (\mathbf{I} - \mathbf{A}(\theta, \mathbf{n}))\underline{B_0B_s}$$
$$= \mathbf{A}(\theta, \mathbf{n})(\underline{B_0B_s} - \underline{A_0A_s}) + (\mathbf{I} - \mathbf{A}(\theta, \mathbf{n}))\underline{B_0B_s} = \mathbf{0}$$

using (5.23) with $i = 0$. Thus the points B_i^1 and B_i coincide.

Example 5.6

Three particles of a body are initially situated at the points A_0, A_1, A_2 whose coordinates referred to axes Ox, Oy, Oz are $[-1, -1, -2]$, $[4, 9, -1]$ and $[6, 5, -1]$. Subsequently the body is moved so that the three particles go to points B_0, B_1, B_2 whose coordinates are $[0, 1, 1]$, $[-6, 4, 10]$ and $[-2, 2, 10]$. Verify that the displacement of the particles is consistent with the statement that the body is rigid. Determine a translation and rotation which combine to give the displacement of the three particles.

A fourth particle of the body (assumed to be rigid) is initially situated at the point A_3 whose coordinates are $[8, 1, -1]$. Determine by two methods the position of the particle after the rigid body displacement.

> *Solution* Straightforward calculation yields the following:

$$\underline{A_0A_1} = [5, 10, 1], \qquad \underline{A_0A_2} = [7, 6, 1], \qquad \underline{A_1A_2} = [2, -4, 0]$$
$$\underline{B_0B_1} = [-6, 3, 9], \qquad \underline{B_0B_2} = [-2, 1, 9], \qquad \underline{B_1B_2} = [4, -2, 0]$$

Thus $|\underline{A_0A_1}| = |\underline{B_0B_1}|$, $|\underline{A_0A_2}| = |\underline{B_0B_2}|$, $|\underline{A_1A_2}| = |\underline{B_1B_2}|$, which implies consistency with the statement that the body is rigid. Further calculation gives

$$\mathbf{d}_0 = [1, 2, 3], \qquad \mathbf{d}_1 = [-10, -5, +11], \qquad \mathbf{d}_2 = [-8, -3, 11]$$

$$2\mathbf{m}_0 = [-1, 0, -1], \qquad 2\mathbf{m}_1 = [-2, 13, 9], \qquad 2\mathbf{m}_2 = [4, 7, 9]$$

so that

$$\mathbf{d}_1 - \mathbf{d}_0 = [-11, -7, 8], \qquad \mathbf{d}_2 - \mathbf{d}_0 = [-9, -5, 8]$$
$$2(\mathbf{m}_1 - \mathbf{m}_0) = [-1, 13, 10], \qquad 2(\mathbf{m}_2 - \mathbf{m}_0) = [5, 7, 10]$$

Taking $\mathbf{d}_0 = [1, 2, 3]$ as the translation, the equations to be solved to determine the rotation are

$$\mathbf{a} \times (\mathbf{m}_1 - \mathbf{m}_0) = \mathbf{d}_1 - \mathbf{d}_0, \qquad \mathbf{a} \times (\mathbf{m}_2 - \mathbf{m}_0) = \mathbf{d}_2 - \mathbf{d}_0$$

To obtain their solution we could use the expression for $\mathbf{a}$ contained in (5.11). An alternative method is to observe that $\mathbf{a}$ must be of the form $\alpha(\mathbf{d}_1 - \mathbf{d}_0) \times (\mathbf{d}_2 - \mathbf{d}_0)$ since it is perpendicular to both $\mathbf{d}_1 - \mathbf{d}_0$ and $\mathbf{d}_2 - \mathbf{d}_0$. Now

$$(\mathbf{d}_1 - \mathbf{d}_0) \times (\mathbf{d}_2 - \mathbf{d}_0) = \begin{vmatrix} \mathbf{i} & \mathbf{j} & \mathbf{k} \\ -11 & -7 & 8 \\ -9 & -5 & 8 \end{vmatrix} = -8[2, -2, 1]$$

Putting $\mathbf{a} = \beta[2, -2, 1]$ we obtain the value of β by substituting for $\mathbf{a}$ in the equation $\mathbf{a} \times (\mathbf{m}_1 - \mathbf{m}_0) = \mathbf{d}_1 - \mathbf{d}_0$. This yields

$$\beta \begin{vmatrix} \mathbf{i} & \mathbf{j} & \mathbf{k} \\ 2 & -2 & 1 \\ -1 & 13 & 10 \end{vmatrix} = 2[-11, -7, 8]$$

or

$$\beta[-33, -21, 24] = 2[-11, -7, 8]$$

so that $\beta = 2/3$. Hence

$$\mathbf{a} = 2\tan\frac{\theta}{2}\mathbf{n} = \frac{2}{3}[2, -2, 1]$$

which gives $\tan\theta/2 = 1$, and therefore $\theta = \pi/2$ while $\mathbf{n} = (1/3)[2, -2, 1]$.

Recalling (5.13) we have that $\mathbf{A}(\pi/2, \mathbf{n})$ referred to $\mathbf{i}, \mathbf{j}, \mathbf{k}$ is given by

$$9\mathbf{A} = \begin{bmatrix} 4 & -7 & -4 \\ -1 & 4 & -8 \\ 8 & 4 & 1 \end{bmatrix}$$

Let the fourth particle go from A_3 to B_3 as a result of the rigid body displacement. Noting (5.23) and putting $i = 0$ and $s = 3$ we have

$$\underline{B_0B_3} = \mathbf{A}(\pi/2, \mathbf{n})\underline{A_0A_3}$$
$$= \frac{1}{9}\begin{bmatrix} 4 & -7 & -4 \\ -1 & 4 & -8 \\ 8 & 4 & 1 \end{bmatrix}\begin{bmatrix} 9 \\ 2 \\ 1 \end{bmatrix} = [2, -1, 9]^T$$

Thus

$$\underline{OB_3} = \underline{OB_0} + \underline{B_0B_3} = [2,\ 0,\ 10]$$

The position of the fourth particle after the displacement can be determined by the use of body coordinates. Writing

$$\underline{A_0A_3} = \alpha\underline{A_0A_1} + \beta\underline{A_0A_2} + \mu\underline{A_0A_1} \times \underline{A_0A_2}$$

We see that α, β, μ satisfy the vector equation

$$[9,\ 2,\ 1] = \alpha[5,\ 10,\ 1] + \beta[7,\ 6,\ 1] + \mu[4,\ 2,\ -40]$$

Elementary algebra yields $\alpha = -1$, $\beta = 2$, $\mu = 0$. Hence

$$\begin{aligned}\underline{B_0B_3} &= \alpha\underline{B_0B_1} + \beta\underline{B_0B_2} + \mu\underline{B_0B_1} \times \underline{B_0B_2} \\ &= [6,\ -3,\ -9] + [-4,\ 2,\ 18] = [2,\ -1,\ 9]\end{aligned}$$

as before.

Infinitesimal rotations

When the angle of rotation is infinitesimal, that is equal to $\delta\theta$, the rotation is said to be an infinitesimal rotation. An infinitesimal rotation will produce only a small change in the position vector $\mathbf{r}$, and hence writing $\mathbf{r} + \delta\mathbf{r}$ for the vector $\mathbf{R}$ in the relation (5.4) we have

$$\mathbf{r} + \delta\mathbf{r} = \cos\delta\theta\mathbf{r} + (1 - \cos\delta\theta)(\mathbf{n} \,.\, \mathbf{r})\mathbf{n} + \sin\delta\theta\mathbf{n} \times \mathbf{r}$$

To the first order of small quantities $\cos\delta\theta = 1$, $\sin\delta\theta = \delta\theta$, so that working to the first order we have

$$\delta\mathbf{r} = \delta\theta\mathbf{n} \times \mathbf{r} \tag{5.25}$$

while the vector $\mathbf{a}$ which defines the rotation (see (5.3)) is equal to $\delta\theta\mathbf{n}$.

Consider now two infinitesimal rotations about intersecting axes through the origin O. The angles turned through are denoted by $\delta\theta_1$ and $\delta\theta_2$ while the directions of the associated axes are given by the unit vectors $\mathbf{n}_1$ and $\mathbf{n}_2$. If $\delta\mathbf{r}_1$ is the change in $\mathbf{r}$ as a result of the first rotation then we have, using (5.25),

$$\delta\mathbf{r}_1 = \delta\theta_1\mathbf{n}_1 \times \mathbf{r}$$

In like manner for the second rotation we have, noting that $\mathbf{r}$ must be replaced by $\mathbf{r} + \delta\mathbf{r}_1$, that the further change $\delta\mathbf{r}_2$ is given by

$$\delta\mathbf{r}_2 = \delta\theta_2\mathbf{n}_2 \times (\mathbf{r} + \delta\mathbf{r}_1)$$

Thus to the first order of small quantities

$$\delta\mathbf{r}_2 = \delta\theta_2\mathbf{n}_2 \times \mathbf{r}$$

and hence the total change $\delta\mathbf{r}$ in $\mathbf{r}$ is (again to the first order of small quantities) given by

$$\delta\mathbf{r} = \delta\mathbf{r}_1 + \delta\mathbf{r}_2 = (\delta\theta_1\mathbf{n}_1 + \delta\theta_2\mathbf{n}_2) \times \mathbf{r}$$

The expression for $\delta\mathbf{r}$ is unaltered if the rotation about the axis in the direction of the vector $\mathbf{n}_2$ is performed first and the rotation about the axis in the direction $\mathbf{n}_1$ second. It is evident for several infinitesimal rotations defined by the vectors $\delta\theta_i\mathbf{n}_i$ ($i = 1, 2, \ldots, p$) that the total change $\delta\mathbf{r}$ in $\mathbf{r}$ is given by

$$\delta\mathbf{r} = \left(\sum_{r=1}^{p} \delta\theta_i\mathbf{n}_i\right) \times \mathbf{r}$$

so that the order in which the rotations are performed does not affect the outcome. We shall make extensive use of this result when deriving the components of the angular vector of a rotating frame of reference.

Angular velocity vector of a rigid body

Any infinitesimal displacement of a rigid body is by virtue of Chasles's theorem equivalent to an infinitesimal translation together with an infinitesimal rotation. Let the rotation, which is independent of the choice of translation, be defined by the vector $\delta\theta\mathbf{n}$. Replacing $\mathbf{R}_i$ in the proof of Chasles's theorem by $\mathbf{r}_i + \delta\mathbf{r}_i$ we have

$$\mathbf{r}_i + \delta\mathbf{r}_i = \underline{\mathrm{OB}_i} = \underline{\mathrm{OC}_i} + \underline{\mathrm{C}_i\mathrm{B}_i}$$

which becomes, noting (5.25) and taking the axis of rotation to pass through B_0,

$$\mathbf{r}_i + \delta\mathbf{r}_i = \mathbf{r}_i + \delta\mathbf{r}_0 + \delta\theta\mathbf{n} \times \underline{\mathrm{B}_0\mathrm{C}_i}$$

Now

$$\underline{\mathrm{B}_0\mathrm{C}_i} = \underline{\mathrm{OC}_i} - \underline{\mathrm{OB}_0} = \underline{\mathrm{OA}_i} + \mathbf{d}_0 - \underline{\mathrm{OB}_0} = \underline{\mathrm{OA}_i} - \underline{\mathrm{OA}_0} = \mathbf{r}_i - \mathbf{r}_0$$

and hence

$$\delta\mathbf{r}_i = \delta\mathbf{r}_0 + \delta\theta\mathbf{n} \times (\mathbf{r}_i - \mathbf{r}_0)$$

If the displacement takes place in time δt then dividing both sides of this relation by δt and letting $\delta t \to 0$ we have

$$\dot{\mathbf{r}}_i = \dot{\mathbf{r}}_0 + \dot{\theta}\mathbf{n} \times (\mathbf{r}_i - \mathbf{r}_0), \qquad i = 0, 1, 2, \ldots, N \tag{5.26}$$

In particular putting $i = j$

$$\dot{\mathbf{r}}_j = \dot{\mathbf{r}}_0 + \dot{\theta}\mathbf{n} \times (\mathbf{r}_j - \mathbf{r}_0)$$

Subtracting corresponding sides of this relation and (5.26) leads to

$$\dot{\mathbf{r}}_j = \dot{\mathbf{r}}_i + \dot{\theta}\mathbf{n} \times (\mathbf{r}_j - \mathbf{r}_i)$$

Writing (as in Chapter 4) $\mathbf{v}_i$ and $\mathbf{v}_j$ for the absolute velocities of the particles P_i and P_j we have

$$\mathbf{v}_j = \mathbf{v}_i + \boldsymbol{\omega} \times (\mathbf{r}_j - \mathbf{r}_i) \tag{5.27}$$

where

$$\boldsymbol{\omega} = \dot{\theta}\mathbf{n} \tag{5.28}$$

The pair of relations (5.27) and (5.28) can be compared directly with the pair (4.19) and (4.18). The vector $\dot{\theta}\mathbf{n}$ is called the angular velocity vector of the rigid body; the vector $\dot{\theta}\mathbf{k}$ that arises in the two-dimensional motion is a particular example of it.

When the particle P_0 is fixed and its position is taken as the origin for position vectors, the relation (5.26) has the particular form

$$\mathbf{v}_i = \dot{\mathbf{r}}_i = \boldsymbol{\omega} \times \mathbf{r}_i \tag{5.29}$$

Rotating systems of unit vectors

It will be seen in Chapter 6 that it is frequently preferable both mathematically and physically to define a vector in terms of a rotating set of unit vectors rather than the fixed set **i**, **j**, **k**. We are concerned here with two problems. The first is the determination of the rate of change of a vector whose components are defined with respect to a rotating set of unit vectors. The second problem is to determine expressions for the components of the angular velocity of this set in terms of certain angular coordinates.

Let $\mathbf{a}(t)$, $\mathbf{b}(t)$, $\mathbf{c}(t)$, subsequently denoted by **a**, **b**, **c**, be a set of mutually perpendicular right-handed unit vectors that depend on the time t. We take the directed lines Oa, Ob, Oc, respectively parallel to and in the same sense as **a**, **b**, **c**, to be an associated set of rotating axes, the origin O being at rest with respect to an inertial frame of reference. As we have already pointed out, a set of axes such as Oa, Ob, Oc can be regarded as a rigid body. In particular the moving point P, whose position vector with respect to O at any time t is $\mathbf{a}(t)$, is fixed with respect to these axes, being one unit of length measured from O along Oa. If $\boldsymbol{\omega}$ is the angular velocity of the axes, then using (5.29) we have

$$\dot{\mathbf{a}} = \boldsymbol{\omega} \times \mathbf{a}, \qquad \dot{\mathbf{b}} = \boldsymbol{\omega} \times \mathbf{b}, \qquad \dot{\mathbf{c}} = \boldsymbol{\omega} \times \mathbf{c} \tag{5.30}$$

The relations in (5.30) can be put in an alternative (matrix) form. If

$$\boldsymbol{\omega} = \omega_1\mathbf{a} + \omega_2\mathbf{b} + \omega_3\mathbf{c}$$

then

$$\boldsymbol{\omega} \times \mathbf{a} = [0, \omega_3, -\omega_2], \qquad \boldsymbol{\omega} \times \mathbf{b} = [-\omega_3, 0, \omega_1], \qquad \boldsymbol{\omega} \times \mathbf{c} = [\omega_2, -\omega_1, 0]$$

referred to the unit vectors **a**, **b**, **c** so that

$$\begin{bmatrix} \dot{\mathbf{a}} \\ \dot{\mathbf{b}} \\ \dot{\mathbf{c}} \end{bmatrix} = \begin{bmatrix} 0 & \omega_3 & -\omega_2 \\ -\omega_3 & 0 & \omega_1 \\ \omega_3 & -\omega_1 & 0 \end{bmatrix} \begin{bmatrix} \mathbf{a} \\ \mathbf{b} \\ \mathbf{c} \end{bmatrix} \tag{5.31}$$

If **A** is a vector represented in terms of **a**, **b**, **c** by

$$\mathbf{A} = A_1\mathbf{a} + A_2\mathbf{b} + A_3\mathbf{c}$$

where A_1, A_2, A_3 may be functions of time, then

$$\begin{aligned} \dot{\mathbf{A}} &= \dot{A}_1\mathbf{a} + A_1\dot{\mathbf{a}} + \dot{A}_2\mathbf{b} + A_2\dot{\mathbf{b}} + \dot{A}_3\mathbf{c} + A_3\dot{\mathbf{c}} \\ &= \dot{A}_1\mathbf{a} + \dot{A}_2\mathbf{b} + \dot{A}_3\mathbf{c} + A_1\boldsymbol{\omega} \times \mathbf{a} + A_2\boldsymbol{\omega} \times \mathbf{b} + A_3\boldsymbol{\omega} \times \mathbf{c} \\ &= \dot{A}_1\mathbf{a} + \dot{A}_2\mathbf{b} + \dot{A}_3\mathbf{c} + \boldsymbol{\omega} \times (A_1\mathbf{a} + A_2\mathbf{b} + A_3\mathbf{c}) \\ &= \dot{A}_1\mathbf{a} + \dot{A}_2\mathbf{b} + \dot{A}_3\mathbf{c} + \boldsymbol{\omega} \times \mathbf{A} \end{aligned} \tag{5.32}$$

We note the particular result obtained by putting $\mathbf{A} = \boldsymbol{\omega}$, namely

$$\dot{\boldsymbol{\omega}} = \dot{\omega}_1\mathbf{a} + \dot{\omega}_2\mathbf{b} + \dot{\omega}_3\mathbf{c} \tag{5.33}$$

The relation in (5.32) can be differentiated a second time to find $\ddot{\mathbf{A}}$. We have

$$\begin{aligned} \ddot{\mathbf{A}} &= \ddot{A}_1\mathbf{a} + \dot{A}_1\dot{\mathbf{a}} + \ddot{A}_2\mathbf{b} + \dot{A}_2\dot{\mathbf{b}} + \ddot{A}_3\mathbf{c} + \dot{A}_3\dot{\mathbf{c}} + \dot{\boldsymbol{\omega}} \times \mathbf{A} + \boldsymbol{\omega} \times \dot{\mathbf{A}} \\ &= \ddot{A}_1\mathbf{a} + \ddot{A}_2\mathbf{b} + \ddot{A}_3\mathbf{c} + \mathbf{2}\boldsymbol{\omega} \times (\dot{A}_1\mathbf{a} + \dot{A}_2\mathbf{b} + \dot{A}_3\mathbf{c}) \\ &\quad + \dot{\boldsymbol{\omega}} \times \mathbf{A} + \boldsymbol{\omega} \times (\boldsymbol{\omega} \times \mathbf{A}) \end{aligned} \tag{5.34}$$

using (5.30) and (5.32).

We now turn to the second problem of finding the components of $\boldsymbol{\omega}$ in terms of the angular coordinates known as Euler's angles.

Euler's angles

We begin by defining the position of the axes Oa, Ob, Oc with respect to the inertial set of axes Ox, Oy, Oz (see Fig. 5.4). The angle between the unit vector **c** and **k** is denoted by θ, where $0 \leq \theta \leq \pi$, while ϕ ($0 \leq \phi \leq 2\pi$) is the angle between Od, the projection of Oc onto the plane xOy, and Ox. The unit vector perpendicular to **c** that lies in the plane of **c** and **k**, and whose sense is such that the angle between it and **k** is equal to $\pi/2 + \theta$ when $0 \leq \theta \leq \pi/2$ and equal to $3\pi/2 - \theta$ when $\pi/2 \leq \theta \leq \pi$, is labelled **s**. The further vector **t** is defined by $\mathbf{t} = \mathbf{c} \times \mathbf{s}$. Taking O$s$, O$t$ to be directed

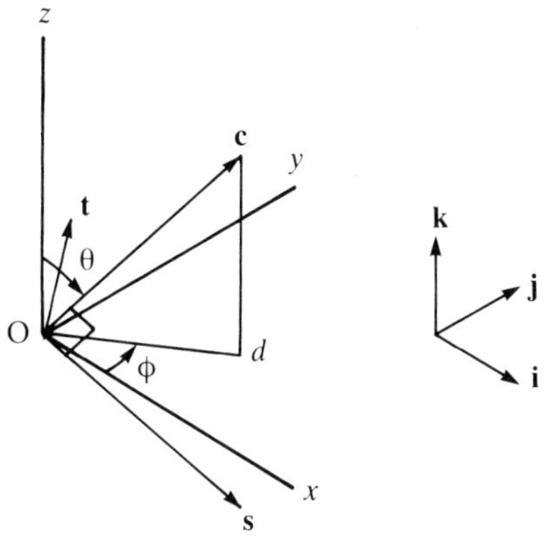

Fig. 5.4

lines in the directions of **s** and **t**, it follows that Os, Ot, Oa, Ob all lie in a plane with Oc as normal. The axis Oa is inclined to Os at an angle ψ measured counter-clockwise from Os about Oc as axis (see Fig. 5.5). The three angles θ, ϕ, ψ are known as Euler's angles: they define the orientation of the axes Oa, Ob, Oc with respect to the axes Ox, Oy, Oz. We now obtain expressions for **a**, **b**, **c** in terms of **i**, **j**, **k** and θ, ϕ, ψ.

The unit vector **c** (see Fig. 5.4) can be resolved into components $\cos\theta$ in the direction of the unit vector **k** and $\sin\theta$ in the direction of Od. The component $\sin\theta$ can, in turn, be resolved into components $\sin\theta\cos\phi$ in the direction of **i** and $\sin\theta\sin\phi$ in the direction of **j**. Thus

$$\mathbf{c} = \sin\theta\cos\phi\mathbf{i} + \sin\theta\sin\phi\mathbf{j} + \cos\theta\mathbf{k}$$

The components for **s** are found by observing that we need only replace θ in the expression for **c** by $\theta + \pi/2$. Hence

$$\mathbf{s} = \cos\theta\cos\phi\mathbf{i} + \cos\theta\sin\phi\mathbf{j} - \sin\theta\mathbf{k}$$

Then $\mathbf{t} = \mathbf{c} \times \mathbf{s}$ is found after some manipulation to be given by

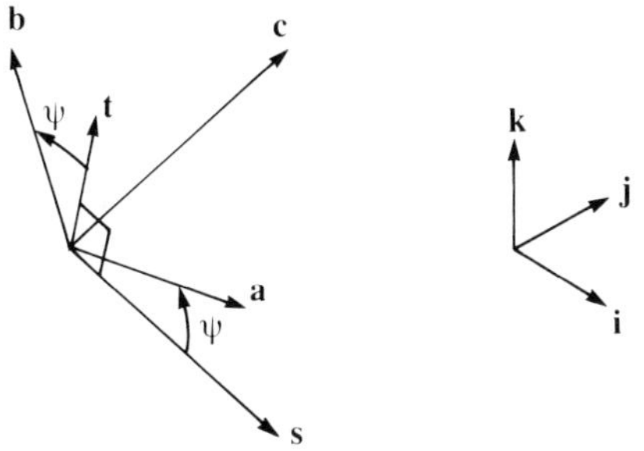

Fig. 5.5

$$\mathbf{t} = -\sin\phi\mathbf{i} + \cos\phi\mathbf{j}$$

We can write these results in the matrix form

$$\begin{bmatrix} \mathbf{s} \\ \mathbf{t} \\ \mathbf{c} \end{bmatrix} = \begin{bmatrix} \cos\theta\cos\phi & \cos\theta\sin\phi & -\sin\theta \\ -\sin\phi & \cos\phi & 0 \\ \sin\theta\cos\phi & \sin\theta\sin\phi & \cos\theta \end{bmatrix} \begin{bmatrix} \mathbf{i} \\ \mathbf{j} \\ \mathbf{k} \end{bmatrix} \tag{5.35}$$

It is readily verified that the relation (5.35) can be written as

$$\begin{bmatrix} \mathbf{s} \\ \mathbf{t} \\ \mathbf{c} \end{bmatrix} = \begin{bmatrix} \cos\theta & 0 & -\sin\theta \\ 0 & 1 & 0 \\ \sin\theta & 0 & \cos\theta \end{bmatrix} \begin{bmatrix} \cos\phi & \sin\phi & 0 \\ -\sin\phi & \cos\phi & 0 \\ 0 & 0 & 1 \end{bmatrix} \begin{bmatrix} \mathbf{i} \\ \mathbf{j} \\ \mathbf{k} \end{bmatrix} \tag{5.36}$$

Resolving (see Fig. 5.5) we have

$$\mathbf{a} = \cos\psi\mathbf{s} + \sin\psi\mathbf{t}, \qquad \mathbf{b} = -\sin\psi\mathbf{s} + \cos\psi\mathbf{t}, \qquad \mathbf{c} = \mathbf{c}$$

which becomes in matrix form

$$\begin{bmatrix} \mathbf{a} \\ \mathbf{b} \\ \mathbf{c} \end{bmatrix} = \begin{bmatrix} \cos\psi & \sin\psi & 0 \\ -\sin\psi & \cos\psi & 0 \\ 0 & 0 & 1 \end{bmatrix} \begin{bmatrix} \mathbf{s} \\ \mathbf{t} \\ \mathbf{c} \end{bmatrix}$$

Writing $\mathbf{A}_1$, $\mathbf{A}_2$, $\mathbf{A}_3$ for the matrices $\mathbf{A}(\psi, \mathbf{k})$, $\mathbf{A}(\theta, \mathbf{j})$, $\mathbf{A}(\phi, \mathbf{k})$ when $\mathbf{i}$, $\mathbf{j}$, $\mathbf{k}$ are taken as the reference set of unit vectors, we obtain, with the help of (5.36) and results contained in Example 5.2,

$$\begin{bmatrix} \mathbf{a} \\ \mathbf{b} \\ \mathbf{c} \end{bmatrix} = \mathbf{A}_1^{\mathrm{T}} \mathbf{A}_2^{\mathrm{T}} \mathbf{A}_3^{\mathrm{T}} \begin{bmatrix} \mathbf{i} \\ \mathbf{j} \\ \mathbf{k} \end{bmatrix} \tag{5.37}$$

The inverse of the relation (5.37) is given by

$$\begin{bmatrix} \mathbf{i} \\ \mathbf{j} \\ \mathbf{k} \end{bmatrix} = \mathbf{A}_3 \mathbf{A}_2 \mathbf{A}_1 \begin{bmatrix} \mathbf{a} \\ \mathbf{b} \\ \mathbf{c} \end{bmatrix} \tag{5.38}$$

as $\mathbf{A}\mathbf{A}^{\mathrm{T}} = \mathbf{I}$ for any rotation matrix.

Components of angular velocity vector of rotating axes

We proceed now to obtain the components of the angular velocity vector $\boldsymbol{\omega}$ of the rotating axes Oa, Ob, Oc in terms of the angles θ, ϕ, ψ and their derivatives. The method that is employed here makes use of the theory of infinitesimal rotations.

The angle ϕ is formed by the lines Ox and Od which have Oz as normal. Therefore in order to change ϕ by a small amount $\delta\phi$ the axes

Oa, Ob, Oc have to be rotated about Oz. This infinitesimal rotation is represented by the vector $\delta\phi\mathbf{k}$. The angle θ is formed by Oc and Oz which have, since $\mathbf{t} = \mathbf{c} \times \mathbf{s}$, Ot as normal. To change θ by a small amount $\delta\theta$ we therefore have to rotate the axes about Ot so that the infinitesimal rotation is given by $\delta\theta\mathbf{t}$. In a similar manner $\delta\psi\mathbf{c}$ defines the rotation that changes the angle ψ. We know from the theory relating to infinitesimal rotations that the three vectors $\delta\phi\mathbf{k}$, $\delta\theta\mathbf{t}$, $\delta\psi\mathbf{c}$ can be added together to obtain the vector that defines the single (infinitesimal) rotation equivalent to the combined effect of the three rotations. Thus if $\delta\varepsilon\mathbf{n}$ defines the equivalent rotation then

$$\delta\varepsilon\mathbf{n} = \delta\phi\mathbf{k} + \delta\theta\mathbf{t} + \delta\psi\mathbf{c}$$

If the equivalent rotation takes place in time δt then dividing both sides of this relation by δt and letting $\delta t \rightarrow 0$ we obtain

$$\boldsymbol{\omega} = \dot{\varepsilon}\mathbf{n} = \dot{\phi}\mathbf{k} + \dot{\theta}\mathbf{t} + \dot{\psi}\mathbf{c} \tag{5.39}$$

The relation (5.39) is not suitable for the work of Chapter 6 as $\mathbf{k}$, $\mathbf{t}$, $\mathbf{c}$ are not mutually perpendicular. There are two representations of $\boldsymbol{\omega}$ that prove particularly useful: one refers $\boldsymbol{\omega}$ to the unit vectors $\mathbf{a}$, $\mathbf{b}$, $\mathbf{c}$ and the other to the vectors $\mathbf{s}$, $\mathbf{t}$, $\mathbf{c}$.

It follows from Fig. 5.4 that $\mathbf{k} = -\sin\theta\mathbf{s} + \cos\theta\mathbf{c}$, and therefore

$$\boldsymbol{\omega} = -\dot{\phi}\sin\theta\mathbf{s} + \dot{\theta}\mathbf{t} + (\dot{\phi}\cos\theta + \dot{\psi})\mathbf{c} \tag{5.40}$$

Since $\mathbf{s} = \cos\psi\mathbf{a} - \sin\psi\mathbf{b}$, $\mathbf{t} = \sin\psi\mathbf{a} + \cos\psi\mathbf{b}$, it also follows that

$$\boldsymbol{\omega} = (\dot{\theta}\sin\psi - \dot{\phi}\sin\theta\cos\psi)\mathbf{a} + (\dot{\theta}\cos\psi + \dot{\phi}\sin\psi\sin\theta)\mathbf{b} + (\dot{\phi}\cos\theta + \dot{\psi})\mathbf{c} \tag{5.41}$$

By putting $\psi = \dot{\psi} = 0$ in the relations (5.40) and (5.41) we obtain expressions for the angular velocity vector $\boldsymbol{\Omega}$ of the rotating set of axes Os, Ot, Oc. We have in particular

$$\boldsymbol{\Omega} = -\dot{\phi}\sin\theta\mathbf{s} + \dot{\theta}\mathbf{t} + \dot{\phi}\cos\theta\mathbf{c} \tag{5.42}$$

We now apply some of the results derived in Sections 5.2 and 5.3.

Hooke's joint

Hooke's joint is a device which joins two shafts enabling one shaft, the input shaft, to drive a second shaft, the output shaft (see Fig. 5.6). Revolute joints at A, B, D, E join the shafts to the arms of the cross, represented by the perpendicular lines AB and DE. The arms are the means by which the driving shaft transmits power to the output shaft; the arm AB is connected to the driving shaft and the arm DE to the driven shaft. The axes

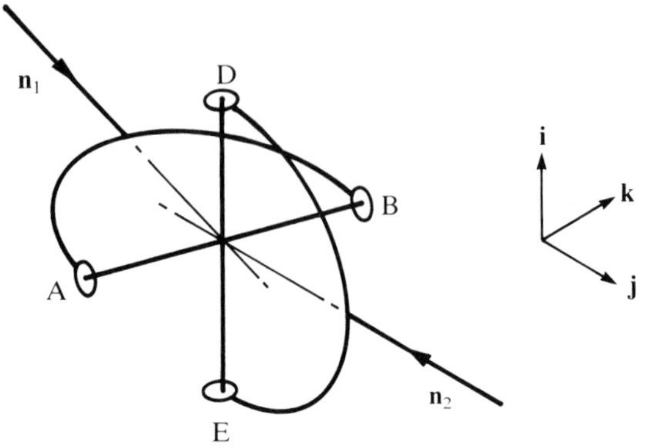

Fig. 5.6

of the driving and driven shafts are in the directions of the fixed unit vectors $\mathbf{n}_1$, $\mathbf{n}_2$, the angle between these vectors being equal to α where $\alpha \neq 0$. The lines of the axes produced intersect at the point O, the common point of the arms AB, DE (see Fig. 5.7). The objective of the analysis is to relate the angular velocity $\boldsymbol{\omega}_2$ of the output shaft to the angular velocity $\boldsymbol{\omega}_1$ of the driving shaft. We also obtain expressions for the components of the angular velocity vector of the cross.

Now AB will move in a fixed plane with $\mathbf{n}_1$ as normal while DE will move in a second fixed plane with $\mathbf{n}_2$ as normal. The line of intersection of these two planes is taken as the x axis, the point O being chosen as origin. The axis Oy is chosen so as to be in the direction of $\mathbf{n}_1$, while Oz forms in the standard manner the third member of the (fixed) right-handed axes Ox, Oy, Oz. At time $t = 0$ the arm DE is taken to coincide with the x axis. Now $\underline{\text{AB}}$ is perpendicular to both $\mathbf{n}_1$ and $\underline{\text{DE}}$ and must therefore lie along the z

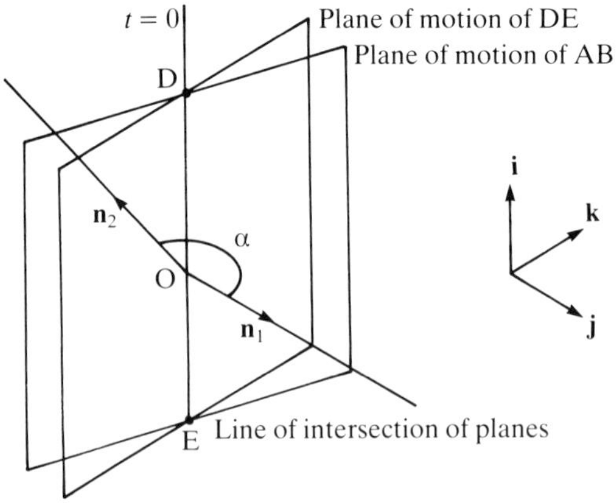

Fig. 5.7

axis at time $t = 0$. Taking $2a$ to be the length of each arm of the cross, the points B, A, D, E of the cross oriented as shown in Fig. 5.7 will, at time $t = 0$, be situated at points in space with coordinates $[0, 0, a]$, $[0, 0, -a]$, $[a, 0, 0]$, $[-a, 0, 0]$ with respect to Ox, Oy, Oz. At time t let θ, ϕ be the angles turned through by AB and DE from their positions at time $t = 0$. Thus the angular velocities of the shafts are given by

$$\boldsymbol{\omega}_1 = \dot{\theta}\mathbf{n}_1 = \dot{\theta}\mathbf{j}, \qquad \boldsymbol{\omega}_2 = \dot{\phi}\mathbf{n}_2 = \dot{\phi}(\cos\alpha\mathbf{j} + \sin\alpha\mathbf{k}) \tag{5.43}$$

The matrix defining the rotation of AB is $\mathbf{A}(\theta, \mathbf{j})$ and that defining the rotation of DE is $\mathbf{A}(\phi, \mathbf{n}_2)$. These matrices are given, using $\mathbf{i}, \mathbf{j}, \mathbf{k}$ as reference vectors, by

$$\mathbf{A}(\theta, \mathbf{j}) = \begin{bmatrix} \cos\theta & 0 & \sin\theta \\ 0 & 1 & 0 \\ -\sin\theta & 0 & \cos\theta \end{bmatrix}$$

$$\mathbf{A}(\phi, \mathbf{n}_2) = \begin{bmatrix} \cos\phi & -\sin\alpha\sin\phi & \cos\alpha\sin\phi \\ \sin\alpha\sin\phi & \cos^2\alpha + \cos\phi\sin^2\alpha & (1 - \cos\phi)\cos\alpha\sin\alpha \\ -\cos\alpha\sin\phi & (1 - \cos\phi)\cos\alpha\sin\alpha & \sin^2\alpha + \cos\phi\cos^2\alpha \end{bmatrix}$$

Hence during time t the points B and D of the cross have moved to the points whose coordinates are given by the elements of the two column vectors $\mathbf{A}(\theta, \mathbf{j})[0, 0, a]^{\mathrm{T}}$ and $\mathbf{A}(\phi, \mathbf{n}^2)[a, 0, 0]^{\mathrm{T}}$, that is by $a[\sin\theta, 0, \cos\theta]^{\mathrm{T}}$ and $a[\cos\phi, \sin\alpha\sin\phi, -\cos\alpha\sin\phi]^{\mathrm{T}}$. But these vectors are perpendicular, and hence taking their scalar product we have

$$\sin\theta\cos\phi - \cos\alpha\cos\theta\sin\phi = 0$$

or

$$\tan\theta = \cos\alpha\tan\phi \tag{5.44}$$

Differentiating both sides of this relation with respect to time leads to the ratio $\dot{\phi}:\dot{\theta}$ of the angular speeds of the shafts. Thus

$$\sec^2\theta\ \dot{\theta} = \cos\alpha\sec^2\phi\ \dot{\phi} = \cos\alpha[1 + \tan^2\phi]\dot{\phi}$$

This becomes, with the help of (5.44),

$$\sec^2\theta\ \dot{\theta} = \cos\alpha[1 + \tan^2\theta\sec^2\alpha]\dot{\phi}$$

which in turn yields, after some elementary manipulation,

$$\dot{\phi}/\dot{\theta} = \cos\alpha/(1 - \sin^2\alpha\cos^2\theta) \tag{5.45}$$

Assuming that $\dot{\theta}$ is constant it is evident that $\dot{\phi}$ varies cyclically between the two values $\dot{\theta}\cos\alpha$ and $\dot{\theta}\sec\alpha$. The former value occurs when θ is an odd multiple of $\pi/2$ and the latter value when θ is a multiple of π. The period of the angular speed $\dot{\phi}$ is $\pi/\dot{\theta}$.

The angular acceleration $\ddot{\phi}\mathbf{n}_2$ is obtained by differentiating both sides of the relation (5.45) with respect to time. The operation yields

$$\ddot{\phi}/\dot{\theta}^2 = -\frac{\cos\alpha\sin^2\alpha\sin 2\theta}{(1-\sin^2\alpha\cos^2\theta)^2} \tag{5.46}$$

Let $[\Omega_1, \Omega_2, \Omega_3]$ be the components of the angular velocity $\mathbf{\Omega}$ of the cross referred to the unit vectors $\mathbf{i}$, $\mathbf{j}$, $\mathbf{k}$. The point A of Hooke's joint can be regarded as belonging to either the driving shaft or the cross. We can therefore equate the two expressions $\boldsymbol{\omega}_1 \times \underline{OA}$ and $\mathbf{\Omega} \times \underline{OA}$ for its (absolute) velocity. Hence

$$a\dot{\theta}\begin{vmatrix} \mathbf{i} & \mathbf{j} & \mathbf{k} \\ 0 & 1 & 0 \\ \sin\theta & 0 & \cos\theta \end{vmatrix} = a\begin{vmatrix} \mathbf{i} & \mathbf{j} & \mathbf{k} \\ \Omega_1 & \Omega_2 & \Omega_3 \\ \sin\theta & 0 & \cos\theta \end{vmatrix}$$

which leads to the following two equations:

$$\Omega_2 = \dot{\theta}, \qquad \Omega_1 = \tan\theta\Omega_3 \tag{5.47}$$

Likewise $\boldsymbol{\omega}_2 \times \underline{OD} = \mathbf{\Omega} \times \underline{OD}$, and therefore

$$\begin{vmatrix} \mathbf{i} & \mathbf{j} & \mathbf{k} \\ 0 & \dot{\phi}\cos\alpha & \dot{\phi}\sin\alpha \\ \cos\phi & \sin\alpha\sin\phi & -\cos\alpha\sin\phi \end{vmatrix} = \begin{vmatrix} \mathbf{i} & \mathbf{j} & \mathbf{k} \\ \Omega_1 & \Omega_2 & \Omega_3 \\ \cos\phi & \sin\alpha\sin\phi & -\cos\alpha\sin\phi \end{vmatrix}$$

We obtain, after some manipulation, the further three equations for Ω_1, Ω_2, Ω_3:

$$\begin{aligned} \cos\alpha\Omega_2 + \sin\alpha\Omega_3 &= \dot{\phi} \\ \cos\alpha\tan\phi\Omega_1 + \Omega_3 &= \dot{\phi}\sin\alpha \\ \sin\alpha\tan\phi\Omega_1 - \Omega_2 &= -\dot{\phi}\cos\alpha \end{aligned} \tag{5.48}$$

Thus there are five equations altogether for the three unknowns Ω_1, Ω_2, Ω_3. Solving for Ω_1, Ω_2, Ω_3 using the first, second and fourth equations yields

$$\Omega_1 = \sin\alpha\sin\theta\cos\theta\,\dot{\phi}, \qquad \Omega_2 = \dot{\theta}, \qquad \Omega_3 = \sin\alpha\cos^2\theta\,\dot{\phi}$$

It is readily verified, with the help of (5.44) and (5.45), that these values satisfy the remaining two equations.

PROBLEMS

Section 5.2

5.1 Show that a rotation of amount π about the axis ON is equivalent to a rotation of amount $-\pi$ about the axis ON.

5.2 Two particles P_1, P_2 undergo a rotation of the same amount about the axis ON. Show if the particles are collinear with the origin O before the rotation that they are collinear with O after the rotation.

5.3 Given that $\theta = \pi/3$ and that $(\surd 3)\mathbf{n} = [-1, 1, 1]$, $\mathbf{r}_1 = [2, 3, -7]$, $\mathbf{r}_2 = [3, 4, -1]$ referred to $\mathbf{i}$, $\mathbf{j}$, $\mathbf{k}$, find using the relation (5.4) the vectors $\mathbf{R}_1$, $\mathbf{R}_2$. Verify that
(a) $\mathbf{R}_1^2 = \mathbf{r}_1^2$, $\mathbf{R}_2^2 = \mathbf{r}_2^2$, $\mathbf{R}_1 \cdot \mathbf{R}_2 = \mathbf{r}_1 \cdot \mathbf{r}_2$
(b) The particle with position vector $\mathbf{r}_1 \times \mathbf{r}_2$ before the rotation is moved to the point with position vector $\mathbf{R}_1 \times \mathbf{R}_2$ by the rotation.
The position vector of the particle P before the rotation is $2\mathbf{r}_1 + \mathbf{r}_2 - \mathbf{r}_1 \times \mathbf{r}_2$. Determine its position vector after the rotation.

5.4 Verify by repeated use of the relation (5.4) that a rotation of amount π about the axis Ox followed by a rotation of amount $\pi/2$ about the axis Oz is equivalent to a rotation of amount $\pi/2$ about Oz followed by a rotation of amount π about Oy.

5.5 The line passing through the points with coordinates [1, 1, 3] and [4, 2, 5] with respect to the axes Ox, Oy, Oz is rotated an amount $\pi/3$ about an axis through O in the direction of the vector $-\mathbf{i} + \mathbf{j} + \mathbf{k}$. Find the vector equation of the line after the rotation. Find also the position before the rotation of the point P of the line whose location after the rotation is the point with coordinates [−3, −3, 3]. Obtain the vector equation of the line after the rotation given that the axis of rotation passes through the point with coordinates [1, 5, 2] instead of the origin O.

5.6 Two particles P_1, P_2 are at time $t = 0$ situated at the points A_1, A_2. Subsequently they are displaced so that at time t they are situated at the points B_1, B_2. Determine, for the following, if there exists a rotation about an axis through O that takes P_1 from A_1 to B_1 and P_2 from A_2 to B_2:
(a) $\underline{OA_1} = [3, 4, 5]$, $\underline{OA_2} = [1, 0, -1]$, $\underline{OB_1} = [5, 4, 3]$, $\underline{OB_2} = [-1, 0, 1]$
(b) $\underline{OA_1} = [5, 7, 9]$, $\underline{OA_2} = [4, -4, 0]$, $\underline{OB_1} = [9, 7, 5]$, $\underline{OB_2} = [0, -4, -4]$
(c) $\underline{OA_1} = [1, 1, 0]$, $\underline{OA_2} = [2, 3, 6]$, $\underline{OB_1} = [-1, 0, 1]$, $\underline{OB_2} = [-3, 6, -2]$
(d) $\underline{OA_1} = [7, 7, 7]$, $\underline{OA_2} = [-1, 1, 1]$, $\underline{OB_1} = [11, -5, -1]$, $\underline{OB_2} = [1, 1, 1]$.
Find, when the rotation exists, the angle of rotation θ and a vector in the direction of the axis of rotation.

5.7 A plane passes through the three points A, B, C whose coordinates with respect to the axes Ox, Oy, Oz are [6, 2, −1], [1, 1, 9], [8, 1, −4]. It is rotated an amount $\pi/2$ about an axis passing through O in the direction of the vector $\mathbf{i} + 2\mathbf{j} - 2\mathbf{k}$. Determine the cartesian equations of the plane before and after the rotation. Determine also the position before the rotation of the point P of the plane whose location after the rotation is the point with coordinates [4, −2, −3].

Obtain the cartesian equation of the plane after the rotation given that the axis of rotation passes through the point with coordinates [−1, 2, 3] instead of the origin O.

5.8 $\underline{O_1A}$ and $\underline{O_2B}$ are two axes parallel to the unit vector **n**. The vector $\underline{O_1O_2}$ is equal to **a** and is perpendicular to **n**. A rigid body is rotated through an angle π about $\underline{O_1A}$ and then through an angle π about $\underline{O_2B}$. Show that the combined effect of the two rotations is equal to a translation of amount 2**a**.

5.9 Given that $\mathbf{R} = \cos\theta\mathbf{r} + (1 - \cos\theta)(\mathbf{n}\,.\,\mathbf{r})\mathbf{n} + \sin\theta\mathbf{n} \times \mathbf{r}$, where **n** is a unit vector, show, without using any results on rotation theory, that
(a) $\mathbf{R} - \mathbf{r} = \tan(\theta/2)\mathbf{n} \times (\mathbf{R} + \mathbf{r}),\ \theta \neq \pi$
(b) $\mathbf{r} = \cos\theta\mathbf{R} + (1 - \cos\theta)(\mathbf{n}\,.\,\mathbf{R})\mathbf{n} - \sin\theta\mathbf{n} \times \mathbf{R}$.

5.10 Find the rotation matrices, referred to the unit vectors **i**, **j**, **k**, for the rotations defined by
(a) $\theta = \pi/2$, $\quad 3\mathbf{n} = [2, 1, 2]$
(b) $\theta = \pi/3$, $\quad (\sqrt{3})\mathbf{n} = [1, 1, -1]$
(c) $\theta = \pi$, $\quad 9\mathbf{n} = [7, 4, 4]$
(d) $\theta = 2\pi/3$, $\quad (\sqrt{3})\mathbf{n} = [1, -1, 1]$
(e) $\theta = \pi$ $\quad (\sqrt{6})\mathbf{n} = [2, 1, 1]$.
Verify for the matrix obtained in (c) properties P5 and P6 of rotation matrices.

5.11 Determine which of the following matrices are rotation matrices:

$$\text{(a)} \begin{bmatrix} 0 & -1 & 0 \\ 0 & 0 & 1 \\ 1 & 0 & 0 \end{bmatrix} \qquad \text{(b)} \frac{1}{7}\begin{bmatrix} 2 & 3 & 6 \\ -6 & -2 & 3 \\ 3 & -6 & 2 \end{bmatrix}$$

$$\text{(c)} \frac{1}{9}\begin{bmatrix} -7 & 4 & 4 \\ 4 & -1 & 8 \\ 4 & 8 & -1 \end{bmatrix} \qquad \text{(d)} \frac{1}{11}\begin{bmatrix} -9 & -2 & -6 \\ -2 & -9 & 6 \\ -6 & 6 & 7 \end{bmatrix}$$

$$\text{(e)} \frac{1}{17}\begin{bmatrix} -8 & -9 & -12 \\ -9 & -8 & 12 \\ -12 & -12 & -1 \end{bmatrix} \qquad \text{(f)} \frac{1}{23}\begin{bmatrix} 3 & -6 & 22 \\ -18 & 13 & 6 \\ 14 & 18 & 3 \end{bmatrix}.$$

Find for those matrices that are rotation matrices the angle of rotation and a vector in the direction of the axis of rotation.

5.12 The matrix **A** is defined by

$$\mathbf{A} = \begin{bmatrix} 2l_1^2 - 1 & 2l_1l_2 & 2l_1l_3 \\ 2l_2l_1 & 2l_2^2 - 1 & 2l_2l_3 \\ 2l_3l_1 & 2l_3l_2 & 2l_3^2 - 1 \end{bmatrix}$$

Show $\det \mathbf{A} = 2(l_1^2 + l_2^2 + l_3^2) - 1$. Hence deduce that a necessary condition for $\mathbf{A}$ to be a rotation matrix is $l_1^2 + l_2^2 + l_3^2 = 1$. Explain why this condition is sufficient to ensure that $\mathbf{A}$ is a rotation matrix. What is the associated angle of rotation?

Show that there are an infinity of matrices with the property $\mathbf{A}^2 = \mathbf{I}$ and find three of them.

5.13 Show the following:

(a) If $\theta = k\pi$, where k is a positive or negative integer, then the rotation matrix $\mathbf{A}(\theta, \mathbf{n})$ is a symmetric matrix.

(b) If $\mathbf{A}(\theta, \mathbf{n})$ is a symmetric matrix then $\theta = k\pi$, where k is a positive or negative integer.

5.14 The point P lies on the axis of rotation which passes through the origin O. Show that $\mathbf{r} = \underline{\mathrm{OP}}$ satisfies the equation $\mathbf{Ar} = \mathbf{Ir}$, where $\mathbf{A}$ is the matrix of the rotation.

Use this result to determine a vector in the direction of the axis of rotation for rotations defined by the following matrices:

$$\text{(a)} \begin{bmatrix} 0 & -1 & 0 \\ 0 & 0 & -1 \\ 1 & 0 & 0 \end{bmatrix} \quad \text{(b)} \frac{1}{3}\begin{bmatrix} 2 & -2 & 1 \\ 1 & 2 & 2 \\ -2 & -1 & 2 \end{bmatrix} \quad \text{(c)} \frac{1}{9}\begin{bmatrix} 1 & 8 & 4 \\ -4 & 4 & -7 \\ -8 & -1 & 4 \end{bmatrix}.$$

5.15 Two particles P_1, P_2 are rotated through the same angle about an axis ON. Their positions before the rotation are given by A_1, A_2 while their positions after the rotation are B_1, B_2. Find, for the following, the angle of rotation θ and a vector in the direction of the axis of rotation:

(a) $\underline{\mathrm{OA}_1} = [-1, 2, 9]$, $\underline{\mathrm{OA}_2} = [0, 1, -4]$, $\underline{\mathrm{OB}_1} = [-6, -7, 1]$, $\underline{\mathrm{OB}_2} = [1, 4, 0]$

(b) $\underline{\mathrm{OA}_1} = [3, 3, -3]$, $\underline{\mathrm{OA}_2} = [4, 1, -1]$, $\underline{\mathrm{OB}_1} = [1, 1, 5]$, $\underline{\mathrm{OB}_2} = [0, 3, 3]$

(c) $\underline{\mathrm{OA}_1} = [3, 2, 2]$, $\underline{\mathrm{OA}_2} = [3, -1, -10]$, $\underline{\mathrm{OB}_1} = [3, 2, 2]$, $\underline{\mathrm{OB}_2} = [-9, 2, 5]$

(d) $\underline{\mathrm{OA}_1} = [0, 2, 4]$, $\underline{\mathrm{OA}_2} = [2, 0, 4]$, $\underline{\mathrm{OB}_1} = [4, 2, 0]$, $\underline{\mathrm{OB}_2} = [4, 0, 2]$.

Find the rotation matrix for each of the rotations. Verify for each of (a), (b), (c), (d) that the particle with the position vector $\underline{\mathrm{OA}_1} \times \underline{\mathrm{OA}_2}$ before the rotation moves to the point with the position vector $\underline{\mathrm{OB}_1} \times \underline{\mathrm{OB}_2}$ as a result of the rotation.

5.16 In the analysis leading to the derivation of the relations (5.11) and (5.12) it is implicitly assumed that neither of $\mathbf{m}_1$ or $\mathbf{m}_2$ is equal to the null vector. When one of these vectors is the null vector the analysis must be revised and the following provides a basis for the revision. Show when $\mathbf{m}_1 = 0$ and $\mathbf{m}_2 \neq 0$ that

(a) $\mathbf{R}_1 = -\mathbf{r}_1$, $\mathbf{d}_1 = -2\mathbf{r}_1$

(b) $\mathbf{m}_2 \,.\, \mathbf{r}_2 = \mathbf{m}_2 \,.\, \mathbf{R}_2$, $\mathbf{m}_2 \,.\, \mathbf{r}_2 = |\mathbf{m}_2|^2$, $\mathbf{m}_2 \,.\, \mathbf{r}_1 = 0$
(c) $\mathbf{R}_1 = -\mathbf{r}_1 + 2(\mathbf{N} \,.\, \mathbf{r}_1)\mathbf{N}$, $\mathbf{R}_2 = -\mathbf{r}_2 + 2(\mathbf{N} \,.\, \mathbf{r}_2)\mathbf{N}$, where $|\mathbf{m}_2|\mathbf{N} = \mathbf{m}_2$.
Deduce that the rotation is of amount π about an axis ON in the direction of the vector $\mathbf{m}_2$. If $\mathbf{m}_2 = \mathbf{m}_1 = \mathbf{0}$, how can (c) be modified to show that the rotation is of amount π?

Find a vector in the direction of the axis of rotation when $\underline{\mathrm{OA}_1} = [-2, -1, 1]$, $\underline{\mathrm{OA}_2} = [3, 1, -2]$, $\underline{\mathrm{OB}_1} = [2, 1, -1]$, $\underline{\mathrm{OB}_2} = [-3, -1, 2]$, the notation being the same as in Problem 5.15.

5.17 When $\mathbf{m}_1$, $\mathbf{m}_2$ are not parallel the vector $\mathbf{a}$ that defines the rotation is given by the relation (5.11). There are alternative expressions for $\mathbf{a}$ dependent on whether or not $\mathbf{d}_1$ and $\mathbf{d}_2$ are parallel. The following form the basis for the derivation of these expressions:
(a) Show, if $\mathbf{d}_1$ and $\mathbf{d}_2$ are not parallel, that $\mathbf{a}$ must be of the form $\alpha\mathbf{d}_1 \times \mathbf{d}_2$, where α is a scalar to be determined. By substituting this expression for $\mathbf{a}$ into either of the equations $\mathbf{a} \times \mathbf{m}_1 = \mathbf{d}_1$, $\mathbf{a} \times \mathbf{m}_2 = \mathbf{d}_2$, deduce that $(\mathbf{d}_1 \,.\, \mathbf{m}_2)\alpha = 1$.
(b) Show if $\mathbf{d}_1 = \alpha\mathbf{d}_2$ then $\mathbf{a}$ is given by $\mathbf{a} = \beta(\mathbf{m}_1 - \alpha\mathbf{m}_2)$, where β is defined by $\mathbf{d}_2 = \beta\mathbf{m}_1 \times \mathbf{m}_2$. *Hint*: eliminate $\mathbf{d}_1$, $\mathbf{d}_2$ from the equations $\mathbf{a} \times \mathbf{m}_1 = \mathbf{d}_1$ and $\mathbf{a} \times \mathbf{m}_2 = \mathbf{d}_2$ to show that $\mathbf{a}$ is parallel to the vector $\mathbf{m}_1 - \alpha\mathbf{m}_2$.

Use the expression for $\mathbf{a}$ in (b) to find the angle of rotation and the direction of the axis of rotation when $\underline{\mathrm{OA}_1} = [0, 0, 6]$, $\underline{\mathrm{OA}_2} = [-2, 0, 1]$, $\underline{\mathrm{OB}_1} = [4, 4, 2]$, $\underline{\mathrm{OB}_2} = [0, 2, -1]$.

5.18 A particle P undergoes a rotation of amount α about the axis Ox and then a rotation of amount β about the axis Oy. Show that the combined rotations are equivalent to a single rotation of amount $2\arccos\{\cos(\alpha/2)\cos(\beta/2)\}$ about an axis whose direction cosines are in the ratios $\tan(\alpha/2){:}\tan(\beta/2){:}-\tan(\alpha/2)\tan(\beta/2)$. It may be assumed that $0 < \alpha < \pi$ and $0 < \beta < \pi$.

5.19 A particle is first rotated about an axis ON through an angle π and then about an intersecting axis OM also through an angle π. Show that the combined rotations are equivalent to a single rotation about an axis through O perpendicular to ON and OM, the angle of this rotation being equal to twice the angle between ON and OM.
Hint: take $\mathbf{i}$ in the direction of $\underline{\mathrm{ON}}$ and $\mathbf{j}$ to lie in the plane of ON and OM, and then determine the rotation matrix for each rotation.

5.20 A plane whose normal is in the direction of the vector with components [1, 1, 1] referred to the unit vectors $\mathbf{i}$, $\mathbf{j}$, $\mathbf{k}$ passes through the point with coordinates [2, 3, −2] with respect to the axes Ox, Oy, Oz. It is rotated through an angle of $2\pi/3$ about an axis that passes through the point [3, 3, 3] and is in the direction of the vector with components [−1, −1, 1], and then through an angle $\pi/3$ about an axis that passes through the point [4, 5, 7] and is in the direction of the vector [1, −1, 1]. Find the cartesian equation of the plane after the two rotations. Determine a translation and a rotation which combine to produce the same displacement of the plane as the two rotations.

5.21 Referred to axes Ox, Oy, Oz the equations of the plane faces of a cube are given by $x = \pm a$, $y = \pm a$, $z = \pm a$. The cube is rotated counter-clockwise through an angle equal to $\pi/4$ about Oz and then clockwise through an angle equal to arc sin $1/\sqrt{3}$ about Ox. Show that after the combined rotations one of the diagonals of length $(2\sqrt{3})a$ lies along Oy.

5.22 If **A** and **B** are rotation matrices, what can be inferred from the result trace(**AB**) = trace(**BA**)? (See Problem 1.47 for properties of the trace of a matrix).

It has already been stated (without proof) in the course of this chapter that any matrix having the properties $\mathbf{AA}^{\mathrm{T}} = \mathbf{I}$ and $\det \mathbf{A} = 1$ defines a rotation. Problems 5.23 to 5.27 provide the basis of a proof of this result. Each problem is preceded by a short statement setting out the (mathematical) objective of the problem.

5.23 The objective of this problem is to show that any element of a 3×3 matrix **A** with the properties $\mathbf{AA}^{\mathrm{T}} = \mathbf{I}$ and $\det\mathbf{A} = 1$ is equal to its cofactor in $\det\mathbf{A}$ (see Problem 1.41 for the definition of the term 'cofactor').

By equating each element in the first column of **I** to the corresponding element in the matrix product $\mathbf{AA}^{\mathrm{T}}$, show that

$$a_{11}^2 + a_{12}^2 + a_{13}^2 = 1$$

$$a_{21}a_{11} + a_{22}a_{12} + a_{23}a_{13} = 0$$

$$a_{31}a_{11} + a_{32}a_{12} + a_{33\,13} = 0$$

Noting the method of analysis that leads to the relations (1.27), show further that

$$a_{11} = c(a_{22}a_{33} - a_{23}a_{32})$$
$$a_{12} = c(a_{23}a_{31} - a_{21}a_{33})$$
$$a_{13} = c(a_{21}a_{32} - a_{22}a_{31})$$

where c is a scalar to be determined.

Deduce with the help of the property $\det\mathbf{A} = 1$ that $c = 1$. Thus each element of the first row of **A** is equal to its cofactor in $\det\mathbf{A}$. It follows in a like manner that each element of the second and third rows of **A** is equal to its cofactor in det**A**.

5.24 The purpose of the problem is to show that $-1 \leq \text{trace}\,\mathbf{A} \leq 3$, where **A** again satisfies $\mathbf{AA}^{\mathrm{T}} = \mathbf{I}$ and $\det\mathbf{A} = 1$. When trace $\mathbf{A} = -1$ or 3 the matrix **A** is symmetric but not otherwise.

In the course of the problem the following elementary properties of the trace of a square matrix are required:

$$\text{trace}\,\mathbf{C}^{\mathrm{T}} = \text{trace}\,\mathbf{C}$$

$$\text{trace}(\mathbf{C} + \mathbf{D}) = \text{trace}\,\mathbf{C} + \text{trace}\,\mathbf{D}$$

The quantity α is defined by

$$\alpha + (1/\alpha) + 1 = \text{trace}\,\mathbf{A}$$

Show that

(a) α is real if trace $\mathbf{A} \geq 3$ or trace $\mathbf{A} \leq -1$

(b) When α is complex, $|\alpha| = 1$ and hence $\alpha = e^{i\theta}$, where $\theta = \arg\alpha$

(c) trace $\mathbf{A}^2 = a_{11}^2 + a_{22}^2 + a_{33}^2 + 2(a_{12}a_{21} + a_{13}a_{31} + a_{23}a_{32})$ $= (\text{trace}\,\mathbf{A})^2 - 2\text{trace}\,\mathbf{A}$

(d) trace $\mathbf{A}^2 = \alpha^2 + (1/\alpha^2) + 1$

(e) trace $(\mathbf{A} - \mathbf{A}^{\mathrm{T}})^2 = \text{trace}\,\mathbf{A}^2 + \text{trace}(\mathbf{A}^{\mathrm{T}})^2 - 2\text{trace}\mathbf{I}$ $= 2(\alpha^2 - 1)^2/\alpha^2$

(f) trace $(\mathbf{A} - \mathbf{A}^{\mathrm{T}})^2 = -2\{(a_{12} - a_{21})^2 + (a_{13} - a_{31})^2 + (a_{23} - a_{32})^2\}$.

Result (f) implies trace $(\mathbf{A} - \mathbf{A}^{\mathrm{T}})^2 \leq 0$, equality occurring when $a_{12} = a_{21}$, $a_{13} = a_{31}$, $a_{23} = a_{32}$, that is when **A** is symmetric.

Deduce from (e) that $\alpha = \pm 1$ or is complex.

With the help of (b) we can now put trace $\mathbf{A} = 1 + 2\cos\theta$, where $\theta = 0$ when $\alpha = 1$ and $\theta = \pi$ when $\alpha = -1$.

5.25 The objective of this problem is to show, when trace $\mathbf{A} = -1$ or 3, that the matrix **A** is a rotation matrix. When trace $\mathbf{A} = -1$ the angle of rotation is π and when trace $\mathbf{A} = 3$ it is zero.

Show that:

(a) $|a_{ij}| \leq 1$ where a_{ij} is the *ij*th element of **A**

(b) $\mathbf{A} = \mathbf{I}$ when trace $\mathbf{A} = 3$. (**I** is a rotation matrix with $\theta = 0$ and **n** arbitrary.)

The matrix **B** is defined by $2\mathbf{B} = \mathbf{A} + \mathbf{I}$, where trace $\mathbf{A} = -1$, and the quantities L_i $(i = 1, 2, 3)$ by $L_i = \surd\{(1 + a_{ii})/2\}$. Show that

(c) $b_{11} = L_1^2$, $b_{22} = L_2^2$, $b_{33} = L_3^2$, where $L_1^2 + L_2^2 + L_3^2 = 1$

(d) $b_{23}L_1^2 = b_{12}b_{13}$, $b_{13}L_2^2 = b_{12}b_{23}$, $b_{12}L_3^2 = b_{13}b_{23}$ (*Hint*: express each side in terms of the elements of **A**, noting that **A** is symmetric and trace $\mathbf{A} = -1$)

(e) $b_{12}b_{13}b_{23} = L_1^2L_2^2L_3^2 \geq 0$.

Deduce from (d) and (e) that the values of b_{12}, b_{13}, b_{23} are obtained from the relations $b_{12} = \pm L_1 L_2$, $b_{13} = \pm L_1 L_3$, $b_{23} = \pm L_2 L_3$ by taking three plus signs or one plus sign and two negative ones. When three plus signs apply we put $l_1 = L_1$, $l_2 = L_2$, $l_3 = L_3$, so that direct comparison with (5.13) shows $\mathbf{A} = -\mathbf{I} + 2\mathbf{B}$ is a rotation matrix with $\theta = \pi$. When $b_{12} = L_1 L_2$, $b_{13} = -L_1 L_3$, $b_{23} = -L_2 L_3$ we define l_1, l_2, l_3 by $l_1 = L_1$, $l_2 = L_2$, $l_3 = -L_3$, so that $\mathbf{A} = -\mathbf{I} + 2\mathbf{B}$, where $\mathbf{B} = [l_i l_j]$. (Alternatively we could take $l_1 = -L_1$, $l_2 = -L_2$, $l_3 = L_3$.) The proofs for the other cases in which the expressions for b_{pq} $(p \neq q)$ contain two minus signs and one plus sign may be completed in a similar fashion.

It is a consequence of the proof that all symmetric rotation matrices must either be equal to the unit matrix or be of the form

$$\begin{bmatrix} 2l_1^2 - 1 & 2l_1 l_2 & 2l_1 l_3 \\ 2l_1 l_2 & 2l_2^2 - 1 & 2l_2 l_3 \\ 2l_1 l_3 & 2l_2 l_3 & 2l_3^2 - 1 \end{bmatrix}$$

where $l_1^2 + l_2^2 + l_3^2 = 1$.

5.26 The aim of this problem is to establish an algebraic result required in the course of Problem 5.27.

The matrix $\mathbf{E} = [e_{ij}]$ $(i, j = 1, 2, 3)$ is defined by

$$\mathbf{E} = \mathbf{A}^2 - (\text{trace}\,\mathbf{A})\mathbf{A} + (\text{trace}\,\mathbf{A})\mathbf{I} - \mathbf{A}^{\mathrm{T}}$$

Show that

(a) $e_{11} = a_{12}a_{21} + a_{13}a_{31} - a_{11}a_{22} - a_{33}a_{11} + a_{22} + a_{33} = 0$
(b) $e_{12} = a_{13}a_{32} - a_{33}a_{12} - a_{21} = 0$
(c) $\mathbf{E} = \mathbf{0} = \mathbf{E}^{\mathrm{T}}$
(d) $\mathbf{A}^2 + (\mathbf{A}^{\mathrm{T}})^2 = (\text{trace}\,\mathbf{A} + 1)(\mathbf{A} + \mathbf{A}^{\mathrm{T}}) - (2\text{trace}\,\mathbf{A})\mathbf{I}$ (*Hint*: add $\mathbf{E}$ and $\mathbf{E}^{\mathrm{T}}$)
(e) $(\mathbf{A} - \mathbf{A}^{\mathrm{T}})^2 + 4\sin^2\theta\mathbf{I} = 2(1 + \cos\theta)(\mathbf{A} + \mathbf{A}^{\mathrm{T}} - 2\cos\theta\mathbf{I})$, where θ has the meaning attributed to it in Problem 5.24(b).

5.27 The steps in this problem complete the proof of the result that $\mathbf{A}$ represents a rotation if $\mathbf{A}\mathbf{A}^{\mathrm{T}} = \mathbf{I}$ and $\det\mathbf{A} = 1$.

The quantities l_1, l_2, l_3 are defined by

$$2\sin\theta\begin{bmatrix} 0 & -l_3 & l_2 \\ l_3 & 0 & -l_1 \\ -l_2 & l_1 & 0 \end{bmatrix} = \mathbf{A} - \mathbf{A}^{\mathrm{T}} \neq \mathbf{0}$$

Show that

(a) $\alpha - 1/\alpha = 2i\sin\theta$
(b) $\text{trace}\,(\mathbf{A} - \mathbf{A}^{\mathrm{T}})^2 = -8(l_1^2 + l_2^2 + l_3^2)\sin^2\theta$
(c) $l_1^2 + l_2^2 + l_3^2 = 1$ (*Hint*: use (e) of Problem 5.24 together with (a) and (b))

(d) $(\mathbf{A} - \mathbf{A}^{\mathrm{T}})^2 + 4\sin^2\theta\mathbf{I} = 4\sin^2\theta\begin{bmatrix} l_1^2 & l_1 l_2 & l_1 l_3 \\ l_2 l_1 & l_2^2 & l_2 l_3 \\ l_1 l_3 & l_2 l_3 & l_3^2 \end{bmatrix}$.

By equating the right-hand side of (d) to the right-hand side of (e) in Problem 5.26, deduce that

$$\mathbf{A} + \mathbf{A}^{\mathrm{T}} - 2\cos\theta\mathbf{I} = 2(1 - \cos\theta)\begin{bmatrix} l_1^2 & l_1 l_2 & l_1 l_3 \\ l_2 l_1 & l_2^2 & l_2 l_3 \\ l_1 l_3 & l_2 l_3 & l_3^2 \end{bmatrix}$$

The addition of the corresponding sides of this relation and the relation

$$\mathbf{A} - \mathbf{A}^{\mathrm{T}} = 2\sin\theta\begin{bmatrix} 0 & -l_3 & l_2 \\ l_3 & 0 & -l_1 \\ -l_2 & l_1 & 0 \end{bmatrix}$$

yields

$$\mathbf{A} = \cos\theta\mathbf{I} + (1 - \cos\theta)\mathbf{B} + \sin\theta\mathbf{C}$$

where **B** and **C** are as defined in (5.11).

5.28 (a) Show if $a_{22} = 1$ that the rotation matrix **A** must be of the form

$$\begin{bmatrix} \cos\phi & 0 & \sin\phi \\ 0 & 1 & 0 \\ -\sin\phi & 0 & \cos\phi \end{bmatrix}$$

(b) Show if $a_{22} = -1$ that the rotation matrix **A** must be of the form

$$\begin{bmatrix} \cos\phi & 0 & \sin\phi \\ 0 & -1 & 0 \\ \sin\phi & 0 & -\cos\phi \end{bmatrix}$$

Hint: in (a) and (b) use the cofactor property of the elements of **A** and also the property that the sum of the squares of the elements of any row (or column) is equal to one.

Section 5.3

5.29 Three particles of a rigid body are initially situated at the points A_1, A_2, A_3. During a time t the body is displaced so that the particles move to the points B_1, B_2, B_3. Determine for each of the displacements, defined by the following, an equivalent translation and rotation:

(a) $\underline{OA_1} = [-5, 5, 3]$, $\underline{OA_2} = [-2, -3, 3]$, $\underline{OA_3} = [6, 4, 7]$
$\underline{OB_1} = [5, 3, 7]$, $\underline{OB_2} = [4, 9, 1]$, $\underline{OB_3} = [-6, 4, 3]$

(b) $\underline{OA_1} = [2, 1, 5]$, $\underline{OA_2} = [3, -4, 6]$, $\underline{OA_3} = [-3, 6, -5]$
$\underline{OB_1} = [3, 1, 2]$, $\underline{OB_2} = [4, 0, -3]$, $\underline{OB_3} = [-7, 6, 7]$.
Verify, numerically, that the rotation may be performed first provided the axis of rotation passes through one of the three points A_1, A_2, A_3.

5.30 A rigid body moves so that one point O of itself remains fixed. The velocities of two particles P_1, P_2 of it are respectively equal to $\mathbf{v}_1$, $\mathbf{v}_2$. If $\mathbf{v}_1 = [2, 4, 4]$, $\mathbf{v}_2 = [8, 7, -2]$ referred to the unit vectors $\mathbf{i}, \mathbf{j}, \mathbf{k}$ when the particles are situated at the points whose position vectors with respect to O are $[2, 0, -1]$ and $[3, -4, -2]$, find the angular velocity $\boldsymbol{\omega}$ of the rigid body.

5.31 The velocities of three particles P_1, P_2, P_3 of a rigid body are equal to $\mathbf{v}_1$, $\mathbf{v}_2$, $\mathbf{v}_3$. When the particles are respectively situated at the points whose position vectors with respect to the origin O are given by $[2, 2, 2]$, $[3, 4, 5]$, $[6, 6, 8]$, then $\mathbf{v}_1 = [1, 3, -2]$, $\mathbf{v}_2 = [2, 1, -1]$, $\mathbf{v}_3 = [3, 1, -2]$. Find the angular velocity $\boldsymbol{\omega}$ of the body.

5.32 The angular velocity $\boldsymbol{\omega}$ and angular acceleration $\dot{\boldsymbol{\omega}}$ of a rigid body moving so that one point O of it remains fixed are given at a certain instant of time by $\boldsymbol{\omega} = [1, 1, 1]$, $\dot{\boldsymbol{\omega}} = [2, 3, 1]$ referred to the unit vectors $\mathbf{i}, \mathbf{j}, \mathbf{k}$. Determine the magnitude of the acceleration of the particle of the body whose position vector with respect to O at this instant is equal to the vector $[2, 6, 3]$.

5.33 The angular velocity $\boldsymbol{\omega}$ of the mutually perpendicular right-handed set of rotating unit vectors $\mathbf{a}, \mathbf{b}, \mathbf{c}$ is given by $\boldsymbol{\omega} = \omega\mathbf{c}$, where ω is a constant.

Find an expression for $\dot{\mathbf{A}}$ in terms of $\mathbf{a}, \mathbf{b}, \mathbf{c}$ when
(a) $\mathbf{A} = r\mathbf{a}$, where r is a function of time
(b) $\mathbf{A} = -\cos\omega t\mathbf{a} + \sin\omega t\mathbf{b}$.
Find $\mathbf{a}$ and $\mathbf{b}$ in terms of $\mathbf{i}, \mathbf{j}$ and t given that $\mathbf{a} = \mathbf{i}$, $\mathbf{b} = \mathbf{j}$ when $t = 0$.
Hint: show $\ddot{\mathbf{a}} = \omega\dot{\mathbf{b}} = -\omega^2\mathbf{a}$.

5.34 The angular velocity $\boldsymbol{\omega}$ of the mutually perpendicular right-handed set of rotating unit vectors $\mathbf{a}, \mathbf{b}, \mathbf{c}$ is given by

$$\boldsymbol{\omega} = -\sin t\mathbf{a} + \mathbf{b} + \cos t\mathbf{c}$$

Find expressions for $\dot{\mathbf{A}}$ and $\ddot{\mathbf{A}}$ in terms of $\mathbf{a}, \mathbf{b}, \mathbf{c}$ when
(a) $\mathbf{A} = \cos^2 t\mathbf{a} - \sin t\mathbf{b} + \cos t\sin t\mathbf{c}$
(b) $\mathbf{A} = -\sin t\mathbf{a} + \cos t\mathbf{c}$.

5.35 The velocities of the particles P_i $(i = 0, 1, 2 \ldots)$ of a rigid body are denoted by $\mathbf{v}_i$. Show that the points Q_i $(i = 0, 1, 2, \ldots)$ with position vectors $\underline{OQ_i}$ given by $\underline{OQ_i} = \mathbf{v}_i$ lie in a plane that passes through the point Q_0 and has its normal parallel to $\boldsymbol{\omega}$, the angular velocity of the body.

Deduce that the speed of the slowest particle is equal to $|\mathbf{v}_0 \cdot \boldsymbol{\omega}|/|\boldsymbol{\omega}|$.

5.36 Using the notation of Problem 5.35, show that

(a) $\mathbf{v}_i \cdot \boldsymbol{\omega} = \mathbf{v}_0 \cdot \boldsymbol{\omega}$ $(i = 1, 2, \ldots)$

(b) The velocities of the particles whose position vectors with respect to P_0 are of the form $(\boldsymbol{\omega} \times \mathbf{v}_0)/\omega^2 + \alpha\boldsymbol{\omega}$ (α arbitrary) are parallel to $\boldsymbol{\omega}$.

The objective of the next problem is to provide an alternative derivation of the relation (5.11). The method of analysis is an extension of that contained in Section 4.3.

5.37 The three vectors **a**, **b**, **c** form a mutually perpendicular right-handed set of unit vectors. Each is a function of the time t. Show that

(a) $\mathbf{a} \cdot \dot{\mathbf{a}} = \mathbf{b} \cdot \dot{\mathbf{b}} = 0$

(b) $\mathbf{a} \cdot \dot{\mathbf{b}} + \dot{\mathbf{a}} \cdot \mathbf{b} = 0$.

Writing $\dot{\mathbf{a}} = a_1\mathbf{a} + a_2\mathbf{b} + a_3\mathbf{c}$, $\dot{\mathbf{b}} = b_1\mathbf{a} + b_2\mathbf{b} + b_3\mathbf{c}$, show further that

(c) $a_1 = b_2 = 0$, $a_2 = -b_1$

(d) $\dot{\mathbf{a}} = \boldsymbol{\omega} \times \mathbf{a}$, $\dot{\mathbf{b}} = \boldsymbol{\omega} \times \mathbf{b}$, where $\boldsymbol{\omega} = [b_3, -a_3, a_2]$ referred to **a**, **b**, **c**

(e) $\dot{\mathbf{c}} = \boldsymbol{\omega} \times \mathbf{c}$. *Hint*: differentiate both sides of the relation $\mathbf{c} = \mathbf{a} \times \mathbf{b}$.

If $\mathbf{r}_i$ $(i = 0, 1, 2, \ldots)$ are the position vectors of the particles P_i of a rigid body with respect to a fixed origin O, it follows from the definition of a rigid body that

$$\mathbf{r}_i - \mathbf{r}_0 = \alpha_i\mathbf{a} + \beta_i\mathbf{b} + \mu_i\mathbf{c}$$

where α_i, β_i, μ_i are constants and **a**, **b**, **c** are a set of mutually perpendicular right-handed unit vectors fixed in the body. Differentiating this relation yields

$$\mathbf{v}_i = \mathbf{v}_0 + \boldsymbol{\omega} \times (\mathbf{r}_i - \mathbf{r}_0)$$

The relation (5.27) can now be deduced.

6 Dynamics of systems of particles and rigid bodies

6.1 INTRODUCTION

In this chapter we are concerned with the study of forces that give rise to the motion of systems of particles and rigid bodies (particular examples of particle systems). We define the quantities momentum and moment of momentum of a system of particles, impulse, work, and energy, and derive general principles governing them.

We apply these principles to a variety of problems such as the balancing of rotors, the determination of forces in mechanisms and engines and the use of the flywheel. The motion of tops and rate gyroscopes is also analysed.

6.2 NOTATION

As in Chapter 4 (see Section 4.1) we take the set of fixed axes Ox, Oy, Oz to be a set of axes in which Newton's laws of motion apply. We continue to refer to the particles that form a rigid body and/or a system of particles as P_s ($s = 0, 1, 2, \ldots, N$). The distinction between the particles and their respective positions in space is made by denoting the latter by Q_s ($s = 0, 1, 2, \ldots, N$). We put $\underline{OQ_s} = \mathbf{r}_s$, so that within the context of this chapter $\mathbf{r}_s$ represents an absolute position vector and therefore its first and second derivatives with respect to time represent absolute velocity and acceleration. Thus, using notation previously defined, we have $\dot{\mathbf{r}}_s = \mathbf{v}_s$ and $\ddot{\mathbf{r}}_s = \mathbf{f}_s$. The mass of the particle P_s is denoted by m_s and the total mass (of the system) $\Sigma_s m_s$ by M.

6.3 INTERNAL FORCES AND EXTERNAL FORCES

When we consider the motion of a system of particles we divide the forces acting into two groups: internal and external forces. Those forces that occur in equal and opposite pairs within the system are called internal forces. For example, in the motion of two particles connected by a spring the force exerted by the spring on one particle is equal and opposite to that exerted by it on the other, and these forces are therefore, by definition, internal forces. External forces also occur, by virtue of Newton's third law, in equal and opposite pairs. However, only one of the pair acts upon the given system; the other acts upon a body which is exterior to the system. Since the earth is external to the dynamical systems we consider, gravity is an external force.

The resultant of the external forces acting on the particle P_s is denoted by $\mathbf{F}_s$ and the resultant of the internal forces by $\mathbf{F}'_s$ (see Fig. 6.1).

6.4 CENTRE OF MASS OF A SYSTEM OF PARTICLES

The centre of mass of a system of particles is the point G whose position vector $\mathbf{r}_G$ with respect to O is defined by

$$M\mathbf{r}_G = \sum_{s=0}^{N} m_s \mathbf{r}_s \tag{6.1}$$

We establish two results concerning the centre of mass:

(a) $\Sigma_s m_s \boldsymbol{\sigma}_s = \mathbf{0}$, where $\underline{GQ_s} = \boldsymbol{\sigma}_s$ $(s = 0, 1, 2, \ldots, N)$.
(b) The centre of mass of a rigid body is fixed relative to the body.

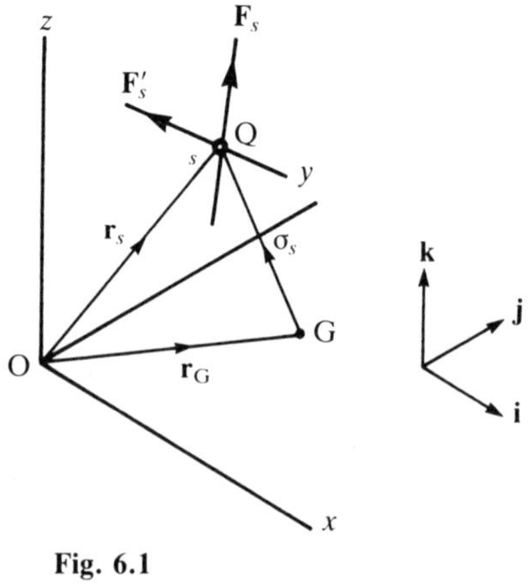

Fig. 6.1

Proof

(a) We have $\underline{OQ_s} = \underline{OG} + \underline{GQ_s}$, so that $\mathbf{r}_s = \mathbf{r}_G + \boldsymbol{\sigma}_s$. Hence

$$\sum_s m_s \mathbf{r}_s = \sum_s m_s(\mathbf{r}_G + \boldsymbol{\sigma}_s)$$
$$= M\mathbf{r}_G + \sum_s m_s \boldsymbol{\sigma}_s$$

It follows immediately from (6.1) that $\Sigma m_s \boldsymbol{\sigma}_s = \mathbf{0}$.

(b) Using the definition of a rigid body we can write

$$\mathbf{r}_s - \mathbf{r}_0 = \underline{Q_0 Q_s} = \alpha_s \mathbf{a} + \beta_s \mathbf{b} + \mu_s \mathbf{c}$$

where α_s, β_s, μ_s are constants and **a**, **b**, **c** are a right-handed set of mutually perpendicular unit vectors whose directions are fixed relative to the rigid body. Thus

$$\sum_s m_s \mathbf{r}_s = \sum_s m_s(\mathbf{r}_0 + \alpha_s \mathbf{a} + \beta_s \mathbf{b} + \mu_s \mathbf{c})$$

and hence, noting (6.1),

$$\mathbf{r}_G = \mathbf{r}_0 + \left\{ \sum_s m_s(\alpha_s \mathbf{a} + \beta_s \mathbf{b} + \mu_s \mathbf{c}) \right\} \Big/ M \tag{6.2}$$

Alternatively

$$\underline{Q_o G} = \frac{1}{M} \sum m_s(\alpha_s \mathbf{a} + \beta_s \mathbf{b} + \mu_s \mathbf{c})$$

which is a vector whose components with respect to the vectors **a**, **b**, **c** are constants. Result (b) is therefore established.

The position of the centre of mass of a rigid body is usually found by a process of integration. A table giving the positions of the centres of mass of a number of bodies is provided in Appendix A.

The centre of mass plays a central role in the discussion of the motion of systems of particles and rigid bodies. We now proceed to derive the equation that governs its motion.

6.5 MOTION OF THE CENTRE OF MASS

It follows from Newton's second law that the equation of motion of the particle P_s is given by

$$m_s \ddot{\mathbf{r}}_s = \mathbf{F}_s + \mathbf{F}'_s$$

where the external and internal forces $\mathbf{F}_s$, $\mathbf{F}'_s$ are defined in Section 6.3. Summing corresponding sides of this equation from $s = 0$ to N we obtain

$$\sum_s m_s \ddot{\mathbf{r}}_s = \sum_s \mathbf{F}_s + \sum_s \mathbf{F}'_s \tag{6.3}$$

Overall the system of forces $\mathbf{F}'_s$ ($s = 0, 1, 2, \ldots, N$) is in equilibrium and therefore $\Sigma_s \mathbf{F}'_s = \mathbf{0}$. Thus the relation (6.3) becomes, with the help of (6.1),

$$M\ddot{\mathbf{r}}_G = \sum_s \mathbf{F}_s \tag{6.4}$$

and this is the equation that governs the motion of the centre of mass. It states that:

The acceleration of the centre of mass multiplied by the total mass of the system is equal to the vector sum of the external forces, these forces being added together as if they were *free* vectors.

The equation (6.4) can be *interpreted* as the equation of motion of a particle of mass M (with absolute position vector $\mathbf{r}_G$) that is acted upon by a force equal to $\Sigma_s \mathbf{F}_s$.

We observe that if $\Sigma \mathbf{F}_s = \mathbf{0}$ then $\dot{\mathbf{r}}_G$, which is the velocity of the centre of mass, must be equal to a constant vector. This result leads to the principle known as the conservation of linear momentum. The linear momentum of a system of particles is denoted by $\mathbf{L}$ and defined by

$$\mathbf{L} = \sum_s m_s \mathbf{v}_s \tag{6.5}$$

Now $\mathbf{v}_s = \dot{\mathbf{r}}_s$ and hence by virtue of (6.1)

$$\mathbf{L} = \Sigma m_s \dot{\mathbf{r}}_s = M\dot{\mathbf{r}}_G = M\mathbf{v}_G \tag{6.6}$$

Thus if $\Sigma_s \mathbf{F}_s = \mathbf{0}$ the linear momentum $\mathbf{L}$ is a constant vector, that is momentum is conserved. In practice $\Sigma_s \mathbf{F}_s$ is not usually equal to the zero vector but it frequently happens that a component of $\Sigma_s \mathbf{F}_s$ in a particular direction is zero and therefore the components of $\mathbf{L}$ and $\dot{\mathbf{r}}_G$ in that direction are constant.

6.6 MOMENT OF MOMENTUM OF A SYSTEM OF PARTICLES ABOUT A POINT

The moment of momentum of a system of particles about the point Q (whether fixed or moving) is denoted by $\mathbf{H}_Q$ and defined by

$$\mathbf{H}_Q = \sum \underline{QQ_s} \times m_s \mathbf{v}_s \tag{6.7}$$

In particular

$$\mathbf{H}_O = \sum \underline{OQ_s} \times m_s \mathbf{v}_s = \sum \mathbf{r}_s \times m_s \mathbf{v}_s \tag{6.8}$$

where O is an example of a fixed point, while

$$\mathbf{H}_G = \sum \underline{GQ_s} \times m_s \mathbf{v}_s = \sum \boldsymbol{\sigma}_s \times m_s \mathbf{v}_s \tag{6.9}$$

where, in general, G will not be a fixed point.

We now derive a number of important results concerning the moment of momentum about a point. They are:

(a) $\mathbf{H}_O = \mathbf{H}_Q + \underline{OQ} \times \mathbf{L}$
(b) $(d/dt)\mathbf{H}_O = \Sigma \mathbf{r}_s \times \mathbf{F}_s$ = sum of the moments of the external forces acting about the fixed point O
(c) $(d/dt)\mathbf{H}_Q = \Sigma \underline{QQ_s} \times \mathbf{F}_s + \mathbf{L} \times \mathbf{v}_Q$
(d) $(d/dt)\mathbf{H}_G = \Sigma \boldsymbol{\sigma}_s \times \mathbf{F}_s$ = sum of the moments of the external forces acting about G.

Proofs

(a) We have

$$\begin{aligned}\mathbf{H}_O - \mathbf{H}_Q &= \sum (\underline{OQ_s} - \underline{QQ_s}) \times m_s \mathbf{v}_s \\ &= \sum \underline{OQ} \times m_s \mathbf{v}_s = \underline{OQ} \times \sum m_s \mathbf{v}_s \\ &= \underline{OQ} \times \mathbf{L}\end{aligned}$$

recalling the definition in (6.5). The result contained in (a) now follows.

(b) From (6.8),

$$\begin{aligned}\frac{d}{dt}\mathbf{H}_O &= \frac{d}{dt}\sum \mathbf{r}_s \times m_s \mathbf{v}_s \\ &= \sum (\dot{\mathbf{r}}_s \times m_s \mathbf{v}_s + \mathbf{r}_s \times m_s \dot{\mathbf{v}}_s)\end{aligned}$$

which reduces to

$$\frac{d}{dt}\mathbf{H}_O = \sum \mathbf{r}_s \times (\mathbf{F}_s + \mathbf{F}'_s)$$

as $\dot{\mathbf{r}}_s$ is equal to $\mathbf{v}_s$ and therefore parallel to $m_s \mathbf{v}_s$. As the system of forces $\mathbf{F}'_s$ ($s = 0, 1, 2, \ldots, N$) is in equilibrium, it follows that $\Sigma \mathbf{r}_s \times \mathbf{F}'_s$, the sum of their moments about the point O, is zero. Hence result (b) is established.

(c) Differentiating both sides of the relation in (a) yields

$$\frac{d}{dt}\mathbf{H}_O = \frac{d}{dt}\mathbf{H}_Q + \underline{\dot{OQ}} \times \mathbf{L} + \underline{OQ} \times \dot{\mathbf{L}}$$

which becomes, noting (6.6) and (6.4),

$$\sum \mathbf{r}_s \times \mathbf{F}_s = \frac{d}{dt}\mathbf{H}_Q + \mathbf{v}_Q \times \mathbf{L} + \underline{OQ} \times \left(\sum \mathbf{F}_s\right)$$

This reduces to

$$\frac{d}{dt}\mathbf{H}_Q = \sum \underline{QQ_s} \times \mathbf{F}_s + \mathbf{L} \times \mathbf{v}_Q$$

(d) If we take Q to coincide with G we have, by (6.6),

$$\mathbf{L} \times \mathbf{v}_Q = M\mathbf{v}_G \times \mathbf{v}_G = \mathbf{0}$$

so that (d) follows directly from (c)

In most problems we will either take moments about a fixed point and therefore apply result (b), or take moments about G when result (d) will be used. However, before we can do this it is necessary to obtain expressions for $\mathbf{H}_O$ and $\mathbf{H}_G$ for a rigid body in terms of its angular velocity vector and further quantities known as moments and products of inertia.

6.7 DETERMINATION OF MOMENT OF MOMENTUM FOR A RIGID BODY

One point of rigid body fixed

We begin by considering the particular motion in which one particle P_0 of the rigid body is fixed. There is no loss of generality in taking the position of P_0 to coincide with the fixed origin O. Recalling (5.29) we have that the velocity of the particle P_s is given by

$$\mathbf{v}_s = \boldsymbol{\omega} \times \mathbf{r}_s$$

where $\boldsymbol{\omega}$ is the angular velocity of the rigid body and $\mathbf{r}_s$ is the position vector of the particle P_s with respect to O. Using the definition in (6.8) it follows that

$$\mathbf{H}_O = \sum m_s \mathbf{r}_s \times (\boldsymbol{\omega} \times \mathbf{r}_s) \tag{6.10}$$

At this stage there are two ways in which we can proceed. All vectors can be expressed in terms of either the fixed unit vectors **i**, **j**, **k** or a set of right-handed mutually perpendicular unit vectors **a**, **b**, **c** fixed relative to the rigid body, the associated axes being Oa, Ob, Oc. We put

$$\mathbf{r}_s = x_s\mathbf{i} + y_s\mathbf{j} + z_s\mathbf{k} = \alpha_s\mathbf{a} + \beta_s\mathbf{b} + \mu_s\mathbf{c} \tag{6.11}$$

where α_s, β_s, μ_s are a set of body coordinates for the particle P_s and thus are constants. The quantities x_s, y_s, z_s will in general depend upon time and are not constants. In addition we put

$$\boldsymbol{\omega} = \omega_x\mathbf{i} + \omega_y\mathbf{j} + \omega_z\mathbf{k} = \omega_1\mathbf{a} + \omega_2\mathbf{b} + \omega_3\mathbf{c} \tag{6.12}$$

$$\mathbf{H}_O = H_x\mathbf{i} + H_y\mathbf{j} + H_z\mathbf{k} = \mathrm{H}_1\mathbf{a} + \mathrm{H}_2\mathbf{b} + \mathrm{H}_3\mathbf{c} \tag{6.13}$$

We proceed by working in terms of the unit vectors **a**, **b**, **c**. The reason for this choice will become clear later. Expanding the triple-vector product in (6.10) yields

$$\mathbf{H}_O = \sum m_s(\mathbf{r}_s^2\boldsymbol{\omega} - (\mathbf{r}_s \,.\, \boldsymbol{\omega})\mathbf{r}_s)$$

which becomes, with the help of (6.11) and (6.12),

$$\begin{aligned}\mathbf{H}_O = \sum m_s\{&(\alpha_s^2 + \beta_s^2 + \mu_s^2)(\omega_1\mathbf{a} + \omega_2\mathbf{b} + \omega_3\mathbf{c})\\ &-(\omega_1\alpha_s + \omega_2\beta_s + \omega_3\mu_s)(\alpha_s\mathbf{a} + \beta_s\mathbf{b} + \mu_s\mathbf{c})\}\end{aligned}$$

Noting (6.13), it is found after some manipulation that

$$\begin{aligned}H_1 &= I_{11}\omega_1 - I_{12}\omega_2 - I_{13}\omega_3\\ H_2 &= -I_{21}\omega_1 + I_{22}\omega_2 - I_{23}\omega_3\\ H_3 &= -I_{31}\omega_1 - I_{32}\omega_2 + I_{33}\omega_3\end{aligned} \tag{6.14}$$

where

$$I_{11} = \sum m_s(\beta_s^2 + \mu_s^2), \qquad I_{22} = \sum m_s(\mu_s^2 + \alpha_s^2), \qquad I_{33} = \sum m_s(\alpha_s^2 + \beta_s^2) \tag{6.15}$$

and

$$I_{12} = I_{21} = \sum m_s\alpha_s\beta_s, \qquad I_{13} = I_{31} = \sum m_s\alpha_s\mu_s, \qquad I_{23} = I_{32} = \sum m_s\beta_s\mu_s \tag{6.16}$$

The quantity I_{11} is called the moment of inertia of the rigid body about the axis Oa; I_{22} is the moment of inertia about Ob, and I_{33} that about Oc. The quantity $\beta_s^2 + \mu_s^2$ is equal to the square of the perpendicular distance from the position of the particle P_s to the axis Oa. Similarly $\mu_s^2 + \alpha_s^2$ and $\alpha_s^2 + \beta_s^2$ represent the squares of perpendicular distances from P_s to the axes Ob and Oc.

The quantities I_{12}, I_{13}, I_{23} are called products of inertia, I_{12} being the product of inertia of the body with respect to the axes Oa and Ob, I_{13} that with respect to Oa and Oc, and I_{23} that with respect to Ob and Oc. For a body of a given shape and mass distribution the moments and products of inertia are fixed once the positions of the axes Oa, Ob, Oc within the rigid body have been specified. Values of moments and products of inertia have been determined by a number of methods for a whole variety of bodies and axes and it will be sufficient for our purposes to quote these values. As the reader will see, considerations of symmetry often enable us to show that a product of inertia is zero. Some useful theorems concerning moments and products of inertia are derived in Section 6.8. A table of moments of inertia for various bodies and axes is provided in Appendix B.

The relations in (6.14) can be expressed in the form

$$\begin{bmatrix} H_1 \\ H_2 \\ H_3 \end{bmatrix} = \mathbf{I}_\mathrm{O} \begin{bmatrix} \omega_1 \\ \omega_2 \\ \omega_3 \end{bmatrix} \tag{6.17}$$

where

$$\mathbf{I}_\mathrm{O} = \begin{bmatrix} I_{11} & -I_{12} & -I_{13} \\ -I_{21} & I_{22} & -I_{23} \\ -I_{31} & -I_{32} & I_{33} \end{bmatrix} \tag{6.18}$$

is called the inertia matrix at O with respect to the axes Oa, Ob, Oc.

If we had worked throughout in terms of the unit vectors **i**, **j**, **k** we would have arrived at a relation of the form

$$\begin{bmatrix} H_x \\ H_y \\ H_z \end{bmatrix} = \mathbf{I}'_\mathrm{O} \begin{bmatrix} \omega_x \\ \omega_y \\ \omega_z \end{bmatrix} \tag{6.19}$$

where I'_O is the inertia matrix at O with respect to the axes Ox, Oy, Oz. Noting the form of (6.15), it follows that the moment of inertia about Ox is equal to $\Sigma m_s(y_s^2 + z_s^2)$ and will therefore be, in general, a function of time. In like manner we see that all other elements of $\mathbf{I}'_\mathrm{O}$ are time dependent. The elements of $\mathbf{I}'_\mathrm{O}$ are for the most part not readily found; we are therefore mathematically compelled to work with the matrix $\mathbf{I}_\mathrm{O}$, and this implies that the unit vectors **a**, **b**, **c** must be used as reference vectors.

In two-dimensional motion of a rigid body $\boldsymbol{\omega} = \omega\mathbf{k}$ (where **k** is perpendicular to the plane of motion), so that recalling the definitions of the vectors **a**, **b** given in Section 4.3 it follows that $\mathbf{c} = \mathbf{a} \times \mathbf{b} = \mathbf{k}$ and therefore, by virtue of (6.12), $\omega_3 = \omega$ and $\omega_1 = \omega_2 = 0$. Hence we have, noting (6.17),

$$\mathbf{H}_\mathrm{O} = \omega(-I_{13}\mathbf{a} - I_{23}\mathbf{b} + I_{33}\mathbf{k}) \tag{6.20}$$

where

I_{33} is the moment of inertia about the axis of rotation
I_{13} is the product of inertia with respect to axes through O in the direction of **a** and the axis of rotation
I_{23} is the product of inertia with respect to axes through O in the direction of **b** and the axis of rotation.

If the products of inertia are zero then

$$\mathbf{H}_\mathrm{O} = \omega I\mathbf{k} \tag{6.21}$$

where I_{33} is simply denoted by I.

No point of rigid body fixed

When no point of the rigid body is fixed we use result (d) in Section 6.6 to help find forces, and therefore we need to know $\mathbf{H}_G$. As a preliminary to finding $\mathbf{H}_G$ we show that

$$\mathbf{H}_G = \sum m_s \boldsymbol{\sigma}_s \times \dot{\boldsymbol{\sigma}}_s \tag{6.22}$$

where $\boldsymbol{\sigma}_s$ is the position vector of the particle P_s with respect to the centre of mass G. We have, by virtue of the definition (6.9),

$$\begin{aligned}\mathbf{H}_G &= \sum \boldsymbol{\sigma}_s \times m_s \mathbf{v}_s \\ &= \sum m_s \boldsymbol{\sigma}_s \times (\dot{\mathbf{r}}_G + \dot{\boldsymbol{\sigma}}_s) \\ &= \left(\sum m_s \boldsymbol{\sigma}_s\right) \times \dot{\mathbf{r}}_G + \sum m_s \boldsymbol{\sigma}_s \times \dot{\boldsymbol{\sigma}}_s \\ &= \sum m_s \boldsymbol{\sigma}_s \times \dot{\boldsymbol{\sigma}}_s\end{aligned}$$

using result (a) in Section 6.4. Now G is fixed with respect to the rigid body and therefore

$$\mathbf{v}_s - \mathbf{v}_G = \boldsymbol{\omega} \times \underline{GQ_s}$$

or

$$\dot{\boldsymbol{\sigma}}_s = \boldsymbol{\omega} \times \boldsymbol{\sigma}_s$$

Thus we have

$$\mathbf{H}_G = \sum m_s \boldsymbol{\sigma}_s \times (\boldsymbol{\omega} \times \boldsymbol{\sigma}_s) \tag{6.23}$$

The right-hand side of (6.23) is comparable in form with the right-hand side of (6.10). Since $\mathbf{r}_s$ is the position vector of the particle P_s with respect to O and $\boldsymbol{\sigma}_s$ is its position vector with respect to G, we may infer that

$$\mathbf{H}_G = \mathbf{I}_G \begin{bmatrix} \omega_1 \\ \omega_2 \\ \omega_3 \end{bmatrix} \tag{6.24}$$

where $\mathbf{H}_G$ is to be interpreted as a column vector and $\mathbf{I}_G$ is the inertia matrix at the centre of mass G with respect to axes Ga, Gb, Gc parallel to the unit vectors **a**, **b**, **c**.

In two-dimensional motion $\mathbf{H}_G$ is given by an expression of the form of (6.20) or if the products of intertia are zero by one of the form of (6.21). The quantity I_{33} is the moment of inertia about an axis through G perpendicular to the plane of the motion.

We now have the mathematical tools to solve a wide variety of dynamical problems relating to the motion of rigid bodies. For the most part the approach to problems will follow a standard pattern. We use the relation (6.6) to analyse the motion of the centre of mass of the body, and results (b) and (d) in Section 6.6 coupled with those contained in (6.17) and (6.24) to analyse the angular motion.

Examples 6.1 to 6.5 relate to two-dimensional motion.

Applications

Example 6.1

A rotor of mass M rotates with constant angular velocity $\boldsymbol{\omega} = \omega\mathbf{k}$ about an axis simply supported by fixed bearings at the points D and E (see Fig. 6.2). Obtain expressions for the reactions of the bearings at D and E on the rotor. Neglect the effects of gravity and friction on the bearings.

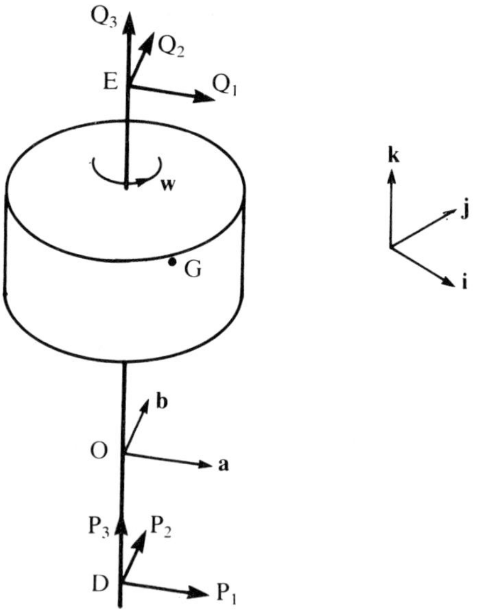

Fig. 6.2

▷ *Solution* Let O be any (fixed) point on the axis of rotation between D and E. The unit vectors **a**, **b** are fixed in the rotor so that they form with **k** a mutually perpendicular right-handed set of unit vectors, **k** being in the direction of $\underline{\text{DE}}$. The associated set of axes at O are labelled Oa, Ob, Oz. We take the reactions of the bearings at D and E to be **P**, **Q**, where

$$\mathbf{P} = P_1\mathbf{a} + P_2\mathbf{b} + P_3\mathbf{k}, \qquad \mathbf{Q} = Q_1\mathbf{a} + Q_2\mathbf{b} + Q_3\mathbf{k}$$

and let

$$\mathbf{r}_G = \underline{OG} = \alpha\mathbf{a} + \beta\mathbf{b} + \mu\mathbf{k}$$

where G is the centre of mass of the rotor.

From standard results we have

$$\dot{\mathbf{a}} = \boldsymbol{\omega} \times \mathbf{a} = \omega\mathbf{k} \times \mathbf{a} = \omega\mathbf{b}, \qquad \dot{\mathbf{b}} = \omega\mathbf{k} \times \mathbf{b} = -\omega\mathbf{a}$$

Hence we obtain, by differentiating the expression for $\underline{OG}$ twice with respect to time,

$$\mathbf{f}_G = -\omega^2(\alpha\mathbf{a} + \beta\mathbf{b})$$

Readers are reminded that α, β, μ are constants as G is fixed within the rotor.

Using (6.4), the equation governing the motion of the centre of mass of a body, yields

$$\mathbf{P} + \mathbf{Q} = -M\omega^2(\alpha\mathbf{a} + \beta\mathbf{b})$$

Equating corresponding components we obtain the equations

$$P_1 + Q_1 = -M\alpha\omega^2, \qquad P_2 + Q_2 = -M\beta\omega^2, \qquad P_3 + Q_3 = 0$$

It follows from (6.20) that

$$\mathbf{H}_O = \omega(-I_{13}\mathbf{a} - I_{23}\mathbf{b} + I_{33}\mathbf{k})$$

where

I_{13} is the product of inertia (of the rotor) with respect to Oa, Oz
I_{23} is the product of inertia with respect to Ob, Oz
I_{33} is the moment of inertia about Oz.

Now

$$\begin{aligned}\dot{\mathbf{H}}_O &= \omega(-I_{13}\dot{\mathbf{a}} - I_{23}\dot{\mathbf{b}}) \\ &= \omega^2(I_{23}\mathbf{a} - I_{13}\mathbf{b})\end{aligned}$$

and therefore, using result (b) of Section 6.6, we have

$$\begin{aligned}\omega^2(I_{23}\mathbf{a} - I_{13}\mathbf{b}) &= \underline{OD} \times \mathbf{P} + \underline{OE} \times \mathbf{Q} \\ &= -\mathrm{OD}\mathbf{k} \times \mathbf{P} + \mathrm{OE}\mathbf{k} \times \mathbf{Q} \\ &= -\mathrm{OD}(P_1\mathbf{b} - P_2\mathbf{a}) + \mathrm{OE}(Q_1\mathbf{b} - Q_2\mathbf{a})\end{aligned}$$

from which we deduce

$$\begin{aligned}\mathrm{OD}P_1 - \mathrm{OE}\,Q_1 &= \omega^2 I_{13} \\ \mathrm{OD}P_2 - \mathrm{OE}\,Q_2 &= \omega^2 I_{23}\end{aligned}$$

Routine manipulation of the equations for the components of **P** and **Q** gives the following:

$$P_1 = \omega^2(I_{13} - M\alpha \text{OE})/\text{DE}$$
$$Q_1 = -\omega^2(I_{13} + M\alpha \text{OD})/\text{DE}$$
$$P_2 = \omega^2(I_{23} - M\beta \text{OE})/\text{DE}$$
$$Q_2 = -\omega^2(I_{23} + M\beta \text{OD})/\text{DE}$$

The forces P_3 and Q_3 cannot be individually determined but $P_3 + Q_3$ is always zero.

Balancing of a rotor

In Example 6.1, if $\alpha = \beta = 0$, that is the centre of mass of the rotor is on the axis of rotation, and the products of inertia I_{13}, I_{23} are both zero, then the forces P_1, Q_1, P_2, Q_2, $P_3 + Q_3$ are all zero. The rotor is said to be balanced. If a rotor is not balanced (this is undesirable because of the consequent rotating forces on the axis of rotation) it can be balanced by attaching masses to it. The masses are chosen so that the centre of mass of the rotor and the added masses are on the axis of rotation and the products of inertia I_{13}, I_{23} for the axes at O (again of rotor plus masses) vanish. The process of adding masses so that the (new) centre of mass lies on the axis of rotation is called static balancing, while fixing masses so that the resulting products of inertia (at O) vanish is called dynamic balancing. It is desirable to balance the rotor overall but it is not uncommon for masses to be added so that the rotor is only statically balanced. This is satisfactory in practice if the rotor is thin relative to the distance between bearings and/or balancing is not critical, as for example in a low-speed rotor.

The terms 'static balancing' and 'dynamic balancing' derive from the fact that the position of the centre of mass of a rotor can be found statically while the quantities I_{13}, I_{23} are determined from dynamical experiments.

We illustrate the process of balancing by means of an example. As a preliminary we define the terms and notation that are used. A transverse plane is a plane whose normal is parallel to the axis of rotation, while a correction plane is a transverse plane in which mass is added or removed so as to help balance the rotor. The position of a transverse plane is specified by giving its perpendicular distance p from a fixed point on the axis of rotation. The position of any mass (added or otherwise) is defined by giving the radial distance (of it) from the axis of rotation and the angle between the radius and a fixed direction in the rotor parallel to a transverse plane. The out-of-balance properties of an actual rotor are frequently measured in terms of a set of masses, the unbalanced rotor being a combination of out-of-balance masses and a perfectly balanced rotor.

Example 6.2

The distribution of out-of-balance masses in a rotor is given as follows. The planes designated A and B are correction planes, X and Y are the planes containing the bearings, while the transverse planes 1, 2, 3 contain the out-of-balance masses. The distances labelled p are measured along the axis of the rotor from the point O, which is the point of intersection of the axis and plane A.

Plane	*Mass*/g	*Radius*/mm	*Angle*/deg	p/mm
A	m_1	50	θ_1	0
1	68.6	40	330	100
2	36.2	65	180	200
3	24.0	65	60	300
B	m_2	50	θ_2	400
X				50
Y				350

Determine the masses m_1, m_2 and the angles θ_1, θ_2 so that the rotor is completely balanced. If no balancing masses are added, determine also the magnitudes of the rotating out-of-balance forces on the bearings when the rotational speed is 2800 rev min^{-1}, assuming the bearings act as simple supports.

> *Solution* We take axes Oa, Ob, Oz at O so that the direction of Oa is parallel to the fixed direction in the rotor from which angles are measured. We require the centre of mass of the five masses to be on the axis of rotation and therefore, recalling the definition of the centre of mass,

$$50(m_1\cos\theta_1 + m_2\cos\theta_2) + 68.6 \times 40\cos330°$$
$$+ 36.2 \times 65\cos180° + 24 \times 65\cos60° = 0$$

Simplifying, we have

$$m_1\cos\theta_1 + m_2\cos\theta_2 = -16.067$$

Similarly

$$m_1\sin\theta_1 + m_2\sin\theta_2 = 0.420$$

The product of inertia of the five masses with respect to Oa, Oz is zero and hence, noting the definition in (6.16),

$$50 \times 400m_2\cos\theta_2 + 68.6 \times 40 \times 100\cos330°$$
$$+ 36.2 \times 65 \times 200\cos180° + 24 \times 65 \times 300\cos60° = 0$$

from which we deduce that $m_2\cos\theta_2 = -0.052$. Thus $m_1\cos\theta_1 = -16.015$. Likewise, since the product of inertia with respect to Ob, Oz is zero, $m_2\sin\theta_2 = -13.405$ and therefore $m_1\sin\theta_1 = 13.825$. Straightforward calculation yields $m_1 = 21.2\,\text{g}$, $m_2 = 13.4\,\text{g}$, $\theta_1 = 139.2°$ and $\theta_2 = 269.8°$.

The quantities $m_1\sin\theta_1$ and $m_1\cos\theta_1$ could have been found directly by taking O at the point of intersection of plane B with the axis of rotation and equating the relevant products of inertia to zero. Thus

$$50 \times 400 m_1\cos\theta_1 + 68.6 \times 40 \times 300\cos 330° + 36.2 \times 65 \times 200\cos 180° + 24 \times 65 \times 100\cos 60° = 0$$

so that $m_1\cos\theta_1 = -16.015$ as before.

To find the magnitudes of the rotating out-of-balance forces on the bearings we use the formulae in Example 6.1 for the forces P_1, Q_1, P_2, Q_2, except that OD must be replaced by $-$OD as the point O is not between the bearings. The quantities I_{13}, I_{23} are the products of inertia with respect to Oa, Oz and Ob, Oz for the three out-of-balance particles. From calculations already carried out we have

$$I_{13} = -50 \times 400 m_2\cos\theta_2\,\text{g mm}^2 = 1.04 \times 10^{-6}\,\text{kg m}^2$$
$$I_{23} = -50 \times 400 m_2\sin\theta_2\,\text{g mm}^2 = 2.681 \times 10^{-4}\,\text{kg m}^2$$
$$M\alpha = -50(m_2\cos\theta_2 + m_1\cos\theta_1)\,\text{g mm} = 8.0335 \times 10^{-4}\,\text{kg m}$$
$$M\beta = -50(m_2\sin\theta_2 + m_1\sin\theta_1)\,\text{g mm} = -2.1 \times 10^{-5}\,\text{kg m}$$
$$\text{OD} = 5 \times 10^{-2}\,\text{m}, \qquad \text{OE} = 3.50 \times 10^{-1}\,\text{m}, \qquad \text{DE} = 3 \times 10^{-1}\,\text{m}$$
$$\omega = 293.22\,\text{rad s}^{-1}$$

Hence

$$P_1 = -80.3\,\text{N}, \qquad Q_1 = 11.2\,\text{N}, \qquad P_2 = 78.9\,\text{N}, \qquad Q_2 = -77.1\,\text{N}$$

The reactions of the out-of-balance masses on the axis of rotation are obtained by reversing the signs in the expressions for P_1, Q_1, P_2, Q_2.

The forces **P** and **Q** acting at the bearings on the rotor may be obtained by an alternative method. It uses the conditions of (statical) equilibrium rather than particular formulae for forces, and for this reason it is often preferred. The force **F** which the (balanced) rotor exerts on an out-of-balance particle of mass m, radius r, angle θ is given in the notation previously defined by

$$\mathbf{F} = -mr\omega^2(\cos\theta\,\mathbf{a} + \sin\theta\,\mathbf{b})$$

and hence the force exerted by the mass on the rotor during motion is $mr\omega^2(\cos\theta\,\mathbf{a} + \sin\theta\,\mathbf{b})$. Thus the rotor moves under the action of the reactions of the bearings and forces of the type $mr\omega^2(\cos\theta\,\mathbf{a} + \sin\theta\,\mathbf{b})$ located at the positions of the out-of-balance masses. Since the rotor is perfectly balanced, the reactions and the forces due to the masses constitute a system of forces in equilibrium. Hence the reactions may be found by equating the vector sum of the forces (overall) and their moments about any point to zero. It is left as an exercise for the reader to formulate and solve the equations for the example under consideration.

Example 6.3

Figure 6.3 shows an RRRR mechanism using the conventions and notation defined in Section 4.4. The lengths of the links are OA = 0.04 m, AB = 0.16 m, BP = 0.08 m and OP = 0.19 m. Each link is made of uniform material of density 7.5 kg per metre length. A torque is applied to link 2 so that it rotates at a uniform angular speed of 25 rad s^{-1} in a counter-clockwise sense. At the instant $\theta_2 = 30°$ the values of θ_3 and θ_4 are 21.91° and 94.96°, while $\dot{\theta}_3 = -5.92$ rad s^{-1}, $\dot{\theta}_4 = 1.84$ rad s^{-1}, $\ddot{\theta}_3 = 78.1$ rad s^{-2} and $\ddot{\theta}_4 = 395.7$ rad s^{-2}. Assuming that the links may be regarded as uniform rods, determine the magnitude of the forces being exerted on link 3 through the revolute joints at A and B when $\theta_2 = 30°$.

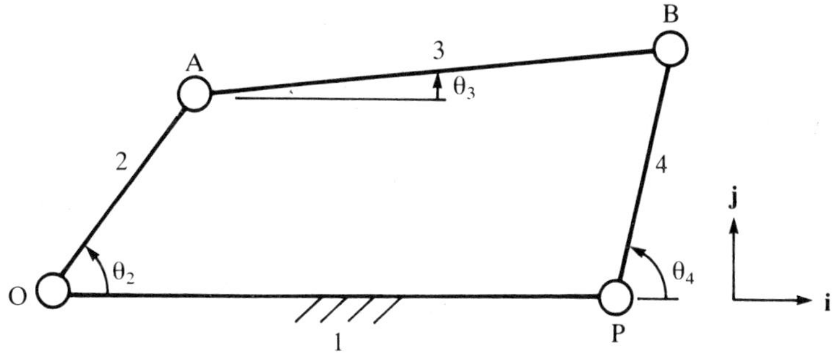

Fig. 6.3

Solution In order to show the form of the relevant equations we do not substitute the given numerical values at the onset. Let M_3, M_4 be the masses of links 3 and 4, and let G_3 be the midpoint of link 3, $\underline{OG_3} = x_3\mathbf{i} + y_3\mathbf{j}$. Let $\mathbf{F}_{ij}$ be the force exerted by link i on link j at a joint, where

$$\mathbf{F}_{12} = [X_1, Y_1, 0], \qquad \mathbf{F}_{23} = [X_2, Y_2, 0]$$
$$\mathbf{F}_{34} = [X_3, Y_3, 0], \qquad \mathbf{F}_{41} = [X_4, Y_4, 0]$$

referred to **i**, **j**, **k**.

The forces acting on the links are shown in the free-body diagram in Fig. 6.4. The couple applied to link 2 is $M\mathbf{k}$. The equation of motion of the centre of mass G_3 of link 3 is given by

$$M_3\underline{\ddot{OG}_3} = \mathbf{F}_{23} + \mathbf{F}_{43}$$

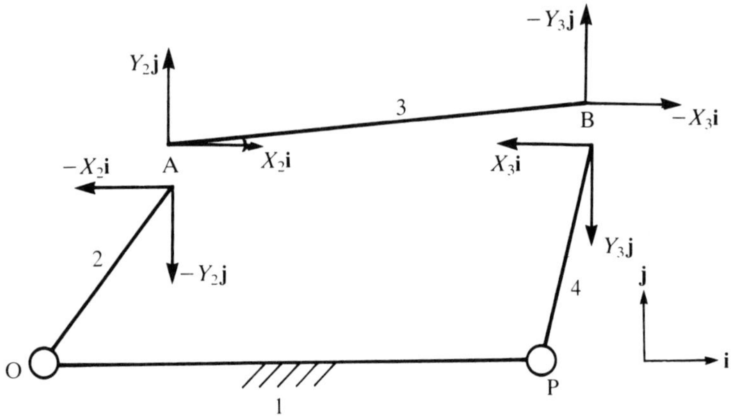

Fig. 6.4

By Newton's third law $\mathbf{F}_{43} = -\mathbf{F}_{34}$, so that

$$M_3\ddot{x}_3 = X_2 - X_3, \qquad M_3\ddot{y}_3 = Y_2 - Y_3$$

Now

$$\underline{G_3A} \times \mathbf{F}_{23} + \underline{G_3B} \times \mathbf{F}_{43} = \frac{r_3}{2}\{(X_2 + X_3)\sin\theta_3 - (Y_2 + Y_3)\cos\theta_3\}\mathbf{k}$$

while the moment of inertia of link 3 about an axis through G_3 perpendicular to its length is $(M_3/12)r_3^2$. Hence, using result (d) of Section 6.6,

$$\frac{M_3}{12}r_3^2\ddot{\theta}_3 = \frac{r_3}{2}\{(X_2 + X_3)\sin\theta_3 - (Y_2 + Y_3)\cos\theta_3\}$$

which may be simplified to give

$$M_3 r_3\ddot{\theta}_3 = 6\{(X_2 + X_3)\sin\theta_3 - (Y_2 + Y_3)\cos\theta_3\}$$

Now $\underline{PB} \times \mathbf{F}_{34} = r_4(Y_3\cos\theta_4 - X_3\sin\theta_4)\mathbf{k}$, while the moment of inertia of link 4 about an axis through P (a fixed point) perpendicular to its length is $M_4 r_4^2/3$. Hence, using result (b) of Section 6.6 with P replacing O,

$$M_4 r_4\ddot{\theta}_4 = 3(Y_3\cos\theta_4 - X_3\sin\theta_4)$$

The four equations derived are sufficient to determine the four quantities X_2, Y_2, X_3, Y_3 and thus the forces exerted on link 3 at the joints A and B can be found. We have

$$\ddot{x}_3 = -r_2\cos\theta_2\,\dot{\theta}_2^2 - \frac{r_3}{2}(\cos\theta_3\,\dot{\theta}_3^2 + \sin\theta_3\,\ddot{\theta}_3)$$
$$= -26.58\,\mathrm{m\,s^{-2}}$$

$$\ddot{y}_3 = -r_2\sin\theta_2\,\dot{\theta}_2^2 - \frac{r_3}{2}(\sin\theta_3\,\dot{\theta}_3^2 - \cos\theta_3\,\ddot{\theta}_3)$$
$$= -7.75\,\mathrm{m\,s^{-2}}$$

using the given numerical values.

The four equations for the forces can be expressed in the form

$$X_2 - X_3 = -31.90, \qquad Y_2 - Y_3 = -9.30$$
$$Y_2 + Y_3 - 0.402(X_2 + X_3) = -2.69, \qquad Y_3 + 11.52X_3 = -73.23$$

so that

$$X_3 = -5.9\,\mathrm{N},\ X_2 = -37.8\,\mathrm{N},\ Y_2 = -14.8\,\mathrm{N},\ Y_3 = -5.5\,\mathrm{N}$$

In practice it is unlikely that the links would be modelled accurately by rods. However the equations of motion would still be formulated in the same way so that, provided the position of the centre of mass G_3 of link 3 and the respective moments of inertia I_2, I_3, I_4 of links 2, 3, 4 about axes parallel to **k** through the points O, G_3, P are known, there would be no difficulty in finding the forces at A and B.

It is evident from the equations we have formulated in Example 6.3 that the forces at all the joints depend linearly on the masses of the links and the quantities I_2, I_3, I_4. Thus if we wish to determine the contribution to the forces at the joints arising from the motion of a particular link, for example link 3, then it is sufficient to solve the relevant equations with M_2, M_4, I_2, I_4 put to zero. We say that we have determined the forces at the joints due to the inertia of link 3.

Example 6.4

A car is initially stationary on level ground. It begins to move directly forward with uniform acceleration of magnitude f with one of its doors open at right angles to the direction of motion. The door, which may be regarded as a uniform plate of mass M, width $2l$ and height $2h$, is hinged to the car at two points A and B. The points A and B are a distance $2d$ apart and symmetrically placed about a horizontal line in the plane of the door that passes through G, its centre of mass. Show, when the door has turned through an angle θ, that

$$4l\ddot{\theta} = 3f\cos\theta$$

The lock of the door will engage if the angular speed of the door is not less than $5\,\mathrm{rad\,s^{-1}}$. Find the minimum value of the acceleration f for the door to close if $l = 0.3\,\mathrm{m}$.

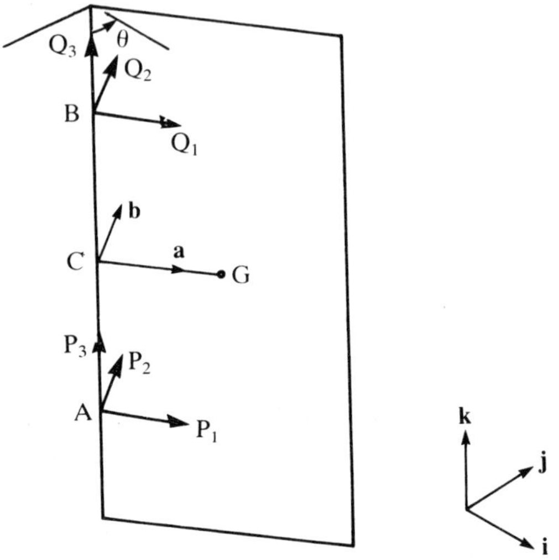

Fig. 6.5

▷ *Solution* Let C be the midpoint of AB. We take the unit vector **a** in the direction of $\overline{CG}$, **k** in the direction of $\overline{AB}$ and **b** so that $\mathbf{k} \times \mathbf{a} = \mathbf{b}$ (see Fig. 6.5). The acceleration of the car is $f\mathbf{i}$, where $\mathbf{i} = -\sin\theta\mathbf{a} - \cos\theta\mathbf{b}$. By standard results,

$$\begin{aligned}\mathbf{f}_G &= \mathbf{f}_C - l\dot{\theta}^2\mathbf{a} + l\ddot{\theta}\mathbf{b} \\ &= -(l\dot{\theta}^2 + f\sin\theta)\mathbf{a} + (l\ddot{\theta} - f\cos\theta)\mathbf{b}\end{aligned}$$

Let the reaction of the hinge on the door at A be **P** and the reaction at B be **Q**, where

$$\mathbf{P} = P_1\mathbf{a} + P_2\mathbf{b} + P_3\mathbf{k}, \qquad \mathbf{Q} = Q_1\mathbf{a} + Q_2\mathbf{b} + Q_3\mathbf{k}$$

The equation governing the motion of the centre of mass G of the door is

$$\mathbf{P} + \mathbf{Q} - Mg\mathbf{k} = M\mathbf{f}_G$$

so that

$$\begin{aligned}P_1 + Q_1 &= -M(l\dot{\theta}^2 + f\sin\theta) \\ P_2 + Q_2 &= M(l\ddot{\theta} - f\cos\theta) \\ P_3 + Q_3 - Mg &= 0\end{aligned}$$

The products of inertia of the door with respect to the axes Ga, Gz and Gb, Gz, where Ga, Gb, Gz are in the directions of **a**, **b**, **k**, are by symmetry (see Section 6.8) zero. Hence

$$\mathbf{H}_G = \frac{M}{3}l^2\dot{\theta}\mathbf{k}$$

Taking moments about G yields

$$\frac{M}{3}l^2\ddot{\theta}\mathbf{k} = \underline{\mathrm{GA}} \times \mathbf{P} + \underline{\mathrm{GB}} \times \mathbf{Q}$$

$$= \begin{vmatrix} \mathbf{a} & \mathbf{b} & \mathbf{k} \\ -l & 0 & -d \\ P_1 & P_2 & P_3 \end{vmatrix} + \begin{vmatrix} \mathbf{a} & \mathbf{b} & \mathbf{k} \\ -l & 0 & d \\ Q_1 & Q_2 & Q_3 \end{vmatrix}$$

$$= [(P_2 - Q_2)d,\ d(Q_1 - P_1) + l(P_3 + Q_3),\ -l(P_2 + Q_2)]$$

referred to **a**, **b**, **k**. Thus

$$P_2 + Q_2 = -\frac{M}{3}l\ddot{\theta}$$

Equating the two expressions for $P_2 + Q_2$ leads to the equation

$$4l\ddot{\theta} = 3f\cos\theta$$

To integrate this equation we multiply both sides by $\dot{\theta}$ giving $4l\dot{\theta}\ddot{\theta} = 3f\cos\theta\,\dot{\theta}$, from which we deduce that

$$\frac{\mathrm{d}}{\mathrm{d}t}(2l\dot{\theta}^2) = \frac{\mathrm{d}}{\mathrm{d}t}(3f\sin\theta)$$

Hence $2l\dot{\theta}^2 = 3f\sin\theta + \text{constant}$. When $\theta = 0$, $\dot{\theta} = 0$, so that the constant is zero. Putting $\theta = \pi/2$ gives $\dot{\theta}^2 = 3f/2l$, and therefore in order for the door to shut we require $3f/2l \geq 25$. Taking $l = 0.3\,\mathrm{m}$, the minimum value of f is found to be $5\,\mathrm{m\,s^{-2}}$.

The crank slider mechanism

Before proceeding to the next example we describe and analyse the crank-slider or RRRP four-link mechanism. It is illustrated diagrammatically in Fig. 6.6. The crank OA is assumed to rotate at a constant angular speed ω_2 and is driven by the action of forces exerted on the piston (link 4) which

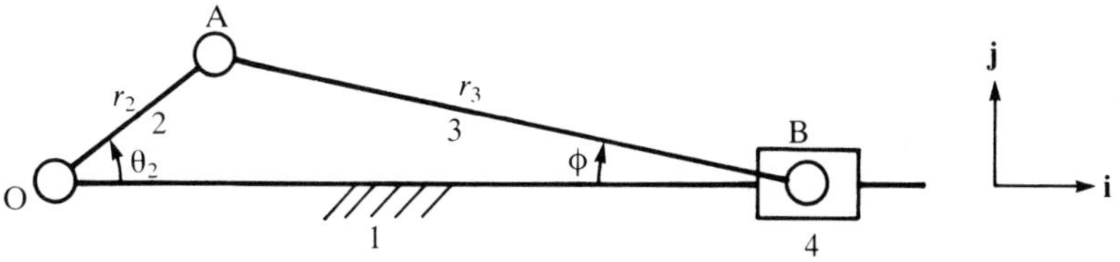

Fig. 6.6

moves along the fixed link. The lengths of links 2 and 3 are denoted by r_2, r_3 and the ratio r_3/r_2 by n, while $|\underline{OB}| = r_4$.

We begin by deriving expressions for the velocity and acceleration of link 4. The derivation is carried out by means of elementary trigonometry in preference to the methods developed in Chapter 4 for the kinematic analysis of mechanisms. Using the notation of Fig. 6.6 we have $r_3\sin\phi = r_2\sin\theta_2$, so that $\cos\phi = \surd(1 - e^2\sin^2\theta_2)$ where $e = 1/n$. Now

$$r_4 = r_2\cos\theta_2 + r_3\cos\phi = r_2\cos\theta_2 + r_3\surd(1 - e^2\sin^2\theta_2)$$

and therefore differentiating with respect to time yields

$$\dot{r}_4 = -\omega_2 r_2\sin\theta_2 - \frac{\omega_2}{2} r_3 e^2\sin 2\theta_2/\surd(1 - e^2\sin^2\theta_2)$$

Hence

$$\mathbf{v}_B = \dot{r}_4\mathbf{i} \approx -\omega_2 r_2(\sin\theta_2 + \frac{e}{2}\sin 2\theta_2)\mathbf{i}$$

the quantity e being assumed to be sufficiently small so that terms containing e^3 and higher powers of e may be neglected. In practice the value of n lies between 3 and 4 so that this is a satisfactory approximation.

Differentiating the expression for r_4 a second time we obtain, after some manipulation, an approximate expression for $\mathbf{f}_B$, namely

$$\mathbf{f}_B \approx -\omega_2^2 r_2\{\cos\theta_2 + (1/n)\cos 2\theta_2\}\mathbf{i}$$

A single cylinder engine may be regarded as being composed of three parts: the housing itself, which contains the piston (acted upon by the pressure of the gases, the force in the connecting rod AB and the reaction of the side wall of the cylinder), and the crank-shaft running in bearings in the housing. It is not difficult to see by consideration of free-body diagrams for the three parts that the frame or foundation to which the housing is attached must bear a load, due to the motion of the piston, equal to the action of a force $-M_4\mathbf{f}_B$ acting at B, where M_4 is the mass of the piston. The force has two constituents: $M_4\omega_2^2 r_2\cos\theta_2$, called the primary force; and $(M_4\omega_2^2 r_2\cos 2\theta)/n$, called the secondary force.

For the present we do not consider the inertia of the motion of the connecting rod. This is done in the discussion of equimomental systems in Section 6.8, where the term 'reciprocating mass' is introduced.

It is convenient at this stage, now the significance of each quantity is appreciated, to drop the suffixes and write θ, ω, r, m for the quantities θ_2, ω_2, r_2, M_4. If $\theta = \alpha$ when $t = 0$ then at time t we have $\theta = \omega t + \alpha$, and therefore the force $-M_4\mathbf{f}_B$ is given by

$$\begin{aligned}-M_4\mathbf{f}_B = m\omega^2 r\{&(\cos\alpha\cos\omega t - \sin\alpha\sin\omega t)\\ &+ (\cos 2\alpha\cos 2\omega t - \sin 2\alpha\sin 2\omega t)/n\}\mathbf{i}\end{aligned}$$

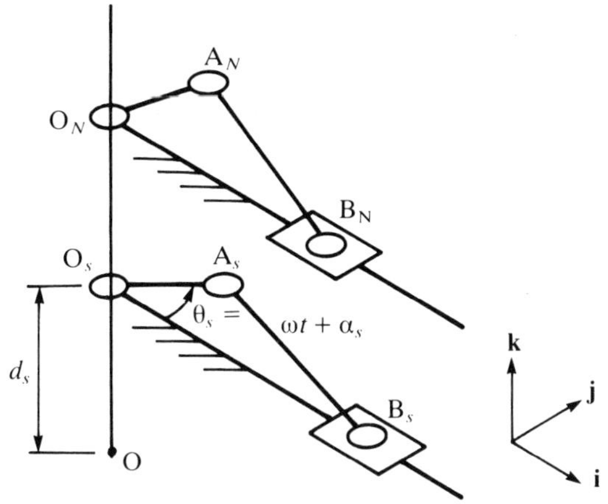

Fig. 6.7

In an engine that is composed of a number of cylinders we seek to minimize the effect of forces of the above type. The process is also called balancing but should not be confused with that described in Example 6.2, where masses were added to a rotor in order to counteract out-of-balance forces on the axis of rotation. The reader will see, however, that the equations that govern the two processes have much in common.

In the analysis that now follows the engine is composed of N cylinders in line. The pistons move in parallel lines and are connected by rods to a common crank-shaft at the points A_s ($s = 1, 2, \ldots, N$) (see Fig. 6.7). The mass of a typical piston is m_s, the length of a typical radius O_sA_s is r_s, while the angle between O_sA_s and the direction of outward motion of a piston is α_s at $t = 0$. The perpendicular distance of the plane of motion of the sth piston measured from the fixed point O on the axis of the crank-shaft is taken to be d_s. The ratio n of the length of the connecting rod to the crank radius is the same for each piston.

A typical primary force is given by

$$m_s\omega^2 r_s(\cos\alpha_s \cos\omega t - \sin\alpha_s \sin\omega t)\mathbf{i}$$

acting at the point B_s, where

$$\underline{OB_s} = \underline{OO_s} + \underline{O_sB_s} = d_s\mathbf{k} + O_sB_s\mathbf{i}$$

it being assumed that the vectors $\underline{OO_s}$ are all in the direction of the unit vector $\mathbf{k}$. Where this is not the case, d_s will have to be replaced by $-d_s$.

It is known from the results contained in Chapter 3 that a number of forces are equivalent to a single force acting at a point together with a couple. The single force is equal to the vector sum of the forces (treated as free vectors). This single force is therefore given for the primary forces by

$$(\omega^2\cos\omega t\sum m_s r_s \cos\alpha_s - \omega^2\sin\omega t\sum m_s r_s \sin\alpha_s)\mathbf{i}$$

and will clearly be zero if m_s, r_s, α_s ($s = 1, 2, \ldots, N$) are such that

$$\sum m_s r_s \cos\alpha_s = 0, \qquad \sum m_s r_s \sin\alpha_s = 0 \tag{6.25}$$

These relations imply that the centre of mass of hypothetical particles of masses m_s situated at the points A_s must lie on the axis of the crank-shaft. Similar conditions were satisfied in the balancing process of Example 6.2.

When the conditions in (6.25) are fulfilled the primary forces reduce to a couple, and from the theory of Chapter 3 we know that the moment of this couple is equal to the sum of the moments of the primary forces about any point. Clearly the most convenient point about which to take moments is the point O. The moment of the couple is obtained by taking the vector product of $\underline{OB_s}$ with the sth primary force and summing. The outcome is the vector

$$\omega^2\left(\cos\omega t\sum_s m_s r_s d_s \cos\alpha_s - \sin\omega t\sum_s m_s r_s d_s \sin\alpha_s\right)\mathbf{j}$$

which will vanish if

$$\sum_s m_s r_s d_s \cos\alpha_s = 0, \qquad \sum_s m_s r_s d_s \sin\alpha_s = 0 \tag{6.26}$$

Clearly the conditions in (6.26) can be interpreted in terms of products of inertia.

In order to determine the single force and couple to which the secondary forces reduce, it is sufficient to replace ωt by $2\omega t$ and α_s by $2\alpha_s$ in the relevant formulae for the primary forces and multiply the outcome by $1/n$. The single force and couple will each be zero if

$$\begin{aligned} &\sum_s m_s r_s \cos 2\alpha_s = 0, \qquad \sum_s m_s r_s \sin 2\alpha_s = 0 \\ &\sum_s m_s r_s d_s \cos 2\alpha_s = 0, \qquad \sum_s m_s r_s d_s \sin 2\alpha_s = 0 \end{aligned} \tag{6.27}$$

When the conditions (6.25)–(6.27) are fulfilled the primary and secondary forces are said to be balanced. In practice very few of the common (engine) configurations satisfy all the conditions, as the following example shows.

Example 6.5

An engine consists of five cylinders in line. The planes of motion of the pistons, each of mass m, are spaced at distances d apart, while the radius of each crank is r. Given that the cranks are spaced successively along the axis of the crank-shaft at angular intervals of 72°, show that the primary and secondary forces are each equivalent to a couple and obtain expressions for the magnitudes of these couples.

> *Solution* Using the notation leading to (6.25) we have $m_s = m$, $r_s = r$, $\alpha_s = (s-1)72°$, $d_s = (s-1)d$ $(s = 1, 2, \ldots, 5)$. Hence

$$\sum m_s r_s \cos\alpha_s = mr(\cos 0° + \cos 72° + \cos 144° + \cos 216° + \cos 288°)$$

which may be shown to be equal to zero by the use of elementary trigonometrical identities (or by a pocket calculator). Likewise

$$\sum m_s r_s \sin\alpha_s = m_r(\sin 0° + \sin 72° + \sin 144° + \sin 216° + \sin 288°) = 0$$

Thus the primary forces reduce to a couple. The moment of the primary couple is given by

$$\begin{aligned}
&mrd\omega^2\{\cos\omega t(\cos 72° + 2\cos 144° + 3\cos 216° + 4\cos 288°)\\
&- \sin\omega t(\sin 72° + 2\sin 144° + 3\sin 216° + 4\sin 288°)\}\mathbf{j}\\
&= mrd\omega^2(3.44095\sin\omega t - 2.5\cos\omega t)\mathbf{j}\\
&= 4.253mrd\omega^2\sin(\omega t - \pi/5)\mathbf{j}\\
&\sum m_s r_s \cos 2\alpha_s = mr(\cos 0° + \cos 144° + \cos 288° + \cos 432° + \cos 576°) = 0\\
&\sum m_s r_s \sin 2\alpha_s = mr(\sin 0° + \sin 144° + \sin 288° + \sin 432° + \sin 576°) = 0
\end{aligned}$$

Thus the secondary forces reduce to a couple. The moment of this couple is found after some elementary manipulation to be

$$\begin{aligned}
&-\{mrd\omega^2(2.5\cos 2\omega t + 0.8123\sin 2\omega t)/n\}\mathbf{j}\\
&= -\{2.629mrd\omega^2/n\sin(2\omega t + 2\pi/5)\}\mathbf{j}
\end{aligned}$$

Hence the magnitudes of the primary and secondary couples have been found.

Before considering examples involving three-dimensional motion we present a short theory of moments and products of inertia. One aspect of this theory has already been anticipated in Example 6.4, namely the use of symmetry to show that a product of inertia with respect to a particular pair of axes is zero. Further results are required in order to study three-dimensional motion.

6.8 THEORY OF MOMENTS OF INERTIA

Principal axes

If the products of inertia I_{12}, I_{13}, I_{23} (see (6.18)) are all zero then we say that the axes Oa, Ob, Oc are principal axes of inertia at O, while the moments of inertia I_{11}, I_{22}, I_{33} are called the principal moments of inertia at O. We now state without proof the following theorem:

At any point O of a mass system it is always possible to find one set of axes Oa, Ob, Oc (say) such that the (three) products of inertia with respect to these axes are all zero.

Where a body has an axis of symmetry or plane of symmetry, principal axes are readily identified. Figure 6.8 shows a plate whose thickness is negligible and whose mass is uniformly distributed. The axis Ob is an axis of symmetry for the (surface) area of the plate. We take Oa to lie in the plane of the plate so that Oc is therefore perpendicular to it. Recalling the notation of (6.11) it follows that μ_s is zero for each elemental particle of the plate and hence (see relations in (6.16)) we have $I_{13} = I_{23} = 0$. Since Ob is an axis of symmetry and the mass of the plate is uniformly distributed it is always possible to pair an elemental particle of mass m_q, coordinates α_q, β_q with respect to Oa, Ob, where $\alpha_q > 0$, with a particle of the same mass but whose coordinates are $-\alpha_q$, β_q. Hence when we sum the quantities $\pm m_s \alpha_s \beta_s$ for all particles of the plate, the outcome is zero. We can, by an extension of this result, show that a set of axes through the centre of mass G of a uniform rectangular block and parallel to the faces of the block is the set of principal axes at G.

We now consider further theorems relating to moments and products of inertia.

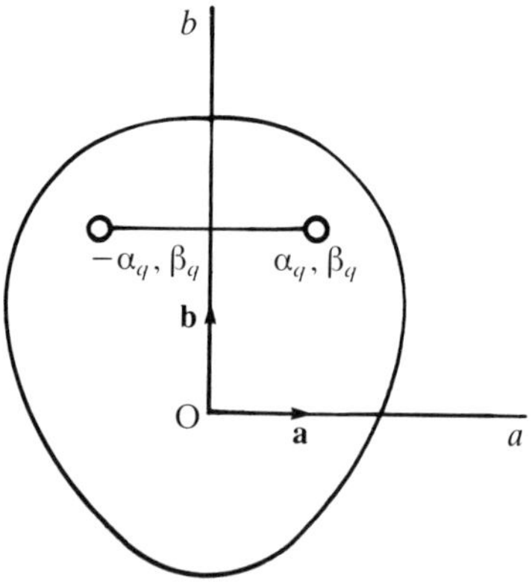

Fig. 6.8

Theorem of perpendicular axes

This theorem states that:

For a body of negligible thickness,

$$I_{33} = I_{11} + I_{22} \tag{6.28}$$

where I_{11}, I_{22} are the (respective) moments of inertia about a pair of perpendicular axes Oa, Ob lying in the plane of the body, and I_{33} is the moment of inertia about an axis through O perpendicular to the plane of the body.

Putting $\mu_s = 0$ in the relations of (6.15) the theorem follows at once.

The theorem is frequently misapplied. It must not be used when the thickness of the body is significant (see Appendix B).

Theorem of parallel axes

This theorem states that:

$$\mathbf{I}_{\mathrm{O}} = \mathbf{I}_{\mathrm{G}} + M\begin{bmatrix} \beta^2+\mu^2 & -\alpha\beta & -\alpha\mu \\ -\alpha\beta & \alpha^2+\beta^2 & -\beta\alpha \\ -\alpha\mu & -\beta\mu & \alpha^2+\beta^2 \end{bmatrix} \tag{6.29}$$

where $\mathbf{I}_{\mathrm{O}}$, $\mathbf{I}_{\mathrm{G}}$ are the inertia matrices at O, G with respect to axes parallel to unit vectors $\mathbf{a}$, $\mathbf{b}$, $\mathbf{c}$, and α, β, μ are defined by $\underline{\mathrm{OG}} = \alpha\mathbf{a} + \beta\mathbf{b} + \mu\mathbf{c}$.

□ *Proof* In order to distinguish between moments and products of inertia referred to the set of axes through O and those referred to axes through G, we will (temporarily) denote the former by $I_{ij}(\mathrm{O})$ $(i, j = 1, 2, 3)$ and the latter by $I_{ij}(\mathrm{G})$.

Now by (6.11)

$$\underline{\mathrm{OP}_s} = \mathbf{r}_s = \alpha_s\mathbf{a} + \beta_s\mathbf{b} + \mu_s\mathbf{c}$$

so it follows from (6.1) that

$$M\underline{\mathrm{OG}} = M(\alpha\mathbf{a} + \beta\mathbf{b} + \mu\mathbf{c}) = \sum m_s(\alpha_s\mathbf{a} + \beta_s\mathbf{b} + \mu_s\mathbf{c})$$

where

$$M\alpha = \sum m_s\alpha_s, \qquad M\beta = \sum m_s\beta_s, \qquad M\mu = \sum m_s\mu_s \tag{6.30}$$

Since

$$\underline{\mathrm{GP}_s} = \underline{\mathrm{OP}_s} - \underline{\mathrm{OG}} = (\alpha_s - \alpha)\mathbf{a} + (\beta_s - \beta)\mathbf{b} + (\mu_s - \mu)\mathbf{c}$$

We have

$$\begin{aligned} I_{11}(\mathrm{G}) &= \sum m_s\{(\beta_s - \beta)^2 + (\mu_s - \mu)^2\} \\ &= \sum m_s(\beta_s^2 + \mu_s^2) + \left(\sum m_s\right)(\beta^2 + \mu^2) - 2\left(\beta\sum m_s\beta_s\right) \\ &\quad - 2\left(\mu\sum m_s\mu_s\right) \end{aligned}$$

$$= I_{11}(\mathrm{O}) - M(\beta^2 + \mu^2)$$

noting (6.30). Also

$$I_{12}(\mathrm{G}) = \sum m_s(\alpha_2 - \alpha)(\beta_s - \beta)$$

$$= \sum m_s\alpha_s\beta_s + \left(\sum m_s\right)\alpha\beta - \alpha\sum m_s\beta_s - \beta\sum m_s\alpha_s$$

$$= I_{12}(\mathrm{O}) - M\alpha\beta$$

The form of the relation between $I_{11}(\mathrm{O})$ and $I_{11}(\mathrm{G})$ and the form of that between $I_{12}(\mathrm{O})$ and $I_{12}(\mathrm{G})$ enable us to assert that the result in (6.29) is true.

If the relation $\underline{\mathrm{OG}} = \alpha\mathbf{a} + \beta\mathbf{b} + \mu\mathbf{c}$ is replaced by $\underline{\mathrm{GO}} = \alpha\mathbf{a} + \beta\mathbf{b} + \mu\mathbf{c}$, result (6.29) is left unaltered. We can therefore think of α, β, μ as the components of the position vector of O with respect to G or vice versa.

Relation between two inertia matrices at O

If $\mathbf{a}'$, $\mathbf{b}'$, $\mathbf{c}'$ is a right-handed set of mutually perpendicular unit vectors related to the set $\mathbf{a}$, $\mathbf{b}$, $\mathbf{c}$ by

$$\begin{bmatrix} \mathbf{a}' \\ \mathbf{b}' \\ \mathbf{c}' \end{bmatrix} = \mathbf{L}\begin{bmatrix} \mathbf{a} \\ \mathbf{b} \\ \mathbf{c} \end{bmatrix}$$

where $\mathbf{L} = [l_{ij}]$, $i, j = 1, 2, 3$, and $\mathbf{I}'_{\mathrm{O}}$ is the inertia matrix with respect to the axes through O in the directions of the vectors $\mathbf{a}'$, $\mathbf{b}'$, $\mathbf{c}'$, then

$$\mathbf{I}'_{\mathrm{O}} = \mathbf{L}\mathbf{I}_{\mathrm{O}}\mathbf{L}^{\mathrm{T}} \tag{6.31}$$

We give a proof of (6.31).

□ *Proof* We can express $\mathbf{I}_{\mathrm{O}}$ in the form

$$\mathbf{I}_{\mathrm{O}} = \sum m_s\begin{bmatrix} \beta_s^2 + \mu_s^2 & -\alpha_s\beta_s & -\alpha_s\mu_s \\ -\alpha_s\beta_s & \alpha_s^2 + \mu_s^2 & -\beta_s\mu_s \\ -\alpha_s\mu_s & -\beta_s\mu_s & \alpha_s^2 + \beta_s^2 \end{bmatrix}$$

$$= \sum m_s(\alpha_s^2 + \beta_s^2 + \mu_s^2)\mathbf{I} - \sum m_s\begin{bmatrix} \alpha_s^2 & \alpha_s\beta_s & \alpha_s\mu_s \\ \alpha_s\beta_s & \beta_s^2 & \beta_s\mu_s \\ \alpha_s\mu_s & \beta_s\mu_s & \mu_s^2 \end{bmatrix}$$

$$= \sum m_s(\alpha_s^2 + \beta_s^2 + \mu_s^2)\mathbf{I} - \sum m_s\begin{bmatrix} \alpha_s \\ \beta_s \\ \mu_s \end{bmatrix}[\alpha_s \quad \beta_s \quad \mu_s] \tag{6.32}$$

Putting $\underline{OP_s} = \alpha_s'\mathbf{a}' + \beta_s'\mathbf{b}' + \mu_s'\mathbf{c}'$, it follows by virtue of (1.33) that

$$\begin{bmatrix} \alpha_s' \\ \beta_s' \\ \mu_s' \end{bmatrix} = \mathbf{L} \begin{bmatrix} \alpha_s \\ \beta_s \\ \mu_s \end{bmatrix}$$

and therefore

$$[\alpha_s', \beta_s', \mu_s'] = [\alpha_s, \beta_s, \mu_s]\mathbf{L}^{\mathrm{T}}$$

Pre-multiplying both sides of (6.32) by $\mathbf{L}$ and then post-multiplying both sides of the result by $\mathbf{L}^{\mathrm{T}}$ yields, since $\mathbf{LL}^{\mathrm{T}} = \mathbf{I}$ and $\alpha_s'^2 + \beta_s'^2 + \mu_s'^2 = \alpha_s^2 + \beta_s^2 + \mu_s^2$,

$$\mathbf{LI}_{\mathrm{O}}\mathbf{L}^{\mathrm{T}} = \sum m_s(\alpha_s'^2 + \beta_s'^2 + \mu_s'^2)\mathbf{I} - \sum m_s \begin{bmatrix} \alpha_s' \\ \beta_s' \\ \mu_s' \end{bmatrix} [\alpha_s', \beta_s', \mu_s'] \tag{6.33}$$

It follows, noting the form of (6.32), that the right-hand side of (6.33) is equal to $\mathbf{I}_{\mathrm{O}}'$, so that (6.31) is proved. The result is called the transformation law for the inertia matrix.

Equimomental mass systems

As a preliminary to defining the term 'equimomental' we consider two mass systems. Figure 6.9(a) shows a uniform (rigid) rod AB of length l and mass M, while Fig. 6.9(b) shows a light rigid rod, also of length l, with a mass $2M/3$ attached to its midpoint and a mass $M/6$ at each of its ends. These two systems have the same mass, while the centre of mass of each is situated at the midpoint of each rod. In Fig. 6.9(a) and (b) the unit vector **a** is along the rod while the vectors **b** and **c** are perpendicular to it.

It is readily seen from the definitions in (6.16) that the products of inertia for either mass system with respect to axes through its centre of

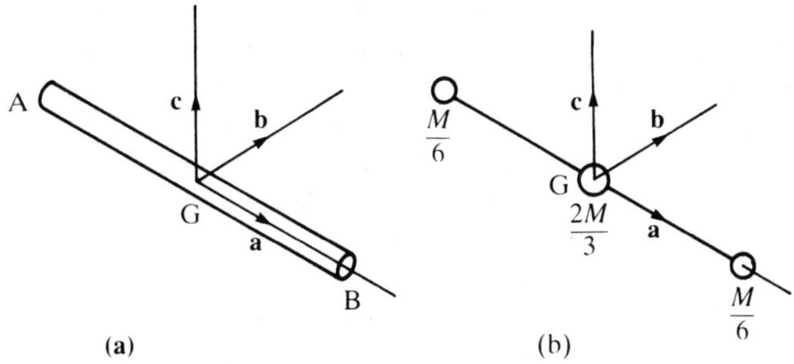

Fig. 6.9

mass in the directions of the unit vectors **a**, **b**, **c** are zero. The moment of inertia of the rod mass M about an axis perpendicular to its length is known to be equal to $Ml^2/12$, and this is seen by calculation to be equal to the moment of inertia (about the comparable axis) for the three particles. Thus the two systems have the same inertia matrix at their centres of mass with respect to axes in the direction of the unit vectors **a**, **b**, **c**. It is given by

$$\frac{Ml^2}{12}\begin{bmatrix} 0 & 0 & 0 \\ 0 & 1 & 0 \\ 0 & 0 & 1 \end{bmatrix}$$

We deduce from the transformation law (6.31) that the inertia matrix at the centre of mass of one system for a given set of axes is equal to the inertia matrix at the centre of mass of the other system for axes whose directions are the same as those of the given set. We can now infer from the theorem of parallel axes that this result holds for all (corresponding) points of the two mass systems and not simply for the centres of mass. The two mass systems are said to be equimomental, the term implying equality of corresponding moments and products of inertia.

The two mass systems may be regarded as being dynamically equivalent. If they were acted upon by equivalent sets of external forces then the motions of the two rods would be identical. Conversely if the motions of the rods were identical then the systems of external forces acting would be equivalent. This follows immediately on recalling (6.16), the vector equation governing the motion of the centre of mass, and result (d) of Section 6.6.

The forces acting on the housing of a crank-slider mechanism due to the motion of the connecting rod are frequently calculated by substituting for the connecting rod a light rod and two masses, one of these being placed on the piston and the second at the other end of the rod. The total mass of the particles is chosen to be equal to the mass of the connecting rod and distributed so that their centre of mass is coincident with that of the connecting rod. Since corresponding moments of inertia are not necessarily equal, it follows that forces determined when the particles are substituted for the connecting rod will only approximate to the actual values (see Problem 6.16). The total mass of the piston and the particle placed at the piston is referred to as a reciprocating mass. We will, in subsequent work, substitute this mass for that of the piston when we determine (approximately) the single force and couple that are equivalent to the primary forces and the single force and couple equivalent to the secondary forces. The reader should note that several mass systems can be equimomental with a given mass system. It is known, for example, that a uniform triangular plate of mass M is equimomental with three particles, each of mass $M/3$, placed at the midpoints of its sides. By subdividing the triangular plate into a number of smaller triangular plates, further equimomental systems may be obtained (see Problem 6.20 for a further example).

It is a straightforward matter, using the theorem of parallel axes, to show that two equimomental systems must have the same total mass.

Radius of gyration

If the total mass of a system is M and I is its moment of inertia about a given axis, then k, its radius of gyration about this axis, is defined by $Mk^2 = I$.

Example 6.6

The axes Oa, Ob, Oc are principal axes of inertia at the point O of a mass system, the principal moments of inertia being A, A, C respectively. Show that any other set of axes Oa', Ob', Oc', where Oc and Oc' coincide (so that Oa', Ob' lie in the plane of Oa, Ob), is also a set of principal axes at O.

Solution The inertia matrix $\mathbf{I}_O$ at O for the axes Oa, Ob, Oc is given by

$$\mathbf{I}_O = \begin{bmatrix} A & 0 & 0 \\ 0 & A & 0 \\ 0 & 0 & C \end{bmatrix}$$

If $\mathbf{a}$, $\mathbf{b}$, $\mathbf{c}$ and $\mathbf{a}'$, $\mathbf{b}'$, $\mathbf{c}'$ are the two sets of unit vectors associated with the axes Oa, Ob, Oc and Oa', Ob', Oc', and the angle between Oa' and Oa is θ (see Fig. 6.10), then

$$\mathbf{a}' = \cos\theta\,\mathbf{a} + \sin\theta\,\mathbf{b}$$

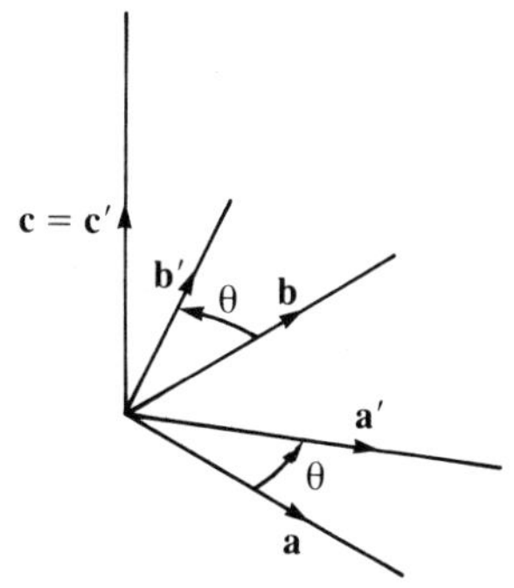

Fig. 6.10

$$\mathbf{b}' = -\sin\theta\mathbf{a} + \cos\theta\mathbf{b}$$

$$\mathbf{c}' = \mathbf{c}$$

so that in the notation of the transformation law (6.31)

$$\mathbf{L} = \begin{bmatrix} \cos\theta & \sin\theta & 0 \\ -\sin\theta & \cos\theta & 0 \\ 0 & 0 & 1 \end{bmatrix}$$

Hence the inertia matrix $\mathbf{I}'_{\mathrm{O}}$ at O for the axes Oa', Ob', Oc' is given by

$$\mathbf{I}'_{\mathrm{O}} = \begin{bmatrix} \cos\theta & \sin\theta & 0 \\ -\sin\theta & \cos\theta & 0 \\ 0 & 0 & 1 \end{bmatrix} \begin{bmatrix} A & 0 & 0 \\ 0 & A & 0 \\ 0 & 0 & C \end{bmatrix} \begin{bmatrix} \cos\theta & -\sin\theta & 0 \\ \sin\theta & \cos\theta & 0 \\ 0 & 0 & 1 \end{bmatrix}$$

from which we deduce (as required) that

$$\mathbf{I}'_{\mathrm{O}} = \begin{bmatrix} A & 0 & 0 \\ 0 & A & 0 \\ 0 & 0 & C \end{bmatrix}$$

Example 6.7

A uniform solid block of metal of mass M is shaped in the form of a cube of side $2a$. Find the inertia matrix $\mathbf{I}_{\mathrm{O}}$ at a vertex O of the cube with respect to axes Oa, Ob, Oc that lie along the three edges of the cube through O. Find also the inertia matrix $\mathbf{I}'_{\mathrm{O}}$ at O for the set of axes Oa', Ob', Oc' whose associated unit vectors $\mathbf{a}'$, $\mathbf{b}'$, $\mathbf{c}'$ are defined in terms of $\mathbf{a}$, $\mathbf{b}$, $\mathbf{c}$ (those of Oa, Ob, Oc) by

$$\sqrt{6}\begin{bmatrix} \mathbf{a}' \\ \mathbf{b}' \\ \mathbf{c}' \end{bmatrix} = \begin{bmatrix} \sqrt{3} & -\sqrt{3} & 0 \\ 1 & 1 & -2 \\ \sqrt{2} & \sqrt{2} & \sqrt{2} \end{bmatrix} \begin{bmatrix} \mathbf{a} \\ \mathbf{b} \\ \mathbf{c} \end{bmatrix}$$

Deduce that any pair of perpendicular lines lying in a plane with the diagonal of the cube through O as normal together with this diagonal form a set of principal axes at O.

▷ *Solution* It follows from the use of symmetry and the results contained in Appendix B that $\mathbf{I}_{\mathrm{G}}$, the inertia matrix at the centre of mass G of the block with respect to axes parallel to Oa, Ob, Oc, is given by

$$\mathbf{I}_{\mathrm{G}} = \frac{2Ma^2}{3}\begin{bmatrix} 1 & 0 & 0 \\ 0 & 1 & 0 \\ 0 & 0 & 1 \end{bmatrix}$$

Hence by the theorem of parallel axes (see (6.29)), as $\underline{\mathrm{OG}} = a(\mathbf{a} + \mathbf{b} + \mathbf{c})$,

$$\mathbf{I}_O = \frac{2Ma^2}{3}\begin{bmatrix} 1 & 0 & 0 \\ 0 & 1 & 0 \\ 0 & 0 & 1 \end{bmatrix} + Ma^2\begin{bmatrix} 2 & -1 & -1 \\ -1 & 2 & -1 \\ -1 & -1 & 2 \end{bmatrix}$$

$$= \frac{Ma^2}{3}\begin{bmatrix} 8 & -3 & -3 \\ -3 & 8 & -3 \\ -3 & -3 & 8 \end{bmatrix}$$

Hence

$$\mathbf{I}'_O = \frac{Ma^2}{18}\begin{bmatrix} \sqrt{3} & -\sqrt{3} & 0 \\ 1 & 1 & -2 \\ \sqrt{2} & \sqrt{2} & \sqrt{2} \end{bmatrix}\begin{bmatrix} 8 & -3 & -3 \\ -3 & 8 & -3 \\ -3 & -3 & 8 \end{bmatrix}\begin{bmatrix} \sqrt{3} & 1 & \sqrt{2} \\ -\sqrt{3} & 1 & \sqrt{2} \\ 0 & -2 & \sqrt{2} \end{bmatrix}$$

This gives, after some manipulation,

$$\mathbf{I}'_O = \frac{Ma^2}{3}\begin{bmatrix} 11 & 0 & 0 \\ 0 & 11 & 0 \\ 0 & 0 & 2 \end{bmatrix}$$

Since $(\sqrt{3})\mathbf{c}' = \mathbf{a} + \mathbf{b} + \mathbf{c}$, the axis O$c'$ is in the direction of the diagonal of the cube through O. Hence using the result contained in Example 6.6 we deduce that this diagonal, together with any pair of perpendicular lines through O lying in a plane with the diagonal as normal, forms a set of principal axes.

Example 6.8

Explain why the inertia matrix for a thin plate for axes Oa, Ob, Oc, where Oc is perpendicular to the plane of the plate, is of the form

$$\begin{bmatrix} A & -H & 0 \\ -H & B & 0 \\ 0 & 0 & A+B \end{bmatrix}$$

Determine the directions of the principal axes at O.

Solution Using the terminology of the definitions (6.16) we have $\mu_s = 0$ and therefore $I_{13} = I_{23} = 0$. Further by the theorem of perpendicular axes $I_{11} + I_{22} = I_{33}$, and hence the inertia matrix for the axes Oa, Ob, Oc takes the given form.

Using the work of Example 6.6 we see that the inertia matrix $\mathbf{I}'_O$ for axes Oa', Ob', Oc is given by

$$\mathbf{I}'_O = \begin{bmatrix} \cos\theta & \sin\theta & 0 \\ -\sin\theta & \cos\theta & 0 \\ 0 & 0 & 1 \end{bmatrix}\begin{bmatrix} A & -H & 0 \\ -H & B & 0 \\ 0 & 0 & A+B \end{bmatrix}\begin{bmatrix} \cos\theta & -\sin\theta & 0 \\ \sin\theta & \cos\theta & 0 \\ 0 & 0 & 1 \end{bmatrix}$$

After some manipulation it is found that

$$\mathbf{I}'_O = \begin{bmatrix} A\cos^2\theta - 2H\sin\theta\cos\theta + B\sin^2\theta & -\{(A-B)\cos\theta\sin\theta + H(\cos^2\theta - \sin^2\theta)\} & 0 \\ -\{(A-B)\cos\theta\sin\theta + H(\cos^2\theta - \sin^2\theta)\} & A\sin^2\theta + 2H\sin\theta\cos\theta + B\cos^2\theta & 0 \\ 0 & 0 & A+B \end{bmatrix}$$

The quantity $(A - B)\cos\theta\sin\theta + H(\cos^2\theta - \sin^2\theta)$ will vanish if θ is chosen so that $\tan 2\theta = 2H/(B - A)$, and hence O$a'$, O$b'$, O$c$ are principal axes at O.

We now consider the first of our examples on three-dimensional motion.

Example 6.9

A turntable is free to rotate about a fixed vertical axis Oz, the plane of the turntable being perpendicular to Oz. A uniform sphere of mass M, radius a and centre G rolls without slipping on the turntable as it rotates. At time t the coordinates of G are $[x, y, a]$ referred to the fixed axes Ox, Oy, Oz. Derive the relations

$$\dot{x} - a\omega_y + \Omega y = 0$$
$$\dot{y} + a\omega_x - \Omega x = 0$$

where $\Omega\mathbf{k}$ and $\boldsymbol{\omega} = \omega_x\mathbf{i} + \omega_y\mathbf{j}$ are the angular velocities of the turntable and sphere respectively.

Formulate the equations of motion of the sphere and obtain the differential equations satisfied by x and y. Verify that their solution is given by

$$\alpha x = u\sin\alpha t, \qquad \alpha y = u(1 - \cos\alpha t)$$

where $\alpha = 2\Omega/7$, if $x = y = \dot{y} = 0$ and $\dot{x} = u$ when $t = 0$.

▷ *Solution* Let A_1 be the label for the point of contact between sphere and turntable when it is regarded as a point of the turntable, and A_2 when it is regarded as a point of the sphere. The suffix is dispensed with when the distinction is not necessary.

We have

$$\mathbf{v}_{A_1} = \Omega\mathbf{k} \times \underline{OA} = \Omega\mathbf{k} \times (x\mathbf{i} + y\mathbf{j}) = -\Omega y\mathbf{i} + \Omega x\mathbf{j}$$
$$\mathbf{v}_G = \dot{x}\mathbf{i} + \dot{y}\mathbf{j}$$

and hence

$$\mathbf{v}_{A_2} = \mathbf{v}_G + \boldsymbol{\omega} \times \underline{GA} = [\dot{x}, \dot{y}, 0] + (\omega_x\mathbf{i} + \omega_y\mathbf{j}) \times (-a\mathbf{k})$$
$$= [\dot{x} - a\omega_y, \dot{y} + a\omega_x]$$

Since the sphere rolls on the turntable there is no relative motion at A and hence $\mathbf{v}_{A_1} = \mathbf{v}_{A_2}$, giving (as required)

$$\dot{x} - \mathrm{a}\omega_y + \Omega y = 0$$
$$\dot{y} + \mathrm{a}\omega_x - \Omega x = 0$$

Let $F_x\mathbf{i} + F_y\mathbf{j} + F_z\mathbf{k}$ be the force that the turntable exerts on the sphere at the point of contact. The equation of motion of its centre of mass G is

$$M[\ddot{x}, \ddot{y}, 0] = [F_x, F_y, F_z - M_g]$$

so that

$$M\ddot{x} = F_x, \qquad M\ddot{y} = F_y, \qquad F_z = M_g$$

The moment of inertia of the sphere about a diameter is $(2/5)Ma^2$ and therefore, recalling (6.19), we see that the moment of momentum of the sphere about its centre of mass G is given by

$$\mathbf{H}_G = \frac{2}{5}Ma^2[\omega_x, \omega_y, 0]$$

The reader should note that contrary to the general procedure used later (see the note subsequent to (6.19)) we do use axes fixed in space and not axes fixed in the sphere. This is only possible because of spherical symmetry.

Equating $\dot{\mathbf{H}}_G$ to the moment of the external forces about G yields

$$\frac{2}{5}Ma^2[\dot{\omega}_x, \dot{\omega}_y, 0] = \underline{GA} \times (F_x\mathbf{i} + F_y\mathbf{j} + F_z\mathbf{k})$$
$$= a[F_y, -F_x, 0]$$

as $\underline{GA} = -a\mathbf{k}$. Hence

$$\frac{2}{5}Ma\dot{\omega}_x = F_y, \qquad \frac{2}{5}Ma\dot{\omega}_y = -F_x$$

We deduce from these equations and those governing the motion of the centre of mass that

$$\ddot{x} = -\frac{2}{5}a\dot{\omega}_y, \qquad \ddot{y} = \frac{2}{5}a\dot{\omega}_x$$

Differentiating the relations

$$\dot{x} - a\omega_y + \Omega y = 0, \qquad \dot{y} + a\omega_x - \Omega x = 0$$

with respect to time and eliminating $\dot{\omega}_x$, $\dot{\omega}_y$ from the results yields, after some manipulation,

$$7\ddot{x} + 2\Omega\dot{y} = 0, \qquad 7\ddot{y} - 2\Omega\dot{x} = 0$$

These are the differential equations satisfied by x and y. The given expressions for x and y are clearly zero when $t = 0$. Now $\dot{x} = u\cos\alpha t$ and $\dot{y} = u\sin\alpha t$, so that $\dot{x} = u$, $\dot{y} = 0$ when $t = 0$ as required. Differentiating once more yields $\ddot{x} = -u\alpha\sin\alpha t$ and $\ddot{y} = u\alpha\cos\alpha t$, so that the differential equations for x and y are also seen to be satisfied.

In the three examples that now follow we are concerned with two sets of rotating axes that have an axis in common. As a preliminary to the examples we derive a result concerning the relative angular velocity of the sets of axes. The result is then related to the work in Chapter 5 on Euler's angles.

6.9 RELATIVE ANGULAR VELOCITY OF TWO SETS OF ROTATING AXES WITH A COMMON AXIS

Let Oa, Ob, Oc and Oa', Ob', Oc' be two sets of rotating axes such that Oc and Oc' coincide. If the associated unit vectors are $\mathbf{a}$, $\mathbf{b}$, $\mathbf{c}$ and $\mathbf{a}'$, $\mathbf{b}'$, $\mathbf{c}'$, and $\boldsymbol{\omega}$, $\boldsymbol{\omega}'$ are their respective angular velocities, then it follows since $\mathbf{c} = \mathbf{c}'$ that

$$\boldsymbol{\omega}' \times \mathbf{c}' = \dot{\mathbf{c}}' = \dot{\mathbf{c}} = \boldsymbol{\omega} \times \mathbf{c}$$

From these relations we deduce that $(\boldsymbol{\omega} - \boldsymbol{\omega}') \times \mathbf{c} = \mathbf{0}$ and hence $\boldsymbol{\omega} - \boldsymbol{\omega}' = \mu\mathbf{c}$, where μ is a scalar that may be zero. Its significance is readily determined. Since Oc and Oc' are coincident the axes Oa, Oa', Ob, Ob' will at any instant lie in a plane. Thus we have (see Fig. 6.11)

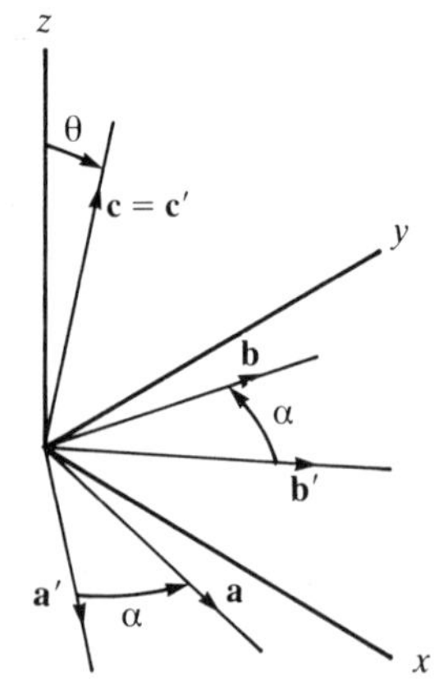

Fig. 6.11

$$\mathbf{a} = \cos\alpha\mathbf{a}' + \sin\alpha\mathbf{b}', \qquad \mathbf{b} = -\sin\alpha\mathbf{a}' + \cos\alpha\mathbf{b}'$$

where α is a variable angle. Differentiating the expression for **a** with respect to time yields

$$\boldsymbol{\omega} \times \mathbf{a} = \dot{\mathbf{a}} = -\dot{\alpha}\sin\alpha\mathbf{a}' + \cos\alpha\dot{\mathbf{a}}' + \dot{\alpha}\cos\alpha\mathbf{b}' + \sin\alpha\dot{\mathbf{b}}'$$

or

$$(\boldsymbol{\omega}' + \mu\mathbf{c}') \times (\cos\alpha\mathbf{a}' + \sin\alpha\mathbf{b}') = \dot{\alpha}(-\sin\alpha\mathbf{a}' + \cos\alpha\mathbf{b}') + \boldsymbol{\omega}' \times (\cos\alpha\mathbf{a}' + \sin\alpha\mathbf{b}')$$

From this relation we deduce, as $\mathbf{c}' \times \mathbf{a}' = \mathbf{b}'$, $\mathbf{c}' \times \mathbf{b}' = -\mathbf{a}'$, that $\mu = \dot{\alpha}$. The vector $\dot{\alpha}\mathbf{c}$ is the relative angular velocity of the axes Oa, Ob, Oc and Oa', Ob', Oc'.

We could have derived the result $\mu = \dot{\alpha}$ by appealing to the work on Euler's angles. The positions of the sets of axes in space can be defined in terms of two sets of Euler's angles θ, ϕ, ψ and θ', ϕ', ψ'. Recalling (5.39) we have

$$\boldsymbol{\omega} = \dot{\phi}\mathbf{k} + \dot{\theta}\mathbf{t} + \dot{\psi}\mathbf{c}$$
$$\boldsymbol{\omega}' = \dot{\phi}'\mathbf{k} + \dot{\theta}'\mathbf{t}' + \dot{\psi}'\mathbf{c}'$$

Now the angles θ, ϕ, θ', ϕ' define the directions of the axes Oc and Oc' in space and the vectors **t** and **t**$'$. But Oc, Oc' coincide and hence $\theta = \theta'$, $\phi = \phi'$, $\mathbf{t} = \mathbf{t}'$, so that

$$\boldsymbol{\omega} - \boldsymbol{\omega}' = (\dot{\psi} - \dot{\psi}')\mathbf{c}$$

giving $\dot{\alpha} = \dot{\psi} - \dot{\psi}'$. If the angular velocity $\boldsymbol{\omega}'$ of Oa', Ob', Oc' has a fixed direction we follow the practice adopted in Chapter 4 and take this direction to be that of **k**. Hence

$$\boldsymbol{\omega}' = \omega'\mathbf{k} = \dot{\phi}'\mathbf{k} + \dot{\theta}'\mathbf{t}' + \dot{\psi}'\mathbf{c}'$$

from which we deduce, since **k**, **t**, **c** are three non-coplanar directions, $\dot{\phi}' = \omega'$, $\dot{\theta}' = \dot{\psi}' = 0$. Thus θ' and ψ' have constant values. It follows that the axes Os, Ot, Oc and Oa', Ob', Oc' are fixed relative to one another, and if they are located in a given rigid body it is convenient to take them as being coincident ($\psi' = 0$). The axes will be referred to as Os, Ot, Oc.

Example 6.10

Figure 6.12 shows a symmetrical rotor, with its centre of mass at G, carried in bearings D and E attached to a circular frame of radius r, which rotates about a fixed axis Gz. The angular velocity $\boldsymbol{\omega}$ of the frame is $\omega\mathbf{k}$ where ω is a variable, while the angular velocity of the rotor relative to that of the frame is $n\mathbf{c}$, **c** being a unit vector in the direction of $\underline{DE}$ and n a constant.

The moment of inertia of the rotor about its axis of symmetry, which is coincident with the line DE, is C, and that about any axis perpendicular to the axis of symmetry is A. The angle between GE and Gz is α. Determine the forces, other than those due to gravity, that act on the bearings at D and E.

▷ *Solution* The axes Gs, Gt, Gc are fixed in the circular frame, the associated unit vectors **s**, **t**, **c** being defined in the usual way by means of the angles θ and ϕ. From the given information it follows that $\dot{\phi} = \omega$ and $\theta = \alpha$. The axes Ga, Gb, Gc (not shown in Fig. 6.12) are fixed in the rotor.

Let $\mathbf{\Omega}$ be the angular velocity of the rotor. Hence $\mathbf{\Omega} - \omega\mathbf{k} = n\mathbf{c}$ or

$$\begin{aligned}\mathbf{\Omega} &= \omega\mathbf{k} + n\mathbf{c}\\ &= \omega(-\sin\alpha\mathbf{s} + \cos\alpha\mathbf{c}) + n\mathbf{c} = -\omega\sin\alpha\mathbf{s} + (\omega\cos\alpha + n)\mathbf{c}\end{aligned}$$

Since the moment of inertia of the rotor about the axis Gs is A and that about Gc is C, the moment of momentum $\mathbf{H}_G$ of the rotor about G is given by

$$\mathbf{H}_G = -A\omega\sin\alpha\mathbf{s} + C(\omega\cos\alpha + n)\mathbf{c}$$

Hence

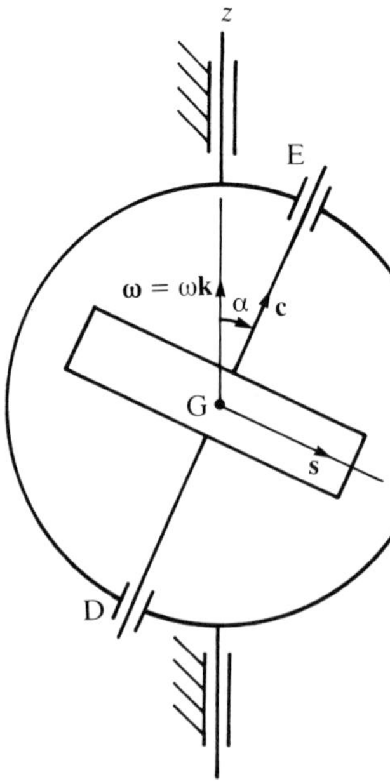

Fig. 6.12

$$\dot{\mathbf{H}}_G = -A\dot{\omega}\sin\alpha\mathbf{s} - A\omega\sin\alpha\dot{\mathbf{s}} + C\dot{\omega}\cos\alpha\mathbf{c} + C(\omega\cos\alpha + n)\dot{\mathbf{c}}$$

But

$$\dot{\mathbf{s}} = \boldsymbol{\omega} \times \mathbf{s} = (-\omega\sin\alpha\mathbf{s} + \omega\cos\alpha\mathbf{c}) \times \mathbf{s} = \omega\cos\alpha\mathbf{t}$$

Likewise

$$\dot{\mathbf{c}} = \boldsymbol{\omega} \times \mathbf{c} = \omega\sin\alpha\mathbf{t}$$

so that

$$\dot{\mathbf{H}}_G = -A\dot{\omega}\sin\alpha\mathbf{s} + \{(C - A)\omega^2\cos\alpha\sin\alpha + C\omega n\sin\alpha\}\mathbf{t} + C\dot{\omega}\cos\alpha\mathbf{c}$$

Let $\mathbf{P}$ and $\mathbf{Q}$ be the reactions of the bearings at D and E on the rotor, where

$$\mathbf{P} = P_1\mathbf{s} + P_2\mathbf{t} + P_3\mathbf{c}, \qquad \mathbf{Q} = Q_1\mathbf{s} + Q_2\mathbf{t} + Q_3\mathbf{c}$$

The sum of the moments of these reactions about G is given by

$$\underline{\mathrm{GD}} \times \mathbf{P} + \underline{\mathrm{GE}} \times \mathbf{Q} = -r\mathbf{c} \times \mathbf{P} + r\mathbf{c} \times \mathbf{Q} = r\mathbf{c} \times (\mathbf{Q} - \mathbf{P})$$

Substituting for $\mathbf{P}$ and $\mathbf{Q}$ in terms of their components yields

$$r\mathbf{c} \times (\mathbf{Q} - \mathbf{P}) = r(P_2 - Q_2)\mathbf{s} + r(Q_1 - P_1)\mathbf{t}$$

Now the sum of the moments about G of the forces acting on the rotor is equal to $\dot{\mathbf{H}}_G$. Hence

$$r(P_2 - Q_2) = -A\dot{\omega}\sin\alpha$$
$$r(Q_1 - P_1) = (C - A)\omega^2\cos\alpha\sin\alpha + Cn\omega\sin\alpha$$

In addition we see that a couple whose moment is in the direction of the unit vector $\mathbf{c}$ must be applied to the rotor. The moment is given by $C\dot{\omega}\cos\alpha\mathbf{c}$. Since the centre of mass of the rotor is at rest, $\mathbf{P} + \mathbf{Q} = \mathbf{0}$, and therefore

$$P_1 = -Q_1, \qquad P_2 = -Q_2, \qquad P_3 + Q_3 = 0$$

where

$$2rP_1 = (A - C)\omega^2\cos\alpha\sin\alpha - Cn\omega\sin\alpha$$
$$2rP_2 = -A\dot{\omega}\sin\alpha$$

Example 6.11

Axes Oa', Ob', Oc' are fixed to a vehicle such that Oa' is in the direction of forward motion and Ob' is upwards. The vehicle carries a high-speed rotor that can turn about its axis of symmetry, which passes through O and is supported in bearings fixed in the vehicle. The angular velocity of the rotor relative to the vehicle is $p\mathbf{c}$, where p is a constant and the unit vector $\mathbf{c}$ lies in the plane of Ob', Oc' such that the angle between $\mathbf{c}$ and Oc' is β. The centre of mass of the rotor is at O, while its moments of inertia at O are C about the axis of symmetry and A about any axis perpendicular to the axis of symmetry.

If the vehicle turns at a constant rate such that the component of its angular velocity about the axis Ob' is ω_2, and simultaneously rolls about the Oa' axis at an angular rate ω_1 (not constant), derive expressions for the torques applied by the vehicle to the rotor during this motion.

▷ *Solution* In order to facilitate the solution of the problem we define axes Oa, Ob, Oc as follows. The axis Oc is in the direction of the axis of the rotor; Oa coincides with Oa'; and Ob is such that Oa, Ob, Oc form in the usual way a right-handed system of mutually perpendicular axes. The unit vectors associated with the axes are, in our usual notation, **a**, **b**, **c** and **a**$'$, **b**$'$, **c**$'$. Thus

$$\mathbf{a} = \mathbf{a}', \ \mathbf{c} = \sin\beta\,\mathbf{b}' + \cos\beta\,\mathbf{c}', \qquad \mathbf{b} = \mathbf{c} \times \mathbf{a} = \cos\beta\,\mathbf{b}' - \sin\beta\,\mathbf{c}'$$

so that

$$\mathbf{a}' = \mathbf{a}, \qquad \mathbf{b}' = \cos\beta\,\mathbf{b} + \sin\beta\,\mathbf{c}, \qquad \mathbf{c}' = -\sin\beta\,\mathbf{b} + \cos\beta\,\mathbf{c}$$

If $\boldsymbol{\omega}$ is the angular velocity of the vehicle then

$$\begin{aligned}\boldsymbol{\omega} &= \omega_1\mathbf{a}' + \omega_2\mathbf{b}' \\ &= \omega_1\mathbf{a} + \omega_2\cos\beta\,\mathbf{b} + \omega_2\sin\beta\,\mathbf{c}\end{aligned}$$

and since by definition the axes Oa, Ob, Oc are fixed in the vehicle this is also their angular velocity.

The angular velocity of the rotor is equal to

$$\omega_1\mathbf{a} + \omega_2\cos\beta\,\mathbf{b} + (\omega_2\sin\beta + p)\mathbf{c}$$

and therefore $\mathbf{H}_\mathrm{O}$, its moment of momentum about O, is given by

$$\mathbf{H}_\mathrm{O} = A\omega_1\mathbf{a} + A\omega_2\cos\beta\,\mathbf{b} + C(\omega_2\sin\beta + p)\mathbf{c}$$

Hence, recalling (5.32), we deduce that

$$\dot{\mathbf{H}}_\mathrm{O} = A\dot{\omega}_1\mathbf{a} + \begin{bmatrix} \mathbf{a} & \mathbf{b} & \mathbf{c} \\ \omega_1 & \omega_2\cos\beta & \omega_2\sin\beta \\ A\omega_1 & A\omega_2\cos\beta & C(p + \omega_2\sin\beta) \end{bmatrix}$$

Since the centre of mass of the rotor is at O, the force of gravity does not enter into the expressions for the moments about O of the forces acting on the rotor. Thus the components of $\dot{\mathbf{H}}_\mathrm{O}$ are equal to the torques applied to the rotor by the vehicle. The components are, referred to **a**, **b**, **c**,

$$A\dot{\omega}_1 + (C - A)\omega_2^2\sin\beta\cos\beta + Cp\omega_2\cos\beta, \ (A - C)\omega_1\omega_2\sin\beta - Cp\omega_1, \ 0$$

In solving this problem we could have used the axes Oa', Ob', Oc' as the reference axes instead of the axes Oa, Ob, Oc, but this would have required us to determine $\mathbf{H}_O$ referred to $\mathbf{a}'$, $\mathbf{b}'$, $\mathbf{c}'$. It is left as an exercise to the reader, noting

$$\begin{bmatrix}\mathbf{a}'\\ \mathbf{b}'\\ \mathbf{c}'\end{bmatrix} = \begin{bmatrix}1 & 0 & 0\\ 0 & \cos\beta & \sin\beta\\ 0 & -\sin\beta & \cos\beta\end{bmatrix}\begin{bmatrix}\mathbf{a}\\ \mathbf{b}\\ \mathbf{c}\end{bmatrix}$$

to show that the inertia matrix at O referred to Oa', Ob', Oc' is

$$\begin{bmatrix}A & 0 & 0\\ 0 & A\cos^2\beta + C\sin^2\beta & (C-A)\sin\beta\cos\beta\\ 0 & (C-A)\sin\beta\cos\beta & A\sin^2\beta + C\cos^2\beta\end{bmatrix}$$

and to obtain $\dot{\mathbf{H}}_O$ and therefore the torques in terms of $\mathbf{a}'$, $\mathbf{b}'$, $\mathbf{c}'$.

The top

The next example concerns the motion of a top. It is convenient at this stage, before working through the example, to define the terminology that is standard in respect of the motion of a top. Figure 6.13 shows a top whose axis of symmetry is in the direction Oc. The axis Oz is vertically upwards, while the directions of Os, Ot, Oc are defined, as in the discussion of Euler's angles, by means of the angles θ, ϕ. If Oa, Ob, Oc are axes fixed in the top with the angle between Ot and Oa equal to the Euler angle, ψ, then the angular velocity $\boldsymbol{\omega}$ of the top is given by

$$\begin{aligned}\boldsymbol{\omega} &= \dot{\phi}\mathbf{k} + \dot{\theta}\mathbf{t} + \dot{\psi}\mathbf{c}\\ &= -\dot{\phi}\sin\theta\mathbf{s} + \dot{\theta}\mathbf{t} + (\dot{\phi}\cos\theta + \dot{\psi})\mathbf{c}\end{aligned}$$

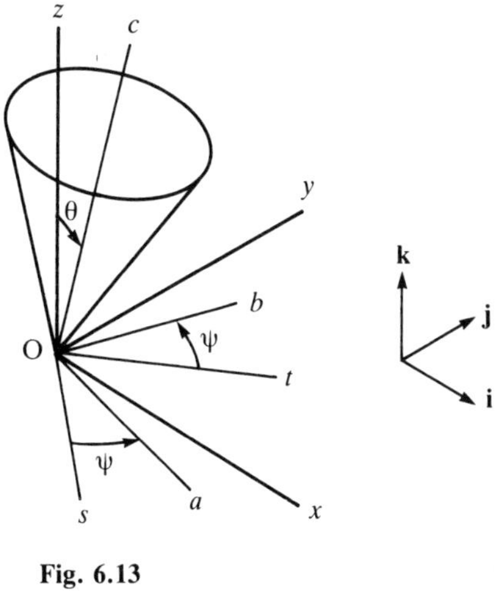

Fig. 6.13

The quantity $\dot{\phi}\cos\theta + \dot{\psi}$ is called the *spin* of the top (about its axis of symmetry) and is usually denoted by n. If the surface with which the point O is in contact is sufficiently rough to prevent the top slipping on it and the only other force acting is gravity, then it can be shown that the spin n is constant (see Problem 6.29). The quantity $\dot{\phi}$ is called the (angular) *speed of precession* of the axis of the top about Oz, while $\dot{\theta}$ defines the angular rate of the *rise and fall* of the axis of the top. In *steady precessional* motion $\dot{\theta} = 0$ and $\dot{\phi}$ is constant.

Example 6.12

The moment of inertia of a top about its axis of symmetry is C and that about any line perpendicular to this axis through the end O of it is A. If the axis of the top is precessing steadily about the vertical axis Oz (O being fixed) with angular speed Ω, show that Ω satisfies the quadratic equation

$$A\cos\alpha\Omega^2 - Cn\Omega + Mgh = 0$$

where M is the mass of the top, h is the distance of its centre of mass G from O, and α is the angle between $\underline{\mathrm{OG}}$ and the axis Oz.

▷ *Solution* In terms of the notation already defined we have $\dot{\phi} = \Omega, \theta = \alpha$, $\Omega\cos\alpha + \dot{\psi} = n$, so that the angular velocity of the top is equal to $-\Omega\sin\alpha\mathbf{s} + n\mathbf{c}$. Hence $\mathbf{H}_\mathrm{O}$, its moment of momentum about O, is given by

$$\mathbf{H}_\mathrm{O} = -A\Omega\sin\alpha\mathbf{s} + Cn\mathbf{c}$$

Recalling (5.32) we have

$$\dot{\mathbf{H}}_\mathrm{O} = \begin{vmatrix} \mathbf{s} & \mathbf{t} & \mathbf{c} \\ -\Omega\sin\alpha & 0 & \Omega\cos\alpha \\ -A\Omega\sin\alpha & 0 & Cn \end{vmatrix}$$

$$= (-A\Omega^2\sin\alpha\cos\alpha + Cn\Omega\sin\alpha)\mathbf{t}$$

But

$$\dot{\mathbf{H}}_\mathrm{O} = \underline{\mathrm{OG}} \times (-Mg\mathbf{k})$$
$$= Mgh\mathbf{c} \times (\sin\alpha\mathbf{s} - \cos\alpha\mathbf{c}) = Mgh\sin\alpha\mathbf{t}$$

Equating the two expressions for $\dot{\mathbf{H}}_\mathrm{O}$ and dividing throughout by $\sin\alpha$ (assumed $\neq 0$) yields the equation $A\cos\alpha\Omega^2 - Cn\Omega + Mgh = 0$ as required.

In the solutions of Examples 6.10 to 6.12 we exploited the fact that the rotors and top had axial symmetry by referring the moment of momentum of each body (about O or G) to unit vectors of which only one was fixed relative to it. When a body does not have axial symmetry we are compelled to take axes fixed in the body. The equations that are now derived governing the body's angular motion are due to Euler.

Euler's equation of motion

We begin by considering the motion of a rigid body that moves with one point O of itself fixed. The principal axes of inertia at O are taken to be Oa, Ob, Oc (associated unit vectors $\mathbf{a}$, $\mathbf{b}$, $\mathbf{c}$), the corresponding principal moments of inertia being A, B, C. If $\boldsymbol{\omega}$, the angular velocity of the body, is given by

$$\boldsymbol{\omega} = \omega_1\mathbf{a} + \omega_2\mathbf{b} + \omega_3\mathbf{c}$$

then

$$\mathbf{H}_0 = A\omega_1\mathbf{a} + B\omega_2\mathbf{b} + C\omega_3\mathbf{c}$$

and hence, recalling (5.32)

$$\dot{\mathbf{H}}_\mathrm{O} = [A\dot{\omega}_1, B\dot{\omega}_2, C\dot{\omega}_3] + \begin{vmatrix} \mathbf{a} & \mathbf{b} & \mathbf{c} \\ \omega_1 & \omega_2 & \omega_3 \\ A\omega_1 & B\omega_2 & C\omega_3 \end{vmatrix}$$

$$= [A\dot{\omega}_1 - (B-C)\omega_2\omega_3, B\dot{\omega}_2 - (C-A)\omega_1\omega_3, C\dot{\omega}_3 - (A-B)\omega_1\omega_2]$$

If $\mathbf{M} = M_1\mathbf{a} + M_2\mathbf{b} + M_3\mathbf{c}$ is equal to the sum of the moments about O of the external forces acting, then

$$A\dot{\omega}_1 - (B - C)\omega_2\omega_3 = M_1$$
$$B\dot{\omega}_2 - (C - A)\omega_3\omega_1 = M_2$$
$$C\dot{\omega}_3 - (A - B)\omega_1\omega_2 = M_3$$

These three equations are known as Euler's equations.

When no point of the body is fixed we consider $\mathbf{H}_\mathrm{G}$ and derive a similar set of equations in which A, B, C are now the principal moments of inertia at G, and M_1, M_2, M_3 are the components of the sum of the moments about G of the external forces acting.

6.10 KINETIC ENERGY AND WORK

Kinetic energy of a particle

The kinetic energy of a particle of mass m moving with absolute velocity $\mathbf{v}$ is defined to be the quantity $m\mathbf{v}^2/2$. It is, by definition, a scalar.

Kinetic energy of a system of particles

We denote the kinetic energy of a system of particles by T. The quantity T is defined by

$$2T = \sum_{s=0}^{N} m_s \mathbf{v}_s^2 \tag{6.34}$$

It follows, noting the definition of $\boldsymbol{\sigma}_s$ in (a) of Section 6.4, that

$$\begin{aligned} 2T &= \sum_{s=0}^{N} m_s(\dot{\mathbf{r}}_G + \dot{\boldsymbol{\sigma}}_s)^2 \\ &= \sum m_s \dot{\mathbf{r}}_G^2 + 2\mathbf{r}_G \,.\, \left(\sum m_s \dot{\boldsymbol{\sigma}}_s\right) + \sum m_s \dot{\boldsymbol{\sigma}}_s^2 \\ &= M\dot{\mathbf{r}}_G^2 + \sum m_s \dot{\boldsymbol{\sigma}}_s^2 \end{aligned} \tag{6.35}$$

The first term on the right-hand side of (6.35) is twice the kinetic energy of a particle whose mass is equal to the total mass of the particles and whose velocity is equal to that of their centre of mass. The second term is twice the kinetic energy of the motion of the particles relative to the centre of mass.

Kinetic energy of a rigid body

For a rigid body we have $\dot{\boldsymbol{\sigma}}_s = \boldsymbol{\omega} \times \boldsymbol{\sigma}_s$, so that (6.35) can be put in the form

$$\begin{aligned} 2T &= M\dot{\mathbf{r}}_G^2 + \sum m_s \dot{\boldsymbol{\sigma}}_s \,.\, (\boldsymbol{\omega} \times \boldsymbol{\sigma}_s) \\ &= M\dot{\mathbf{r}}_G^2 + \boldsymbol{\omega} \,.\, \sum (\boldsymbol{\sigma}_s \times m_s \dot{\boldsymbol{\sigma}}_s) \end{aligned}$$

using property (1.20) of triple-scalar products. Hence

$$2T = M\dot{\mathbf{r}}_G^2 + \boldsymbol{\omega} \,.\, \mathbf{H}_G \tag{6.36}$$

recalling (6.22).

When one point of the rigid body is fixed the formula in (6.36) can be simplified. Taking the fixed point of the body to be at O we have

$$\begin{aligned} 2T &= \sum m_s \dot{\mathbf{r}}_s^2 \\ &= \sum m_s \dot{\mathbf{r}}_s \,.\, (\boldsymbol{\omega} \times \mathbf{r}_s) \\ &= \boldsymbol{\omega} \,.\, \sum \mathbf{r}_s \times m_s \dot{\mathbf{r}}_s \\ &= \boldsymbol{\omega} \,.\, \mathbf{H}_O \end{aligned} \tag{6.37}$$

The formulae (6.35) and (6.36) underline once more the importance of the centre of mass. We note that in two-dimensional motion ($\boldsymbol{\omega} = \omega\mathbf{k}$) the formulae assume a particularly simple form.

Recalling (6.20) and (6.21) it is seen that $\boldsymbol{\omega} \cdot \mathbf{H}_O = I\omega^2$. Thus when a body is rotating about a fixed axis,

$$T = \frac{1}{2}I\omega^2 \tag{6.38}$$

where I is the moment of inertia of the body about the axis of rotation.

With the help of (6.24) we see $\boldsymbol{\omega} \cdot \mathbf{H}_G$ is equal to $I_G\omega^2$, where I_G is the moment of inertia of the body about an axis through its centre of mass G in the direction of $\boldsymbol{\omega}$. Hence (6.36) becomes

$$T = \frac{1}{2}M\dot{\mathbf{r}}_G^2 + \frac{1}{2}I_G\omega^2 \tag{6.39}$$

Work

The work done by a constant force $\mathbf{F}$ when its point of application moves along a straight line from A to B is defined to be the quantity $\mathbf{F} \cdot \underline{AB}$. When the force varies in magnitude and/or direction and the point of application is displaced along a curved path, then to determine the work done the curved path is split up into infinitesimal displacements (see Fig. 6.14) starting at A and ending at B. The work done is defined to be

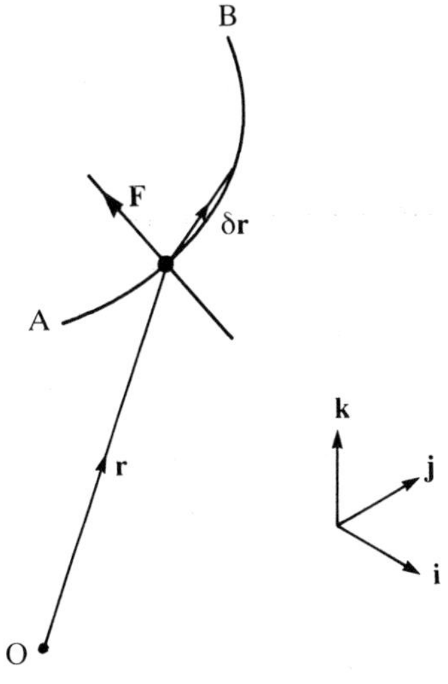

Fig. 6.14

$\Sigma_A^B \mathbf{F} \cdot \delta\mathbf{r}$, the summation, as the notation indicates, extending over all displacements from A to B. If the work done for any pair of points A, B depends not on the path but only on its end points, then the force **F** is said to be a conservative force. Gravitational force is an example of a conservative force.

Work is, by definition, a scalar quantity and may be positive, negative or zero. A useful point to note is that if the force and the vector representing the displacement of its point of application are perpendicular then the work done by the force is zero.

In order to evaluate the sum $\Sigma_A^B \mathbf{F} \cdot \delta\mathbf{r}$ we have to express it as a line integral. If $\mathbf{F} = [X, Y, Z]$ and $\delta\mathbf{r} = [\delta x, \delta y, \delta z]$ referred to the unit vectors **i**, **j**, **k**, then

$$\sum_{A}^{B} \mathbf{F} \cdot \delta\mathbf{r} = \sum_{A}^{B} (X\delta x + Y\delta y + Z\delta z)$$

$$= \int_{A}^{B} (X\mathrm{d}x + Y\mathrm{d}y + Z\mathrm{d}z)$$

The integral is referred to as a line integral. The limits A, B do not stand for numerical values but indicate that the path of integration (which has to be specified except when **F** is a conservative force) starts at the point A and ends at the point B. The evaluation of a line integral is generally carried out by expressing the various quantities in terms of a single parameter t (say). We illustrate the technique by means of the following example.

Example 6.13

A bead is threaded onto a fixed circular wire, radius a, centre O. The bead is acted upon by a force **F** of constant magnitude F in the direction of the tangent to the circle (see Fig. 6.15). Determine the work done by the force when the bead makes a complete circuit of the wire.

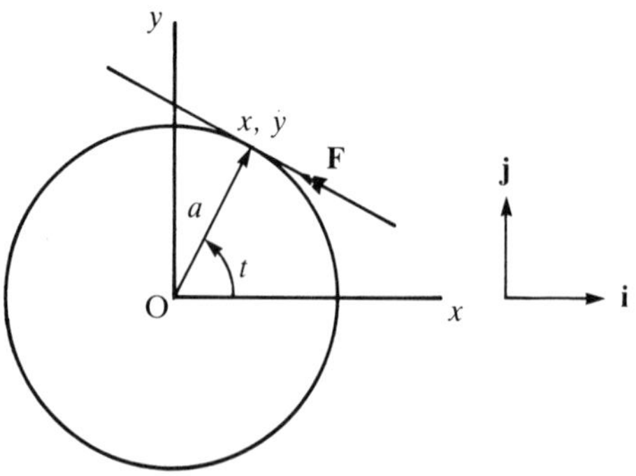

Fig. 6.15

> *Solution* Using the notation of Fig. 6.15 we have $x = a\cos t$, $y = a\sin t$ and $\mathbf{F} = F[-\sin t, \cos t, 0]$ referred to the unit vectors **i**, **j**, **k**. Hence $dx = -a\sin t\,dt$, $dy = a\cos t\,dt$, so that the work done in a complete circuit is equal to

$$\int_0^{2\pi} \{(-F\sin t)(-a\sin t) + (F\cos t)a\cos t\}\,dt = \int_0^{2\pi} aF\,dt = 2\pi aF$$

Work done by gravity

We take the unit vector **k** to be vertically upward and hence the force due to gravity on a particle of mass m is equal to $-mg\mathbf{k}$. If the particle moves from the point A to the point B under the action of gravity (and possibly other forces) then the work W done by gravity on the particle is given by

$$W = \sum_{A}^{B} -mg\delta z = -mg\int_{z_A}^{z_B} dz = mg(z_A - z_B) \tag{6.40}$$

where $\underline{OA} = [x_A, y_A, z_A]$ and $\underline{OB} = [x_B, y_B, z_B]$ referred to **i**, **j**, **k**.

If $z_A - z_B > 0$ the particle has fallen a (vertical) distance $z_A - z_B$ and therefore the work done by gravity is $mg \times$ distance dropped. If $z_A - z_B < 0$ then $z_B - z_A$ is the (vertical) distance risen and work done is $-mg \times$ distance risen.

It is left as an exercise to the reader to show, with the help of (6.40), that the work done by gravity on a system of particles during its motion is equal to the work done by gravity on a particle of mass M located at the centre of mass of the particles.

Relation between work and energy

The equation of motion of the particle P_s (in the notation already defined) is

$$m_s\dot{\mathbf{v}}_s = \mathbf{F}_s + \mathbf{F}'_s$$

Taking the scalar product of both sides of this equation with $\mathbf{v}_s$ yields

$$m_s\mathbf{v}_s \cdot \dot{\mathbf{v}}_s = (\mathbf{F}_s + \mathbf{F}'_s) \cdot \mathbf{v}_s$$

or

$$\frac{d}{dt}\left(\frac{1}{2}m_s\mathbf{v}_s^2\right) = (\mathbf{F}_s + \mathbf{F}'_s) \cdot \mathbf{v}_s \tag{6.41}$$

The expressions $\mathbf{F}_s \cdot \mathbf{v}_s$ and $\mathbf{F}'_s \cdot \mathbf{v}_s$ are the rates of working of the forces $\mathbf{F}_s$

and $\mathbf{F}'_s$ respectively. If we sum corresponding sides of (6.41) from $s = 0$ to $s = N$ we have, since $2T = \Sigma m_s \mathbf{v}_s^2$,

$$\frac{\mathrm{d}T}{\mathrm{d}t} = \sum_{s=0}^{N} \mathbf{F}_s \cdot \mathbf{v}_s + \sum_{s=0}^{N} \mathbf{F}'_s \cdot \mathbf{v}_s \tag{6.42}$$

Integrating both sides of (6.36) with respect to t over the time interval $t = t_0$ to $t = t_1$ leads to

$$T(t_1) - T(t_0) = \sum_{s=0}^{N} \int_{t_0}^{t_1} (\mathbf{F}_s + \mathbf{F}'_s) \cdot \mathbf{v}_s \, \mathrm{d}t$$

Now $\mathbf{v}_s = \mathrm{d}\mathbf{r}_s/\mathrm{d}t$, so that

$$\int_{t_0}^{t_1} \mathbf{F}_s \cdot \mathbf{v}_s \, \mathrm{d}t = \int_{t_0}^{t_1} \left(\mathbf{F}_s \cdot \frac{\mathrm{d}\mathbf{r}_s}{\mathrm{d}t}\right) \mathrm{d}t$$
$$= \int_{\mathrm{A}_s}^{\mathrm{B}_s} \mathbf{F}_s \cdot \mathrm{d}\mathbf{r}_s$$

where A_s, B_s are the positions of the particles P_s at times $t = t_0$ and $t = t_1$ respectively. The limits A_s, B_s indicate that the line integral is to be evaluated along some path starting at A_s and ending at B_s. Thus

$$T(t_1) - T(t_0) = \sum_{s=0}^{N} \int_{\mathrm{A}_s}^{\mathrm{B}_s} (\mathbf{F}_s + \mathbf{F}'_s) \cdot \mathrm{d}\mathbf{r}_s \tag{6.43}$$

so that the change in kinetic energy of the system of particles is equal to the total work done by the external and internal forces.

It should not be assumed that the total work done by the internal forces is zero. To demonstrate this statement we analyse, in simple terms, the effect on the motion of an ice skater induced by the lowering of his (or her) outstretched arms when spinning about a vertical axis. We assume this vertical axis is an axis of symmetry for the skater (arms raised or lowered) and that the moment of inertia of the skater about this axis is I_1 with arms raised and I_2 for arms lowered. It is a consequence of the definition of a moment of inertia that $I_1 > I_2$. Let $\omega_1\mathbf{k}$ and $\omega_2\mathbf{k}$ be the values of the angular velocity of the skater for the two positions of the arms, $\mathbf{k}$ being a unit vector in the direction of the axis of spin. It follows, if the frictional forces between the ice and the skater are negligible, that the moment of momentum about the skater's centre of mass is conserved when his (or her) arms are lowered. Hence, recalling (6.20),

$$I_1\omega_1 = I_2\omega_2$$

The change in the rotational kinetic energy of the skater is given (see (6.39)) by

$$\frac{1}{2}(I_2\omega_2^2 - I_1\omega_1^2) = \frac{I_1}{2I_2}(I_1 - I_2)\omega_1^2 > 0$$

and this is a measure of the work done by the internal forces and gravity.

There is no need to consider the motion of the skater's centre of mass (it has been lowered during the motion) provided it is at rest in the two positions of the arms.

Power

The work done by a force $\mathbf{F}$ in the infinitesimal displacement $\delta\mathbf{r}$ of its point of application is, by definition, $\mathbf{F} \cdot \delta\mathbf{r}$. If the displacement occurs in time δt then

$$\lim_{\delta t \to 0} \mathbf{F} \cdot \frac{\delta\mathbf{r}}{\delta t} = \mathbf{F} \cdot \mathbf{v}$$

where $\mathbf{v}$, the velocity of the point of application, represents the rate at which the force is doing work. The quantity $\mathbf{F} \cdot \mathbf{v}$ is referred to as power. Clearly by integrating $\mathbf{F} \cdot \mathbf{v}$ with respect to time over the interval $t = t_1$ to $t = t_2$ we obtain the work done by $\mathbf{F}$ in the interval.

Work done by a couple

Let the couple be of moment $\mathbf{M}$ and represented by the force pair $\mathbf{F}$, $-\mathbf{F}$ acting at A and B respectively, so that $\mathbf{M} = \underline{\mathrm{BA}} \times \mathbf{F}$. If $\mathbf{v}_\mathrm{A}$ and $\mathbf{v}_\mathrm{B}$ are the velocities of the points of application then $\mathrm{d}W/\mathrm{d}t$, the rate of working of the forces, is given by

$$\frac{\mathrm{d}W}{\mathrm{d}t} = \mathbf{F} \cdot (\mathbf{v}_\mathrm{A} - \mathbf{v}_\mathrm{B})$$

When A and B are points fixed in a rigid body whose angular velocity is $\boldsymbol{\omega}$ then $\mathbf{v}_\mathrm{A} - \mathbf{v}_\mathrm{B} = \boldsymbol{\omega} \times \underline{\mathrm{BA}}$, so that

$$\frac{\mathrm{d}W}{\mathrm{d}t} = \mathbf{F} \cdot (\boldsymbol{\omega} \times \underline{\mathrm{BA}}) = \boldsymbol{\omega} \cdot (\underline{\mathrm{BA}} \times \mathbf{F}) = \boldsymbol{\omega} \cdot \mathbf{M}$$

In two-dimensional motion $\boldsymbol{\omega} = \dot{\theta}\mathbf{k}$, $\mathbf{M} = M\mathbf{k}$, and therefore $\mathrm{d}W/\mathrm{d}t = M\dot{\theta}$. Hence $\mathrm{d}W/\mathrm{d}\theta = M$. If $\theta = \theta_1$ when $t = t_1$ and $\theta = \theta_2$ when $t = t_2$ we deduce that the work W done by the couple in the time interval $t = t_1$ to $t = t_2$ is given by

$$W = \int_{\theta_1}^{\theta_2} M \,\mathrm{d}\theta \qquad (6.44)$$

The scalar quantity M is often referred to as a torque, and when we speak of the torque required to drive a machine or of the torque exerted by a motor it is this quantity we have in mind. The letter T is frequently used as an alternative to M to denote torque. Before considering examples on work and energy we describe the role of the flywheel in a machine.

The flywheel

In simple terms a flywheel may be regarded as a (light) wheel with a heavy rim. The purpose of the heavy rim is to give the flywheel a large axial moment of inertia. This large axial moment of inertia enables the flywheel when fitted to a machine to perform one of two functions. It can control the fluctuations (see Example 6.14) in the rotational speed of the machine, or it can act as a reservoir of mechanical energy (for example in a child's mechanical toy).

Example 6.14

The torque $T(\theta)$ required to drive a punching machine is of period 2π. It is given for $0 \leq \theta \leq 2\pi$ by Fig. 6.16, the quantity θ being the angular displacement about its axis of the shaft that drives the machine. The shaft is connected to a motor that exerts a constant torque M on it. A flywheel is mounted on the shaft such that the axes of the shaft and flywheel are coincident, the axial moment of inertia of the flywheel and shaft being I. Determine the torque M if the average angular speed ω of the shaft in any cycle ($\alpha \leq \theta \leq \alpha + 2\pi$) is constant. Given that $\omega = 100\,\text{rev min}^{-1}$, find I if the variation in $\dot{\theta}$ is not to exceed $\pm 5\%$. If the average speed ω is reduced to $50\,\text{rev min}^{-1}$, find also the variation in rotational speed with this value of I, assuming the torque required to drive the machine is unchanged.

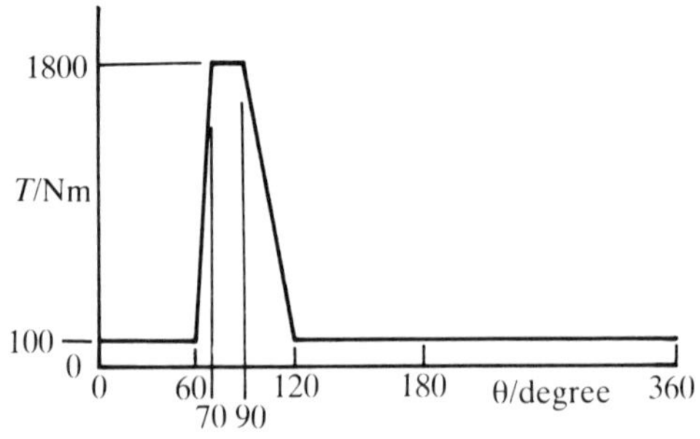

Fig. 6.16

> *Solution* As the average angular speed of the shaft in any cycle is constant there can be no loss or gain of kinetic energy during the cycle. Hence the work done to drive the punching machine must be equal to the work done by the motor in a cycle. Using (6.44) we have

$$2\pi M = \int_0^{2\pi} T(\theta)\,d\theta$$

$$= \left\{2\pi \times 100 + \frac{1700}{2}\left(\frac{\pi}{3} + \frac{\pi}{9}\right)\right\} \text{N m}$$

Hence $M = 288.89$ N m. (Note that in the integration θ must be in radians.)

The torque exerted by the punching machine on the shaft is, by Newton's third law, equal and opposite to that exerted by the shaft on the punching machine. By considering the rate of change of moment of momentum of the flywheel about its centre we see that

$$I\ddot{\theta} = M - T(\theta)$$

Now $\dot{\theta}$ is a maximum or minimum when $\ddot{\theta} = 0$, that is when $T(\theta) = M$. The two solutions of this equation in $0 < \theta < 2\pi$ are given by θ_1, $\theta_2(\theta_1 < \theta_2)$, where

$$100 + \frac{\theta_1 - 60}{10}1700 = 288.89$$

$$1800 - \frac{\theta_2 - 90}{30}1700 = 288.89$$

Thus $\theta_1 = 61.11°$ and $\theta_2 = 116.67°$.

Since $M - T(\theta) > 0$ for $0 > \theta > \theta_1$, it is clear that $\dot{\theta}$ is a maximum when $\theta = \theta_1$ and a minimum when $\theta = \theta_2$. Denoting the maximum and minimum values of $\dot{\theta}$ by ω_1, ω_2 it follows, using (6.38) and (6.44), that

$$\frac{1}{2}I(\omega_2^2 - \omega_1^2) = \int_{\theta_1}^{\theta_2} \{M - T(\theta)\}\,d\theta$$

or

$$\frac{1}{2}(\omega_1 - \omega_2)(\omega_1 + \omega_2)I = \int_{\theta_1}^{\theta_2} \{T(\theta) - M\}\,d\theta$$

Now we require $\omega_1 - \omega_2 = 0.1\omega$. Assuming $\omega_1 + \omega_2 = 2\omega$ it follows that

$$0.1\omega^2 I = \frac{1}{2}(1800 - 288.9)\left(\frac{\pi}{9} + \theta_2 - \theta_1\right) \text{J}$$

from which we deduce, putting $\omega = 100$ rev min^{-1}, that $I = 90.86$ kg m^2.

Let the maximum and minimum values of $\dot{\theta}$ be Ω_1, Ω_2 when $\omega = 50$ rev min^{-1}. Hence

$$\frac{1}{2}I(\omega_1^2 - \omega_2^2) = \frac{1}{2}I(\Omega_1^2 - \Omega_2^2)$$

giving, assuming that $\Omega_1 + \Omega_2 = 100\,\text{rev min}^{-1}$, $\Omega_1 - \Omega_2 = 4(\omega_1 - \omega_2)$. The variation in the angular speed of the shaft is therefore $\pm 20\%$.

Example 6.15

A flat plate of mass M rotates about a horizontal axis, supported in smooth bearings, that is perpendicular to its plane. The point of intersection of the plate and the axis is O while OG $= h$, where G is the centre of mass of the plate. If θ is the angle between OG and the downward vertical (see Fig. 6.17) at time t, show that $I\dot{\theta}^2 - 2\text{Mgh}\cos\theta$ is constant, I being the moment of inertia of the plate about the axis of rotation.

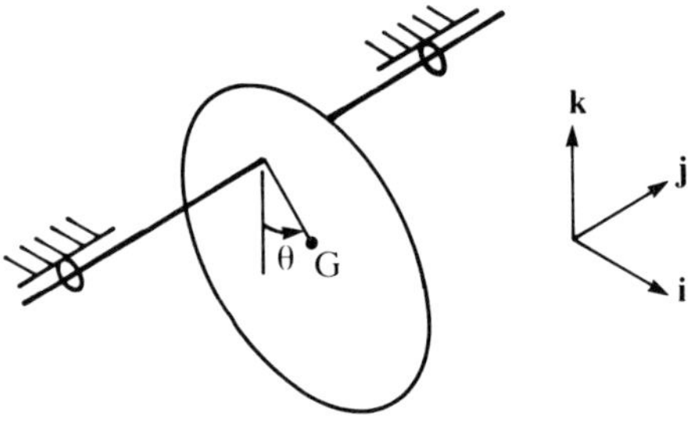

Fig. 6.17

▷ *Solution* If $\dot{\theta} = \Omega$ and $\theta = \alpha$ at time $t = 0$, then using the relations (6.38) and (6.43) we have that $(1/2)I(\dot{\theta}^2 - \Omega^2)$ is the work done by gravity during the time interval t. Now G will have fallen a distance $h(\cos\theta - \cos\alpha)$ in this time interval, and therefore with the help of (6.40) we deduce that

$$\frac{1}{2}I(\dot{\theta}^2 - \Omega^2) = Mgh(\cos\theta - \cos\alpha)$$

or

$$I\dot{\theta}^2 - 2Mgh\cos\theta = I\Omega^2 - 2Mgh\cos\alpha = \text{constant}$$

Differentiating this equation with respect to time yields, after some simplication, the further equation

$$I\ddot{\theta} = -Mgh\sin\theta$$

The equation can be alternatively derived by considering the rate of change of moment of momentum about O.

Example 6.16

A uniform circular disc of mass M and radius a rolls and slips down a line of greatest slope of a plane that is inclined at an angle α to the horizontal. The plane of the disc remains vertical during the motion, which is two-dimensional. Show when the centre C of the disc has moved a distance x parallel to the plane and the disc has turned through an angle θ from rest that

$$M\dot{x}^2 + \frac{1}{2}Ma^2\dot{\theta}^2 = 2Mg\{x\sin\alpha - \mu\cos\alpha(x - a\theta)\}$$

when μ is the coefficient of friction between disc and plane.

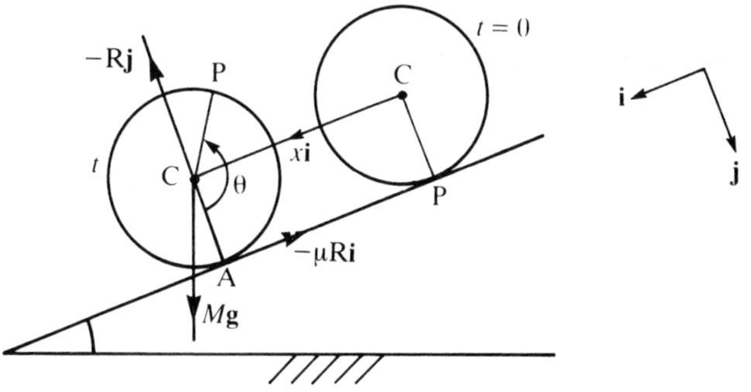

Fig. 6.18

> *Solution* The angular velocity of the disc (see Fig. 6.18) is equal to $\dot{\theta}\mathbf{k}$ while the velocity of its centre of mass is $\dot{x}\mathbf{i}$. Hence using (6.39) we see that the kinetic energy T of the disc is given by

$$2T = M\dot{x}^2 + \frac{1}{2}Ma^2\dot{\theta}^2$$

The forces acting on the disc are the normal reaction of the plane and friction at A, the point of contact of disc and plane, and gravity at C. These forces are given (vectorially) by $-R\mathbf{j} - \mu R\mathbf{i}$ and $Mg(\sin\alpha\mathbf{i} + \cos\alpha\mathbf{j})$. The velocity of the point of the disc in contact with the plane is equal to $\dot{x}\mathbf{i} + \dot{\theta}\mathbf{k} \times \underline{\mathrm{CA}} = (\dot{x} - a\dot{\theta})\mathbf{i}$. Hence applying (6.42) we obtain the equation

$$\frac{dT}{dt} = Mg(\sin\alpha\mathbf{i} + \cos\alpha\mathbf{j}) \cdot \dot{x}\mathbf{i} + (-\mu R\mathbf{i} - R\mathbf{j}) \cdot (\dot{x} - a\dot{\theta})\mathbf{i}$$

Since there is no motion of the disc perpendicular to the plane it follows, using the equation of the motion of the centre of mass C, that $R = Mg\cos\alpha$. Thus

$$\frac{dT}{dt} = \dot{x}Mg\sin\alpha - \mu Mg\cos\alpha(\dot{x} - a\dot{\theta})$$

Integrating with respect to time yields

$$T = Mg\sin\alpha x - \mu Mg(x - a\theta) + \text{constant}$$

Now $x = \theta = 0 = \dot{x} = \dot{\theta}$ at the start of the motion and therefore the constant is zero. The required equation is now derived. The quantity $Mg\sin\alpha x$ represents the work done by gravity during the motion, and $-\mu Mg(x - a\theta)$ that done by friction. The reader should particularly note that the work done by friction is not equal to $-\mu Mgx$. In fact if the motion is one of pure rolling ($\dot{x} - a\dot{\theta} = 0$), friction does no work at all.

6.11 IMPULSE AND MOMENT OF IMPULSE

The equation of motion of a particle of mass m moving with (absolute) velocity $\mathbf{v}$ and acted upon by a force $\mathbf{F}$ whose components referred to $\mathbf{i}, \mathbf{j}, \mathbf{k}$ are $[X, Y, Z]$ is

$$m\frac{d\mathbf{v}}{dt} = \mathbf{F}$$

Integrating both sides of this equation with respect to time from $t = t_0$ to $t = t_1$ yields

$$\int_{t_0}^{t_1} m\frac{d\mathbf{v}}{dt} = \int_{t_0}^{t_1} \mathbf{F}\,dt \tag{6.45}$$

where

$$\int_{t_0}^{t_1} \mathbf{F}\,dt = \left(\int_{t_0}^{t_1} X\,dt\right)\mathbf{i} + \left(\int_{t_0}^{t_1} Y\,dt\right)\mathbf{j} + \left(\int_{t_0}^{t_1} Z\,dt\right)\mathbf{k}$$

Hence

$$m\{\mathbf{v}(t_1) - \mathbf{v}(t_0)\} = \int_{t_0}^{t_1} \mathbf{F}\,dt$$

The integral $\int_{t_0}^{t_1} \mathbf{F}\,dt$ is described as the impulse applied to the particle, in the time interval $t = t_0$ to $t = t_1$, by the force $\mathbf{F}$. Equation (6.45) states that the impulse (applied) is equal to the change in momentum of the particle.

Similarly, integrating both sides of (6.4) with respect to time, again from $t = t_0$ to $t = t_1$, and recalling (6.5) leads to the following for a system of particles:

The vector sum of the impulses due to the external forces is equal to the change in momentum of the system of particles. (6.46)

In general no comparable equation exists between change in moment of momentum and the moments of the impulses. We have, by result (d) of Section 6.6,

$$\frac{\mathrm{d}}{\mathrm{d}t}\mathbf{H}_{\mathrm{G}} = \sum_{s} \boldsymbol{\sigma}_s \times \mathbf{F}_s$$

and hence

$$\int_{t_0}^{t_1} \frac{\mathrm{d}}{\mathrm{d}t}\mathbf{H}_{\mathrm{G}}\,\mathrm{d}t = \int_{t_0}^{t_1} \left(\sum_{s} \boldsymbol{\sigma}_s \times \mathbf{F}_s\right)\mathrm{d}t$$

Now

$$\int_{t_0}^{t_1} (\boldsymbol{\sigma}_s \times \mathbf{F}_s)\,\mathrm{d}t \neq \boldsymbol{\sigma}_s \times \int_{t_0}^{t_1} \mathbf{F}_s\,dt$$

as $\boldsymbol{\sigma}_s$ will, in general, be a function of time. Thus the change in moment of momentum about G cannot be expressed in terms of the moments of the impulses about G. An important exception occurs when the magnitudes $|\mathbf{F}_s|$ of the forces $\mathbf{F}_s$ are very large and the time interval $t_1 - t_0$ is infinitesimal, the impulses $\int_{t_0}^{t_1} \mathbf{F}_s\,\mathrm{d}t$ being finite vectors. We assume that the system of particles does not effectively change its position during the application of the forces, and therefore

$$\begin{aligned}[\mathbf{H}_{\mathrm{G}}]_{t=t_0}^{t=t_1} &= \sum \int_{t_0}^{t_1} (\boldsymbol{\sigma}_s \times \mathbf{F}_s)\,\mathrm{d}t \\ &= \sum \boldsymbol{\sigma}_s \times \int_{t_0}^{t_1} \mathbf{F}_s\,\mathrm{d}t \\ &= \text{sum of moments of external impulses about G} \qquad (6.47)\end{aligned}$$

We observe, noting result (b) of Section 6.6, that G can be replaced by O and $\boldsymbol{\sigma}_s$ by $\mathbf{r}_s$ so that:

The change in moment of momentum about the point O is equal to the sum of the moments of the external impulses about O.

When a large force is applied to a body for a short time we say that the body has been struck by an (impulsive) blow and specify it by giving its magnitude (measured in SI base units $\mathrm{kg\,m\,s^{-1}}$ or SI units N s) and direction.

An application of the laws that relate to impulsive motion is given in the example that now follows.

Example 6.17

A moving rigid body is struck an impulsive blow **J** at the point P. If **u**, **v** are the velocities of P just before and after the blow, show that the change in the kinetic energy of the body is equal to $\mathbf{J} \cdot (\mathbf{u} + \mathbf{v})/2$.

▷ *Solution* Let $\mathbf{v}_1$ be the velocity of the centre of mass G of the body, $\boldsymbol{\omega}_1$ its angular velocity and $\mathbf{H}_1$ its moment of momentum about G before the blow. After the blow let these quantities be equal to $\mathbf{v}_2$, $\boldsymbol{\omega}_2$, $\mathbf{H}_2$. By (6.46) and (6.47)

$$M(\mathbf{v}_2 - \mathbf{v}_1) = \mathbf{J}, \qquad \mathbf{H}_2 - \mathbf{H}_1 = \underline{GP} \times \mathbf{J}$$

The change in kinetic energy is equal to

$$\frac{1}{2}(\boldsymbol{\omega}_2 \cdot \mathbf{H}_2 - \boldsymbol{\omega}_1 \cdot \mathbf{H}_1) + \frac{1}{2}M(\mathbf{v}_2^2 - \mathbf{v}_1^2)$$

By expressing $\boldsymbol{\omega}_1$, $\boldsymbol{\omega}_2$, $\mathbf{H}_1$, $\mathbf{H}_2$ in component form it is seen that $\boldsymbol{\omega}_1 \cdot \mathbf{H}_2 = \boldsymbol{\omega}_2 \cdot \mathbf{H}_1$ and therefore the change in the kinetic energy is given by

$$\frac{1}{2}M(\mathbf{v}_2 - \mathbf{v}_1) \cdot (\mathbf{v}_2 + \mathbf{v}_1) + \frac{1}{2}(\boldsymbol{\omega}_1 + \boldsymbol{\omega}_2) \cdot (\mathbf{H}_2 - \mathbf{H}_1)$$

which is equal to

$$\frac{1}{2}\mathbf{J}(\mathbf{v}_1 + \mathbf{v}_2) + (\boldsymbol{\omega}_1 + \boldsymbol{\omega}_2) \cdot (\underline{GP} \times \mathbf{J})$$

This becomes after some manipulation

$$\frac{1}{2}\mathbf{J}(\mathbf{v}_1 + \boldsymbol{\omega}_1 \times \underline{GP} + \mathbf{v}_2 + \boldsymbol{\omega}_2 \times \underline{GP}) = \frac{1}{2}\mathbf{J} \cdot (\mathbf{u} + \mathbf{v})$$

PROBLEMS

Section 6.7

6.1 Masses of 2.5, 3.5, 6.5 kg are attached to a shaft in a plane perpendicular to its length at radii 8, 11, 6 cm and at angular positions 0°, 50°, 210° with respect to a fixed line in the plane. The shaft is supported in two bearings, one each side of the plane at distances 5 and 5.5 cm from it. Assuming that the shaft is perfectly balanced, find the out-of-balance force on each bearing when the shaft rotates at 30 rad s^{-1}.

Find the magnitude and angular setting of the single mass at 10 cm radius that will balance the shaft and particles.

6.2 A uniform rectangular plate of mass M and sides $2a$, $2b$, where $a > b$, is attached to a shaft so that the axis of the shaft and a diagonal of the plate are coincident. If the shaft rotates with constant angular speed ω, show that the reactions on the bearings of the shaft are equivalent to a couple whose moment is perpendicular to the plane of the plate and is of magnitude $(1/3)abM\omega^2(a^2 - b^2)/(a^2 + b^2)$.
Hint: in order to find products of inertia use the fact that a triangular plate is equimomental with three particles placed at the midpoints of its sides.

6.3 Details of the distribution of out-of-balance masses in a rotor are given as follows. A and B are the correction planes, X and Y are the planes of the bearings and p is measured from plane A along the axis of the rotor. Determine the rotational speed at which the rotating out-of-balance force on bearing X has a magnitude of 45 N and also the magnitude of the rotating out-of-balance force on bearing Y at this speed.

Find the magnitude and angular positions of the masses required at a radius of 40 mm in planes A and B to balance the rotor completely.

Plane	*Mass*/g	*Radius*/mm	*Angle*/degree	p/mm
A				0
X				50
1	36.2	65	90	100
2	49.5	65	30	200
3	68.7	40	180	300
Y				350
B				400

6.4 A balanced shaft runs in bearings X and Y that are 1 m apart. Attached to the shaft are three masses m_i ($i = 1, 2, 3$). The information below gives the distances p_i of the masses, measured along the axis, from bearing X and the products $m_i r_i$, where r_i is the radial distance of the mass m_i from the axis of the shaft. The angular positions of the masses are adjusted to give static balance. Determine the magnitudes of the rotating out-of-balance forces on the bearings when the angular speed of the shaft is 3000 rev min^{-1}. Find the magnitudes and angular positions of the two masses, located at radii of 0.1 m and having distances from bearing X of 0.3 m and 1.1 m, that will balance the shaft and particles.

Mass/kg	p_i/m	$m_i r_i/10^{-4}$ kg m
m_1	0.3	12.0
m_2	0.8	9.6
m_3	1.1	7.2

6.5 A rotor is supported on bearings at X and Y that are 0.41 m apart. The out-of-balance masses in the rotor are equivalent to a mass m_1 at radius r_1 and angular position of 0° in the plane of X, where $m_1 r_1 = 3.61 \times 10^{-4}$ kg m, and a mass m_2 at radius r_2 at 130° in the plane of Y, where $m_2 r_2 = 2.16 \times 10^{-4}$ kg m. The correction planes A and B are 0.25 m apart, with A being 0.08 m from X and B being 0.08 m from Y.

Determine the magnitudes and angular positions of the masses (treated as point masses) that must be removed at a radius of 0.15 m in the planes A and B to balance the rotor completely.

If the balancing mass in plane B only is removed, what are the magnitudes of the rotating out-of-balance forces on the bearings when the rotational speed is 3000 rev min^{-1}, assuming the bearings act as simple supports.

6.6 Determine for the mechanism in Example 6.3 (see Fig. 6.3) the forces exerted on the fixed link at the revolute joints O and P when $\theta_2 = 30°$. Determine also the magnitude of the torque that is applied to link 2.

6.7 In the mechanism shown diagrammatically in Fig. 6.19, link 2 is the driving link. The mass of link 3 is 7.3 kg, its centre of mass is at G_3, and its moment of inertia about the axis through G_3 perpendicular to the plane of the motion is 0.06 kg m^2. The acceleration of G_3 is -450 m s^{-2}**i** $-$ 2245 m s^{-2}**j** and the angular acceleration of link 3 is 4900 rad s^{-2}**k**. Determine the forces at the joints O and P and the

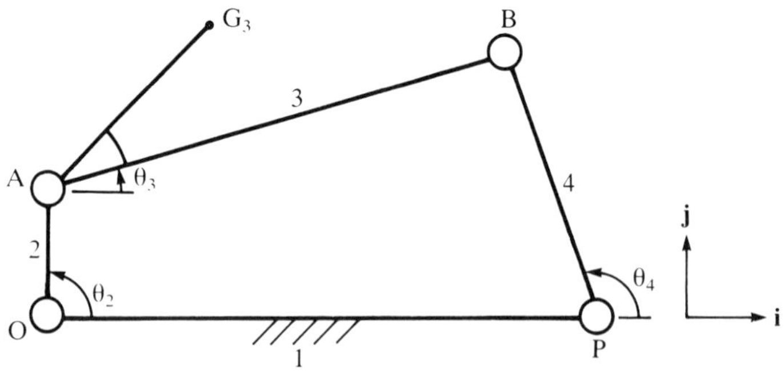

Fig. 6.19

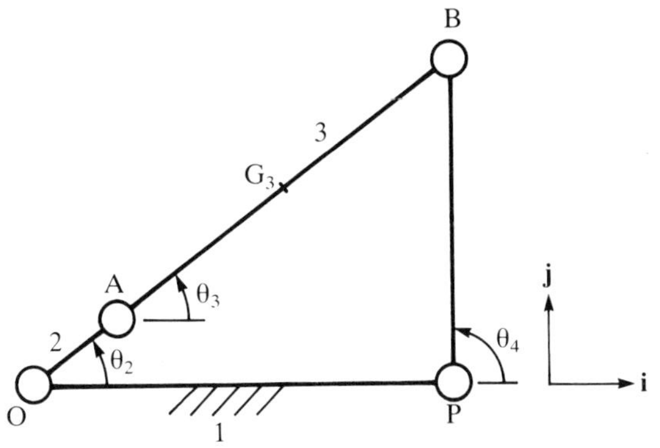

Fig. 6.20

external torque on link 2 due to the inertia of link 3. Neglect friction and gravity forces.
(See note subsequent to Example 6.3 for definition of inertia forces.)

The dimensions of the mechanism are: OP = 0.36 m, OA = 0.08 m, AB = 0.31 m, PB = 0.18 m, AG_3 = 0.16 m, $\angle BAG_3 = 30°$, $\theta_2 = 90°$, $\theta_3 = 11.61°$ and $\theta_4 = 110.47°$.

6.8 In the mechanism shown diagrammatically in Fig. 6.20 link 2 is the driving crank. For the position shown the acceleration of the point A is $-200\,\text{m s}^{-2}\mathbf{i} - 150\,\text{m s}^{-2}\mathbf{j}$ and the acceleration of the point B is $-390.6\,\text{m s}^{-2}\mathbf{i}$. The mass of link 3 is 4 kg, its centre of mass is at G_3, and its moment of inertia about the axis through G_3 perpendicular to the plane of motion is $0.08\,\text{kg m}^2$.

Determine the forces on the fixed link at O and P and the torque on link 2, assuming that link 2 is statically balanced (so that its centre of mass is at O), due to the inertia of links 2 and 3. Neglect gravity and friction forces.

The dimensions of the mechanism are as follows: OP = 0.4 m, OA = 0.1 m, AB = 0.4 m, AG_3 = 0.2 m, PB = 0.3, $\theta_2 = \theta_3 = \arctan 3/4$ and $\theta_4 = 90°$.

6.9 Part of a mechanism is shown in Fig. 6.21. Link 2 is the driving link, while the relative motion of link 3 and link 4 at A is along the line of link 4. The mass of link 4 is 6.4 kg, its centre of mass is at G_4, and its moment of inertia about the axis through G_4 perpendicular to the plane of motion is $0.17\,\text{kg m}^2$. The mass of link 3 is 1 kg, its centre of mass being at A. The moment of inertia of link 3 about an axis through A parallel to the unit vector **k** is negligible. Link 2 is statically balanced (so that its centre of mass is at O) while its moment of inertia about an axis through O perpendicular to the plane of motion is $0.04\,\text{kg m}^2$. Given that the acceleration of G_4 is $-192\,\text{m s}^{-2}\mathbf{i} -$

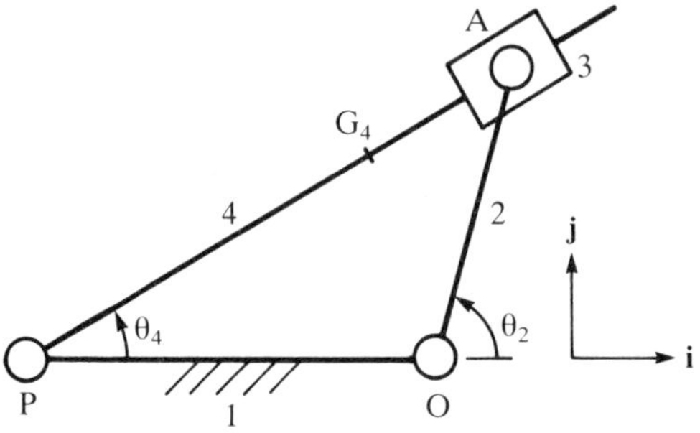

Fig. 6.21

$400\,\text{m}\,\text{s}^{-2}\mathbf{j}$, find the angular accelerations of links 2 and 4. Find also the forces at the joints O and P and the external torque applied to link 2. Neglect forces due to gravity and friction.

The mechanism has the dimensions: $OP = 0.25\,\text{m}$, $OA = 0.18\,\text{m}$, $PG_4 = 0.24\,\text{m}$, $PA = 0.344\,\text{m}$, $\theta_4 = 30.38°$ and $\theta_2 = 75°$

6.10 In the mechanism shown diagrammatically in Fig. 6.22 link 2 is the driving crank. Link 4 has been balanced so that its centre of mass G_4 is coincident with the point P, and its moment of inertia about the axis through G_4 perpendicular to the plane of motion is $0.036\,\text{kg}\,\text{m}^2$. For the position shown the angular acceleration of link 4 is $-915\,\text{rad}\,\text{s}^{-2}\mathbf{k}$. Determine the forces on the bearings at the points O, A, B and P and the torque on link 2 due to the inertia of link 4. Neglect forces due to gravity and friction.

The dimensions are given by: $\underline{OP} = 0.6\,\text{m}\mathbf{i} + 0.24\,\text{m}\mathbf{j}$, $OA = 0.12\,\text{m}$, $AB = 0.48\,\text{m}$, $PB = 0.24\,\text{m}$, $\theta_2 = 45°$, $\theta_3 = 350.3°$ and $\theta_4 = 260.0°$.

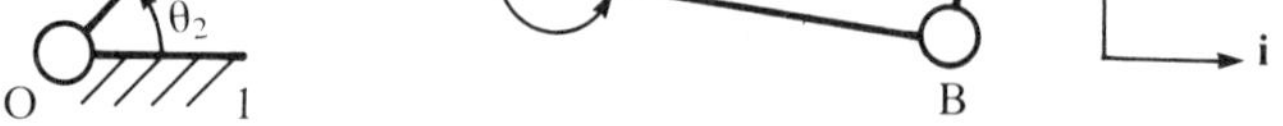

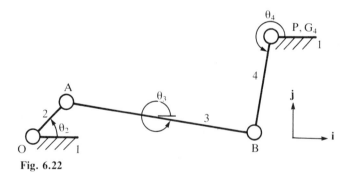

Fig. 6.22

6.11 Show when the disc in Example 6.16 rolls, without slipping, down the plane the frictional force between the disc and the plane is equal to $-(1/3)Mg\sin\alpha\mathbf{i}$. Hence deduce for rolling without slipping that $3\mu \geq \tan\alpha$.

6.12 A uniform rod AB of mass M and length $2l$ can rotate freely in a vertical plane S about a horizontal axis L passing through the end A. The axis is given a constant horizontal accceleration $f\mathbf{i}$ perpendicular to its length (see Fig. 6.23). Initially the rod is vertical and at rest with B vertically below A. At time t the rod is inclined to the downward vertical at an angle θ. The horizontal and vertical components of the force exerted by the axis on the rod are $F_1\mathbf{i}$ and $F_2\mathbf{j}$. Derive the equations governing the motion of the centre of mass G of the rod and its rate of change of moment of momentum about G. Eliminate F_1, F_2 from these equations to show that

$$4l\ddot{\theta} = 3(f\cos\theta - g\sin\theta)$$

Show further that

$$2l\dot{\theta}^2 = 3(f\sin\theta + g\cos\theta - g)$$

Deduce that $\dot{\theta}$ is zero when $\tan\theta/2 = f/g$
(Note in Fig. 6.23 the unit vector $\mathbf{k}$ is into the paper.)

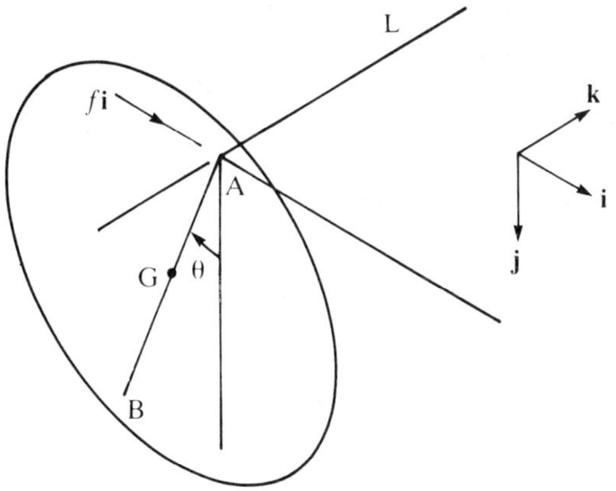

Fig. 6.23

6.13 The arrangement of a four-cylinder radial engine is shown diagrammatically in Fig. 6.24. The hatched lines that denote the centre lines of the cylinders lie in one plane, being spaced at intervals of 90°. The pistons are connected to the crank OA by rods AB, AC, AD, AE of equal length. The ratio of the length of a rod to the crank radius r is n, where $n > 3.5$, while the reciprocating masses are each equal to m. It may be assumed that during motion the crank and connecting rods move in the same plane, the angular velocity of the crank being $\omega\mathbf{k}$, where $\mathbf{k}$ is perpendicular to the plane of the centre lines of the cylinders.

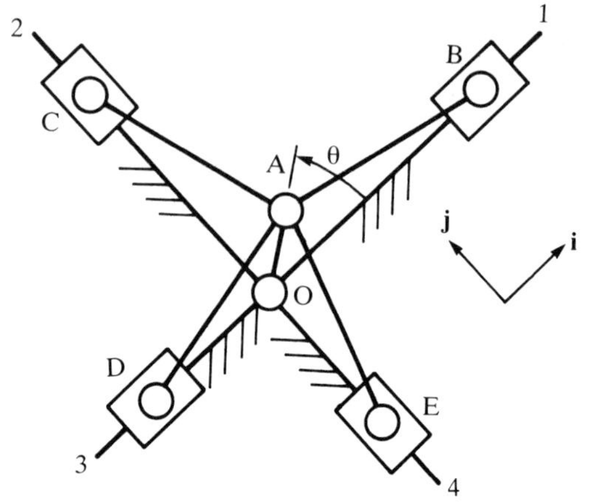

Fig. 6.24

Show that the resultant of the secondary forces is zero and that the resultant of the primary forces may be balanced by the addition of a mass to the crank-shaft opposite to the crank-pin A. Determine the magnitude of the balancing mass if it is added at a radius r.

If the engine were arranged so that pistons 2 and 4 moved in a plane parallel to the plane of the motion of the pistons 1 and 3, being attached to the crank-shaft by a crank parallel to OA, what effect would this have on the balance of the engine?

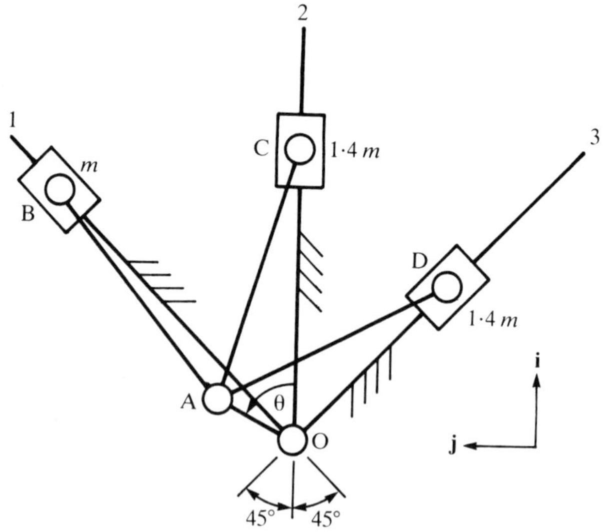

Fig. 6.25

6.14 The arrangement of a three-cylinder reciprocating compressor is shown in Fig. 6.25. The centre lines of the cylinders are in one plane, the pistons being connected to the common crank OA, radius r, by rods AB, AC, AD each of the same length. The ratio of the length of a rod to the radius of the crank is n (>3.5) while the reciprocating masses are m at B and $1.4m$ at C and D. The angular velocity of the crank OA is constant and equal to $\omega\mathbf{k}$, and at time $t = 0$ it lies along the centre line 2. Show that the resultant of the primary forces is equal to

$$m\omega^2 r\{\sqrt{(1.48)}\cos(\omega t + \beta)\mathbf{i} + \sqrt{(6.80)}\cos(\omega t + \delta)\mathbf{j}\}$$

where $\tan\beta = 6$, $13\tan\delta = 1$ and $0 < \delta < \beta < \pi/2$.

A mass αm is attached to the crank (extended) at a point E distance r from O and on the opposite side of O to A. Find the magnitude of the primary forces (including that due to the mass) and deduce that when $\alpha = 1.9$ it is constant. For this value of α show that the resultant of the primary forces is given by

$$m\omega^2 r\sqrt{(0.53)}\{\sin(\omega t + \varepsilon)\mathbf{i} + \cos(\omega t + \varepsilon)\mathbf{j}\}$$

where $7\tan\varepsilon = 2$ and $0 < \varepsilon < \pi/2$.

6.15 Figure 6.26(a) shows diagrammatically the vee-twin engine. It consists of a pair of pistons connected to a crank OA by rods AB, AC of equal length, the angle between the centre lines OR, OL of the cylinders being 90°. The rods and centre lines of the cylinder may be assumed to be coplanar, while the angular velocity of the crank OA is equal to $\omega\mathbf{k}$, where ω is a constant.

A vee-eight engine has eight pistons that are arranged as four pairs,

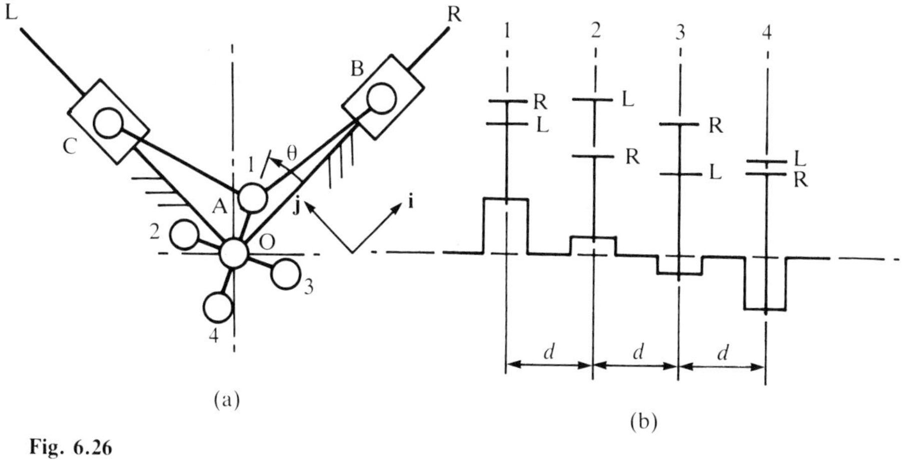

Fig. 6.26

each pair being identical with the vee-twin engine of Fig. 6.26(a). The planes, numbered 1 to 4, containing the centre lines of the pairs of cylinders, are parallel and at distance d apart, while the cranks are spaced at 90°, the spacings being as depicted in Fig. 6.26(b).

If the reciprocating masses are all equal to m, find expressions for the primary and secondary forces and couples in terms of m, ω, r and n (which has its usual meaning).

6.16 In a crank-slider mechanism (see Fig. 6.6) the radius of the crank OA is 31.5 mm. The length of the connecting rod is 105 mm, its centre of mass G_3 being at a distance 19 mm from A. The mass of the connecting rod is 0.617 kg and that of the piston is 0.36 kg, while the moment of inertia of the rod about an axis through G_3 perpendicular to the plane of motion is 1.79×10^{-3} kg m^2. Given that $\theta_2 = 30°$ and the angular speed of the crank is 5500 rev min^{-1}, find by exact means:

(a) The angular velocity and acceleration of link 3 and the acceleration of the piston

(b) The force on the crank bearings due to the inertia of the connecting rod and the piston.

Find by approximate means the acceleration of the piston. Find also the force on the crank bearings, again due to the inertias of link 3 and the piston, by replacing the connecting rod by a light rod and two masses suitably chosen. Use the approximate value of the piston's acceleration.

Section 6.8

6.17 Show that the sum of the three moments of inertia of a rigid body about any set of three mutually perpendicular axes through a point O is constant.

6.18 Show that the moment of inertia of a uniform spherical shell of mass M and radius a about a diameter is $2Ma^2/3$. The mass of the shell may be assumed to be concentrated on the surface of a sphere of radius a. *Hint:* Use the definition of a moment of inertia to show that the sum of the moments of inertia about three mutually perpendicular diameters is $2Ma^2$. Make use of this result for a spherical shell to show, by integration, that the moment of inertia of a uniform sphere of mass M, radius a, about a diameter is $2Ma^2/5$.

6.19 The inertia matrix at O with respect to the axis Ox, Oy, Oz is $A\mathbf{I}$. Show that any set of mutually perpendicular axes through O is a set of principal axes at O.

6.20 A uniform rod of mass M, length l, is equimomental with two particles of masses M_1, M_2 placed at points on the rod distances l_1, l_2 from its midpoint. Show that $12l_1 l_2 = l^2$. What can be inferred from this result?

6.21 Show that the moment of inertia of a uniform circular hoop of mass M, radius a, about an axis through its centre perpendicular to its plane is Ma^2. Use this result to show, by integration, that the moment of inertia of a uniform circular disc of mass M, radius a, about an axis through its centre perpendicular to its plane is $Ma^2/2$. What is the moment of inertia of the disc about a diameter? The disc may be assumed to be of negligible thickness.

6.22 A uniform square plate is of mass M and side $2a$, its centre of mass being the point O. Find the inertia matrix at O for axes Ox, Oy, Oz, where Ox, Oy are parallel to the sides of the plate, using the fact that a triangular plate is equimomental with three particles placed at the midpoints of its sides.

Determine the inertia matrix at the corner A of the plate for mutually perpendicular axes AX, AY, AZ where AX passes through O and AY lies in the plane of the plate by

(a) Using the theorem of parallel axes and then the transformation law for moments of inertia

(b) Using the transformation law and then the theorem of parallel axes.

6.23 A, B, C are the vertices of a uniform triangular plate, the angle at C being a right angle. Find the angles which the principal axes of inertia make at C with the side CA, given that the lengths of the sides CA, CB of the plate are a, b.

6.24 (a) By considering the product $\det\mathbf{L}\det(\mathbf{I}_\mathrm{O} - \alpha\mathbf{I})\det\mathbf{L}^\mathrm{T}$, where the matrix $\mathbf{L}$ has the meaning given to it in the relation (6.31), show that the roots of the two cubic equations (in α) $\det(\mathbf{I}_\mathrm{O} - \alpha\mathbf{I}) = 0$ and $\det(\mathbf{I}'_\mathrm{O} - \alpha\mathbf{I}) = 0$ are identical. Deduce that if principal axes exist at O then the principal moments of inertia must be given by the roots of $\det(\mathbf{I}_\mathrm{O} - \alpha\mathbf{I}) = 0$. *Hint:* take O$a'$, O$b'$, O$c'$ to be principal axes at O and find the roots of $\det(\mathbf{I}'_\mathrm{O} - \alpha\mathbf{I}) = 0$.

(b) The inertia matrix $\mathbf{I}_\mathrm{O}$ is given by

$$\mathbf{I}_\mathrm{O} = Ma^2 \begin{bmatrix} 8 & -3 & -3 \\ -3 & 8 & -3 \\ -3 & -3 & 8 \end{bmatrix}$$

Find the principal moments of inertia at O.

(c) A hollow box of mass M is made up of six thin square plates each of side $2a$. Find the inertia matrix for a set of right-handed axes that coincide with a set of three concurrent edges. Noting the result in part (a), determine the principal moments of inertia at a vertex of the box.

Section 6.9

6.25 A rate gyroscope is shown diagrammatically in Fig. 6.27. The gimbal is mounted in bearings at A and B fixed in the supporting platform so that it can rotate relative to the platform about AB. The midpoint of AB is the point G which is also the position of the centres of mass of the rotor and the gimbal. The rotor is mounted in bearings in the gimbal at C and D so that it can rotate relative to the gimbal about the axis CD which is perpendicular to AB and passes through G. The axis CD is an axis of symmetry for the rotor. The mutually perpendicular right-handed axes Ga, Gb, Gc (see Fig. 6.27) are fixed in the gimbal, being such that Ga is in the direction of $\underline{\text{AB}}$ and Gc is in the direction of $\underline{\text{CD}}$. The axes Ga', Gb', Gc' are fixed in the platform, Ga' and Ga being coincident while the angle between Gc and Gc' is α. The unit vectors associated with the axes are $\mathbf{a}$, $\mathbf{b}$, $\mathbf{c}$ and $\mathbf{a}'$, $\mathbf{b}'$, $\mathbf{c}'$. The angular velocity $\boldsymbol{\omega}$ of the rotor relative to the angular velocity of the gimbal is $n\mathbf{c}$, while the supporting platform rotates about the axis Gb', which is fixed, with constant angular speed p. The rotation of the gimbal relative to the platform is constrained by springs exerting a moment $-k\alpha\mathbf{a}$ and by viscous dampers exerting a moment $-f\dot{\alpha}\mathbf{a}$ on the gimbal.

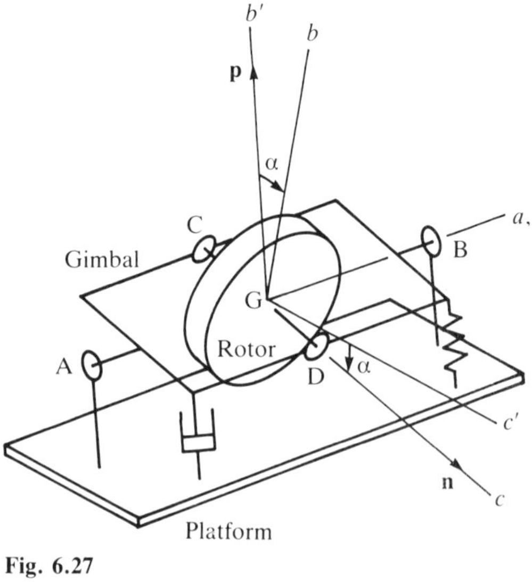

Fig. 6.27

Obtain expressions for the components of the angular velocities of the rotor and gimbal referred to the unit vectors **a**, **b**, **c**, and hence derive the equation

$$A\ddot{\alpha} + f\dot{\alpha} + k\alpha + (B - C)p^2\cos\alpha\sin\alpha + C'np\cos\alpha = 0$$

where A, B, C are the combined moments of inertia of the rotor and gimbal about Ga, Gb, Gc, and C' is the moment of inertia of the rotor about Gc. It may be assumed that Ga, Gb, Gc are principal axes for the gimbal at G.

Show that the bearings at A and B exert a couple on the gimbal and obtain an expression for this couple referred to the unit vectors **b** and **c**. Neglect gravity.

6.26 The axis CD shown in Fig. 6.28 is fixed at right angles to the rotating shaft AB. A uniform circular ring for which CD is a diameter is carried in bearings at C and D that enable it to rotate about the axis CD. The moment of inertia of the ring about a diameter is A and that about an axis through its centre perpendicular to its plane is $2A$. A torque $-k\theta\mathbf{a}$ is applied to the ring, where **a** is a unit vector in the direction of $\underline{\text{CD}}$ and θ is the angle (measured anticlockwise about **a**) between AB and the plane of the ring). If Ω is the constant angular speed of the shaft AB, show that

$$A\ddot{\theta} - A\Omega^2\sin\theta\cos\theta = -k\theta$$

and find the couple that must be applied to the shaft to maintain the motion.

Given that $A\Omega^2 = 2k$, obtain the steady-state values of θ.

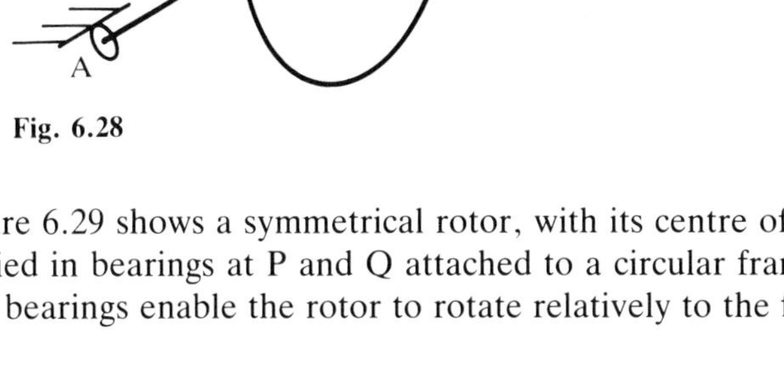

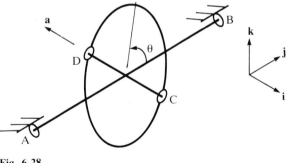

Fig. 6.28

6.27 Figure 6.29 shows a symmetrical rotor, with its centre of mass at G, carried in bearings at P and Q attached to a circular frame radius r. The bearings enable the rotor to rotate relatively to the frame about

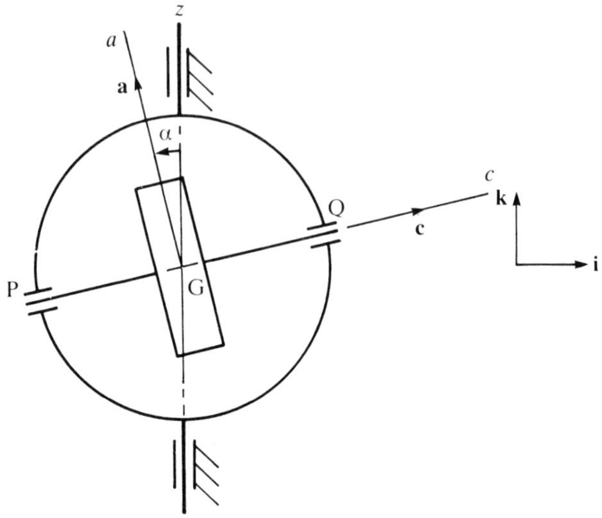

Fig. 6.29

the axis PQ which is the axis of symmetry of the rotor. The circular frame rotates with constant angular velocity $\omega\mathbf{k}$ about the fixed axis Gz. The axes Ga, Gb, Gc shown in Fig. 6.29 are defined as follows: Gc lies along PQ and is in the direction of $\underline{\text{PQ}}$; Ga is perpendicular to Gc and is in the plane defined by Gc and Gz; and Gb is chosen so that the axes are, in the usual way, right-handed and mutually perpendicular. The associated unit vectors are **a**, **b**, **c** while the angle between Ga and Gz is α.

Explain why α is constant, and show that Ga, Gb, Gc are fixed relative to the circular frame. The angular velocity of the rotor relative to the frame is $n\mathbf{c}$, where n is a constant, and the moments of inertia of the rotor about Ga, Gb, Gc are $0.65C$, $0.65C$, C. Find the components, referred to **a**, **b**, **c**, of the forces other than those due to gravity that act on the bearings at P and Q.

6.28 A freely suspended symmetrical rotor is shown in Fig. 6.30. The outer gimbal ring which rotates about the fixed axis Gz carries the inner gimbal in bearings attached to it at E and F. The bearings enable the inner gimbal to rotate relative to the outer about EF. The rotor is carried in bearings attached to the points P, Q of the inner gimbal and can rotate relative to the inner gimbal about PQ. The lines EF and PQ are perpendicular to each other and pass through the centre of mass G of the rotor. The axis EF is perpendicular to Gz, while PQ is the axis of symmetry of the rotor. The right-handed axes Ga, Gb, Gc fixed in the inner gimbal are defined as follows: Gb is in the direction of $\underline{\text{EF}}$, and Gc is in the direction of $\underline{\text{PQ}}$.

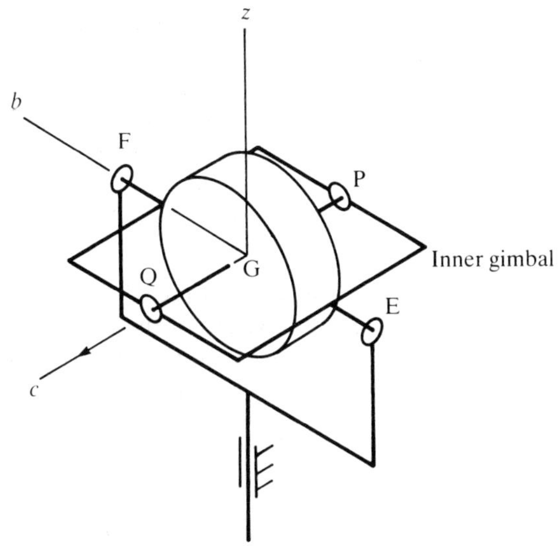

Fig. 6.30

If the angle between Ga and Gz is θ and the angular velocity of the outer gimbal is $\dot{\phi}\mathbf{k}$, show that

$$(A + I_2)\ddot{\theta} + (I_1 - I_3 + A - C)\dot{\phi}^2\sin\theta\cos\theta - Cn\dot{\phi}\cos\theta = 0$$

where A, A, C are the moments of inertia of the rotor about Ga, Gb, Gc; I_1, I_2, I_3 are those of the inner gimbal about Ga, Gb, Gc; and $n\mathbf{c}$ is the angular velocity of the rotor relative to the inner gimbal.

6.29 Using the notation defined for the motion of a top (see Example 6.12) derive the following equations:

$$\dot{\phi}\cos\theta + \dot{\psi} = \text{constant} = n$$
$$A(\ddot{\theta} - \dot{\phi}^2\sin\theta\cos\theta) + Cn\dot{\phi}\sin\theta = Mgh\sin\theta$$
$$A\ddot{\phi}\sin\theta + 2A\dot{\theta}\dot{\phi}\cos\theta - Cn\dot{\theta} = 0$$

6.30 A uniform right-circular cylinder of mass m, radius r and length l is rigidly attached to a rigid shaft of negligible mass supported in short bearings distance l apart. The axis of rotation passes through the centre of mass of the cylinder and the angle between the axis of rotation and the axis of the cylinder is θ. Obtain expressions for the dynamic reactions of the bearings when the angular velocity $\boldsymbol{\omega}$ of the shaft is a constant vector.

Section 6.10

6.31 The torque produced by an engine is $(2100\sin\theta + 900\sin 2\theta)$ N m for $0 \leq \theta \leq \pi$ and $375\sin\theta$ N m for $\pi \leq \theta \leq 2\pi$, where θ is the crank angle. This cycle is then repeated. The load is directly coupled to the engine and may be assumed to exert a constant torque on the engine. The average rotational speed in each cycle is 850 rev min^{-1} and the total moment of inertia of the rotating parts is 270 kg m^2. Determine the maximum instantaneous acceleration of the engine. Show that the minimum rotational speed in the cycle occurs at $\theta = 0.14$ rad and the maximum at $\theta = 2.38$ rad, and determine the percentage fluctuation in the speed of the engine.

6.32 An electric motor exerts a constant torque M on the input shaft of a mechanism. The average angular speed of the shaft during each cycle is ω, which is constant. The torque T required to drive the shaft varies with θ, where $\dot{\theta}$ is the angular speed of the shaft, as follows:

$$T = \begin{cases} A\sin^2(3\theta/4) & 0 \leq \theta \leq 4\pi/3 \\ 0 & 4\pi/3 \leq \theta \leq 2\pi \end{cases}$$

where A is a constant.

Find the torque exerted by the electric motor in terms of A and hence show that the maximum and minimum values of the angular speed of the shaft occur when $\theta = \theta_1$ and $\theta = \theta_2$, where θ_1, θ_2 are roots of $3\cos(3\theta/2) = 1$ and $0 < \theta_1 < \theta_2 < 4\pi/3$.

Determine the total work done by T and M in the interval $\theta = \theta_1$ to $\theta = \theta_2$, and deduce that the difference between the maximum and minimum angular speeds of the shaft is $1.05A/I\omega$, where I is the moment of inertia of the shaft about its axis. It may be assumed that the average of the maximum and minimum values of the angular speed of the shaft is ω.

6.33 A body moves so that one point O of itself is fixed under the action of a couple of moment $\mathbf{M}$. Use Euler's equations to show that $dT/dt = \boldsymbol{\omega} \cdot \mathbf{M}$, where T is the kinetic energy of the body and $\boldsymbol{\omega}$ its angular velocity.

6.34 Show for a top that

$$A(\dot{\theta}^2 + \dot{\phi}^2\sin^2\theta) + Cn^2 + 2Mgh\cos\theta$$

is constant. The symbols have the meanings given in and just prior to Example 6.12.

Section 6.11

6.35 In an impact shear test a heavy pendulum carrying a hammer at its lower end is released from rest at an angle of 60° to the downward vertical. At the bottom of its swing the hammer meets the test piece at a point 1 m below the pivot of the pendulum and, after shearing through it, rises to an angle of 30° to the downward vertical. If the mass of the pendulum and hammer is 25 kg, the distance of its centre of gravity from the pivot is 0.75 m, and its moment of inertia about the pivotal axis is 60 kg m^2, find

(a) The energy dissipated during the impulse

(b) The total impulse of the test piece on the hammer.

6.36 A swing door, of width 0.9 m and mass 30 kg, is hinged about a vertical edge. It is struck horizontally when closed by a blow of magnitude 50 N s at the midpoint of the unhinged side. In the ensuing motion the door swings through 3 radians and strikes an inelastic door-stop at ground level 0.6 m from the line of the hinges. Find the impulsive reaction of the stop, assuming that the motion of the door is resisted by a constant torque of 10 N m. Determine the impulsive reaction on the hinges.

7 Robots

7.1 INTRODUCTION

This chapter provides an introduction to the kinematics and dynamics of robots whose links are connected by revolute joints. The rotation theory of Chapter 5 provides the basis of the work on kinematics while Lagrange's equations are used to obtain the dynamical equations of motion. Lagrange's equations are stated without proof and their use is explained by reference to the motion of a top.

7.2 DEFINITION OF A ROBOT

For our purposes we regard a robot as an assembly of links connected by prismatic and/or revolute joints. The links, in contrast to those of the mechanisms considered in Chapter 4, do not form closed loops and as a consequence more than one independent variable is required to specify the position of a robot at any instant. As in Chapter 4 a revolute joint permits relative rotation, between the two links it connects, about a single axis that is fixed relative to them (see also Section 6.9).

7.3 ROTATED AXIS THEOREM

In a two-dimensional mechanism the axes associated with the revolute joints are parallel to each other. However, in a robot the axes are not all parallel (see Fig. 7.1) and this accounts in some measure for their complexity. For example (see Fig. 7.1 again) if we rotate links 1, 2, 3 about the fixed axis 1, then the directions of axes 2 and 3 are changed and clearly any analysis of subsequent rotations about these two axes must take the changes into consideration. In order to do this we require a further theorem on

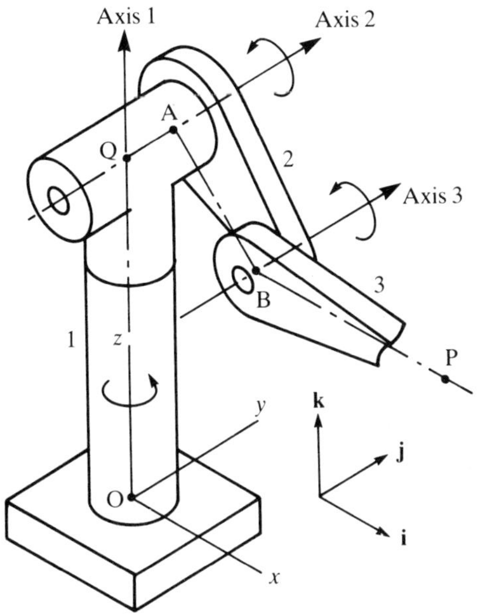

Fig. 7.1

rotations which, for the purposes of reference, we call the rotated axis theorem. The proof of the theorem is not given here, but readers who wish to derive one can do so by working through the various parts of Problem 7.2. Problem 7.1 asks for a proof of a particular application of the theorem.

The rotated axis theorem states that:

If $\mathbf{l}$, $\mathbf{m}$ are unit vectors and $\mathbf{n} = \mathbf{A}(\theta, \mathbf{l})\mathbf{m}$ then

$$\mathbf{A}(\theta, \mathbf{l})\mathbf{A}(\phi, \mathbf{m}) = \mathbf{A}(\phi, \mathbf{n})\mathbf{A}(\theta, \mathbf{l}) \tag{7.1}$$

Alternatively, in words:

A rotation of amount ϕ about an axis OM in the direction of the unit vector $\mathbf{m}$ followed by a rotation of amount θ about the axis OL in the direction of $\mathbf{l}$ (see Fig. 7.2(a)) is equivalent to a rotation of amount θ about OL followed by a rotation of amount ϕ about ON, where ON is obtained by rotating OM about OL through the angle θ (see Fig. 7.2(b) and (c)).

7.4 KINEMATIC ANALYSIS OF ROBOT WITH THREE LINKS CONNECTED BY REVOLUTE JOINTS

Figure 7.3 shows diagrammatically a robot formed of three links and three revolute joints. Link 1 is attached to the fixed point O by means of a

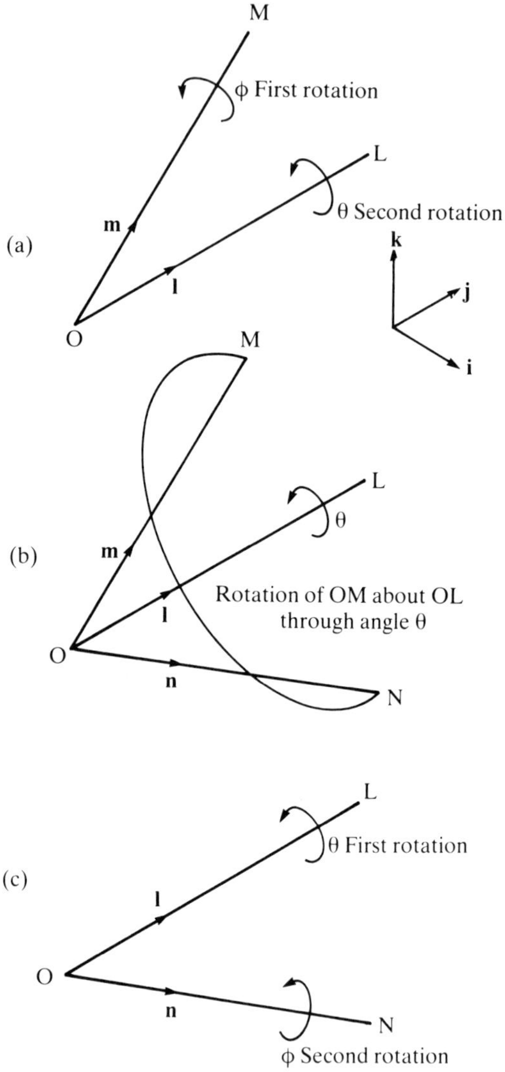

Fig. 7.2

revolute joint and can rotate about the fixed axis 1 which is in the direction of the unit vector $\mathbf{n}_1$. The revolute joint connecting links 1 and 2 is at A and that connecting links 2 and 3 is at B.

We begin our study of the robot's kinematics by considering a particular motion of the robot. Firstly links 1, 2, 3 are rotated about axis 1 an amount θ_1, with no relative rotation at A or B. Next links 2 and 3 are

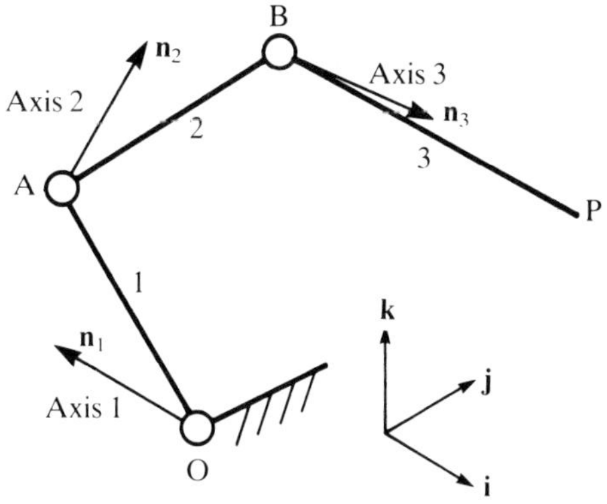

Fig. 7.3

rotated an amount θ_2 about axis 2 (this implies that link 1 is fixed and that there is no relative rotation at B). Finally link 3 is rotated an amount θ_3 about axis 3. Our task is to obtain an expression for the displacement, due to the three rotations, of a point P fixed in link 3. It is essential in our analysis to distinguish between the points O, A, B, P, . . . , fixed in the links of the robot, and the (successive) positions they occupy in space as the robot moves. Let A_0, B_0, P_0 be the positions occupied by A, B, P and let $\mathbf{n}_2$, $\mathbf{n}_3$ be unit vectors in the directions of axes 2 and 3 before the first rotation. As a consequence of the first rotation let A go from A_0 to A_1, B from B_0 to B_1 and P from P_0 to P_1. Using Fig. 7.4(a), the relation (5.15) and the matrix form of (5.7), we have

$$\underline{OA_1} = \mathbf{A}(\theta_1, \mathbf{n}_1)\underline{OA_0}, \qquad \underline{A_1B_1} = \mathbf{A}(\theta_1, \mathbf{n}_1)\underline{A_0B_0}, \qquad \underline{B_1P_1} = \mathbf{A}(\theta_1, \mathbf{n}_1)\underline{B_0P_0} \tag{7.2}$$

The axes 2 and 3 will, by virtue of the relation (5.16), be in the direction of the unit vectors $\mathbf{N}_2$, $\mathbf{L}_3$ after the first rotation, where

$$\mathbf{N}_2 = \mathbf{A}(\theta_1, \mathbf{n}_1)\mathbf{n}_2, \qquad \mathbf{L}_3 = \mathbf{A}(\theta_1, \mathbf{n}_1)\mathbf{n}_3 \tag{7.3}$$

We now let B go from B_1 to B_2 and P from P_1 to P_2 as a consequence of the second rotation (see Fig. 7.4(b)). Using (5.15) and (5.7) once more and noting (7.2) it follows that

$$\underline{A_1B_2} = \mathbf{A}(\theta_2, \mathbf{N}_2)\underline{A_1B_1} = \mathbf{A}(\theta_2, \mathbf{N}_2)\mathbf{A}(\theta_1, \mathbf{n}_1)\underline{A_0B_0} \tag{7.4}$$

$$\underline{B_2P_2} = \mathbf{A}(\theta_2, \mathbf{N}_2)\underline{B_1P_1} = \mathbf{A}(\theta_2, \mathbf{N}_2)\mathbf{A}(\theta_1, \mathbf{n}_1)\underline{B_0P_0} \tag{7.5}$$

Recalling (5.16) it is seen that axis 3 is now in the direction of the unit vector $\mathbf{N}_3$, where

$$\mathbf{N}_3 = \mathbf{A}(\theta_2, \mathbf{N}_2)\mathbf{L}_3 = \mathbf{A}(\theta_2, \mathbf{N}_2)\mathbf{A}(\theta_1, \mathbf{n}_1)\mathbf{n}_3 \tag{7.6}$$

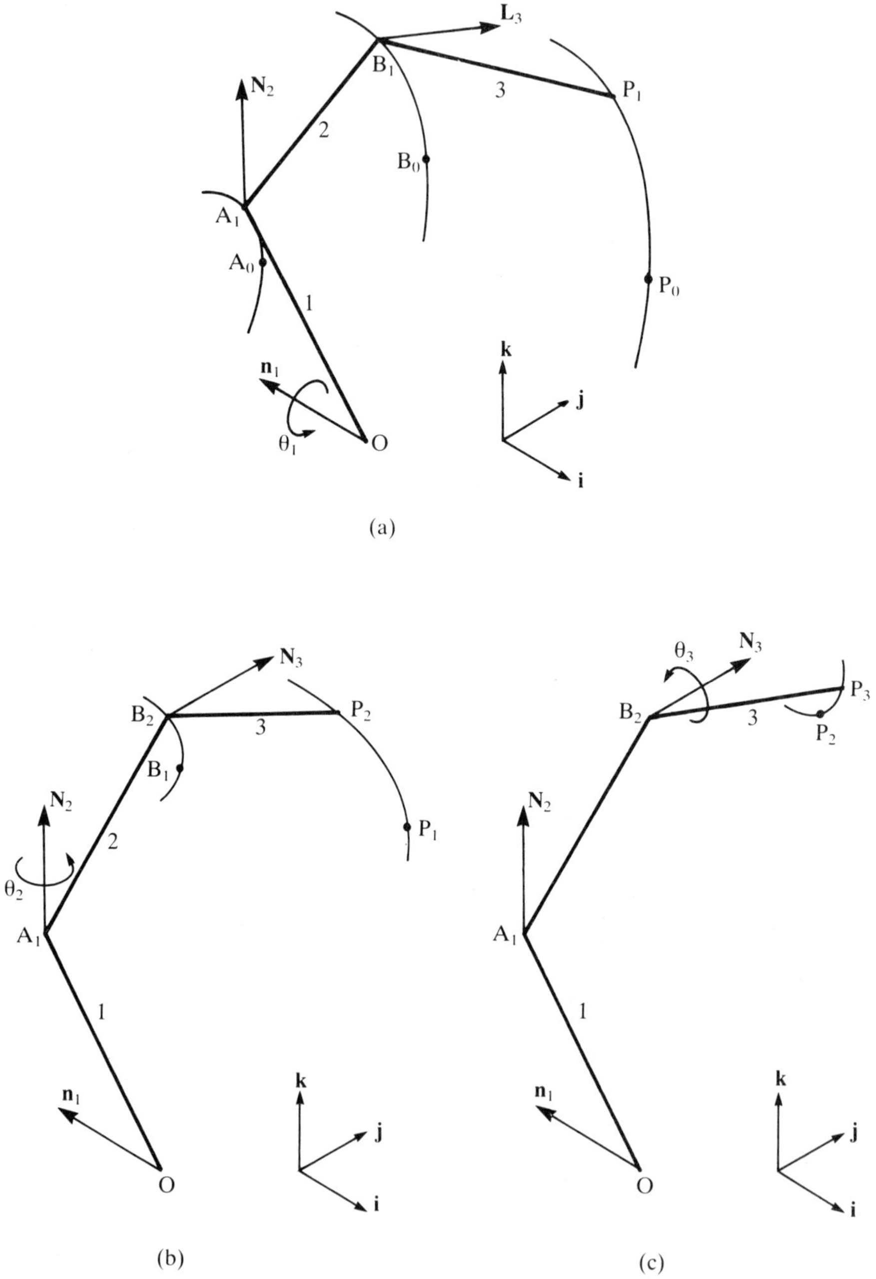
$\mathbf{L}_3$
$\mathbf{N}_2$
B_1
2
3
P_1
B_0
A_1
A_0
1
P_0
$\mathbf{n}_1$
θ_1
O
k
j
i
(a)
$\mathbf{N}_3$
B_2
P_2
3
B_1
$\mathbf{N}_2$
2
P_1
θ_2
A_1
1
$\mathbf{n}_1$
O
k
j
i
(b)
θ_3
$\mathbf{N}_3$
P_3
B_2
3
P_2
$\mathbf{N}_2$
A_1
1
$\mathbf{n}_1$
O
k
j
i
(c)

Fig. 7.4

If P goes from P_2 to P_3 as a result of the rotation about axis 3 (see Fig. 7.4(c)) then

$$\underline{B_2P_3} = \mathbf{A}(\theta_3, \mathbf{N}_3)\underline{B_2P_2}$$

which yields, noting (7.5),

$$\underline{B_2P_3} = \mathbf{A}(\theta_3, \mathbf{N}_3)\mathbf{A}(\theta_2, \mathbf{N}_2)\mathbf{A}(\theta_1, \mathbf{n}_1)\underline{B_0P_0} \tag{7.7}$$

If we put $\theta = \theta_1$, $\mathbf{l} = \mathbf{n}_1$, $\phi = \theta_2$, $\mathbf{m} = \mathbf{n}_2$ in the rotated axis theorem it follows from (7.3) that $\mathbf{n} = \mathbf{N}_2$ and therefore

$$\mathbf{A}(\theta_1, \mathbf{n}_1)\mathbf{A}(\theta_2, \mathbf{n}_2) = \mathbf{A}(\theta_2, \mathbf{N}_2)\mathbf{A}(\theta_1, \mathbf{n}_1) \tag{7.8}$$

Likewise putting $\theta = \theta_2$, $\mathbf{l} = \mathbf{N}_2$, $\phi = \theta_3$, $\mathbf{m} = \mathbf{L}_3$ then by (7.6) we have $\mathbf{n} = \mathbf{N}_3$ and thus

$$\mathbf{A}(\theta_2, \mathbf{N}_2)\mathbf{A}(\theta_3, \mathbf{L}_3) = \mathbf{A}(\theta_3, \mathbf{N}_3)\mathbf{A}(\theta_2, \mathbf{N}_2)$$

Hence

$$\begin{aligned}
&\mathbf{A}(\theta_3, \mathbf{N}_3)\mathbf{A}(\theta_2, \mathbf{N}_2)\mathbf{A}(\theta_1, \mathbf{n}_1)\\
&\quad = \mathbf{A}(\theta_2, \mathbf{N}_2)\mathbf{A}(\theta_3, \mathbf{L}_3)\mathbf{A}(\theta_1, \mathbf{n}_1)\\
&\quad = \mathbf{A}(\theta_2, \mathbf{N}_2)\mathbf{A}(\theta_1, \mathbf{n}_1)\mathbf{A}(\theta_3, \mathbf{n}_3) \qquad \text{using (7.1) and (7.3)}\\
&\quad = \mathbf{A}(\theta_1, \mathbf{n}_1)\mathbf{A}(\theta_2, \mathbf{n}_2)\mathbf{A}(\theta_3, \mathbf{n}_3) \qquad \text{by (7.8)}
\end{aligned} \tag{7.9}$$

Thus we find, with the help of (7.2), (7.4), (7.7), (7.8) and (7.9),

$$\begin{aligned}
\underline{OP_3} &= \underline{OA_1} + \underline{A_1B_2} + \underline{B_2P_3}\\
&= \mathbf{A}(\theta_1, \mathbf{n}_1)\underline{OA_0} + \mathbf{A}(\theta_1, \mathbf{n}_1)\mathbf{A}(\theta_2, \mathbf{n}_2)\underline{A_0B_0}\\
&\quad + \mathbf{A}(\theta_1, \mathbf{n}_1)\mathbf{A}(\theta_2, \mathbf{n}_2)\mathbf{A}(\theta_3, \mathbf{n}_3)\underline{B_0P_0}
\end{aligned} \tag{7.10}$$

The relation (7.10) gives the position vector of the point P of the robot, after the sequence of rotations, in terms of vectors that depend only on the *initial* positions of the points A, B, P in space and rotation matrices that depend on the *initial* directions of the axes of the robot and the angles of rotation. In addition (7.10) is in a form that is not committed to any particular axes of reference, and we are therefore at liberty to choose the set of axes that is most mathematically (and physically) suitable for our purpose.

We now repeat the sequence of rotations, the angles of rotation being θ_1', θ_2', θ_3'. Since the points A, B, P of the robot occupy the positions A_1, B_2, P_3 (before this sequence) and the axes are in the directions of the unit vectors $\mathbf{n}_1$, $\mathbf{N}_2$, $\mathbf{N}_3$ it follows, using (7.10), that if P goes from P_3 to P_3' then

$$\begin{aligned}
\underline{OP_3'} &= \mathbf{A}(\theta_1', \mathbf{n}_1)\underline{OA_1} + \mathbf{A}(\theta_1', \mathbf{n}_1)\mathbf{A}(\theta_2', \mathbf{N}_2)\underline{A_1B_2}\\
&\quad + \mathbf{A}(\theta_1', \mathbf{n}_1)\mathbf{A}(\theta_2', \mathbf{N}_2)\mathbf{A}(\theta_3', \mathbf{N}_3)\underline{B_2P_3}
\end{aligned} \tag{7.11}$$

Using (7.2) we find that

$$\mathbf{A}(\theta_1', \mathbf{n}_1)\underline{\mathrm{OA}_1} = \mathbf{A}(\theta_1', \mathbf{n}_1)\mathbf{A}(\theta_1, \mathbf{n}_1)\underline{\mathrm{OA}_0}$$
$$= \mathbf{A}(\theta_1' + \theta_1, \mathbf{n}_1)\underline{\mathrm{OA}_0}$$

using property P3 of the rotation matrix (see Chapter 5). Likewise by (7.4)

$$\mathbf{A}(\theta_1', \mathbf{n}_1)\mathbf{A}(\theta_2', \mathbf{N}_2)\underline{\mathrm{A}_1\mathrm{B}_2}$$
$$= \mathbf{A}(\theta_1', \mathbf{n}_1)\mathbf{A}(\theta_2', \mathbf{N}_2)\mathbf{A}(\theta_2, \mathbf{N}_2)\mathbf{A}(\theta_1, \mathbf{n}_1)\underline{\mathrm{A}_0\mathrm{B}_0}$$
$$= \mathbf{A}(\theta_1', \mathbf{n}_1)\mathbf{A}(\theta_2' + \theta_2, \mathbf{N}_2)\mathbf{A}(\theta_1, \mathbf{n}_1)\underline{\mathrm{A}_0\mathrm{B}_0}$$
$$= \mathbf{A}(\theta_1', \mathbf{n}_1)\mathbf{A}(\theta_1, \mathbf{n}_1)\mathbf{A}(\theta_2' + \theta_2, \mathbf{n}_2)\underline{\mathrm{A}_0\mathrm{B}_0}$$
$$= \mathbf{A}(\theta_1' + \theta_1, \mathbf{n}_1)\mathbf{A}(\theta_2' + \theta_2, \mathbf{n}_2)\underline{\mathrm{A}_0\mathrm{B}_0}$$

The third term on the right-hand side of (7.11) can be simplified in a similar manner, and hence

$$\mathrm{OP}_3' = \mathbf{A}(\theta_1 + \theta_1', \mathbf{n}_1)\underline{\mathrm{OA}_0} + \mathbf{A}(\theta_1 + \theta_1', \mathbf{n}_1)\mathbf{A}(\theta_2 + \theta_2', \mathbf{n}_2)\underline{\mathrm{A}_0\mathrm{B}_0}$$
$$+ \mathbf{A}(\theta_1 + \theta_1', \mathbf{n}_1)\mathbf{A}(\theta_2 + \theta_2', \mathbf{n}_2)\mathbf{A}(\theta_3 + \theta_3', \mathbf{n}_3)\underline{\mathrm{B}_0\mathrm{P}_0} \quad (7.12)$$

The relation (7.12) enables us to deduce that the order in which the rotations about the axes of the robot are performed is immaterial. Successive applications of (7.12) show that rotations of amounts θ, ϕ, ψ about axes 1, 2, 3 (in that order) are equivalent to rotations of amounts 0, ϕ, 0 about axes 1, 2, 3 (in that order), followed by rotations of amount θ, 0, 0 again about axes 1, 2, 3, and then by rotations of amounts 0, 0, ψ. Since the latter (nine) rotations are identical with rotations of amount ϕ, θ, ψ about axes 2, 1, 3 in that order, we deduce that the order is immaterial.

The results contained in (7.10) and (7.12) are readily extended to robots with more than three links that are connected by revolute joints (see Section 7.8).

We now apply the results just derived to the particular robot in Fig. 7.1. The objective of the analysis that follows is to find the values of θ_1, θ_2, θ_3 when the location in space of the point P, fixed relative to link 3, is specified.

Determination of the angles of rotation

Figure 7.5 shows the initial configuration of the robot. Axes 2 and 3 are parallel to one another and perpendicular to the fixed axis 1. The axes Ox, Oy, Oz are defined as follows: Oz is in the direction of axis 1, Oy is parallel to the common direction of the axes 2 and 3. The initial positions A_0, B_0, P_0 of the points A, B, P of the robot are collinear, the line $\mathrm{A}_0\mathrm{P}_0$ being parallel to Ox. Using the notation of Fig. 7.5 yields

$$\underline{\mathrm{OA}_0} = b\mathbf{j} + a\mathbf{k}, \quad \underline{\mathrm{A}_0\mathrm{B}_0} = c\mathbf{i}, \quad \underline{\mathrm{B}_0\mathrm{P}_0} = d\mathbf{i}, \quad \mathbf{n}_1 = \mathbf{k}, \quad \mathbf{n}_2 = \mathbf{n}_3 = \mathbf{j}$$

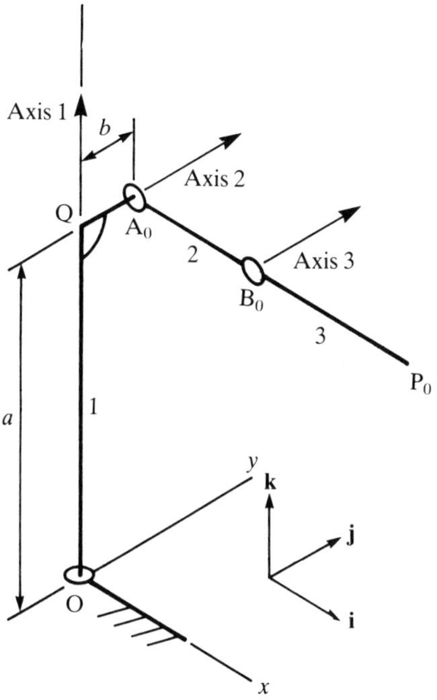

Fig. 7.5

Writing $\underline{OP_3} = [x, y, z]$, the relation (7.10) leads, with the help of properties P2 and P3 of rotation matrices (see Chapter 5), to

$$\mathbf{A}(-\theta_1, \mathbf{k})\begin{bmatrix} x \\ y \\ z \end{bmatrix} = \begin{bmatrix} 0 \\ b \\ a \end{bmatrix} + \mathbf{A}(\theta_2, \mathbf{j})\begin{bmatrix} c \\ 0 \\ 0 \end{bmatrix} + \mathbf{A}(\theta_2 + \theta_3, \mathbf{j})\begin{bmatrix} d \\ 0 \\ 0 \end{bmatrix}$$

Hence using results contained in Example 5.2 we have

$$x\cos\theta_1 + y\sin\theta_1 = c\cos\theta_2 + d\cos(\theta_2 + \theta_3) \tag{7.13}$$

$$-x\sin\theta_1 + y\cos\theta_1 = b \tag{7.14}$$

$$z - a = -c\sin\theta_2 - d\sin(\theta_2 + \theta_3) \tag{7.15}$$

Squaring each side of the three equations and then adding corresponding sides leads, after some simplification, to

$$x^2 + y^2 + (z - a)^2 = b^2 + c^2 + d^2 + 2cd\cos\theta_3$$

Since $-1 \leq \cos\theta_3 \leq 1$ we deduce that the regions defined by

$$x^2 + y^2 + (z - a)^2 > b^2 + (c + d)^2$$

and

$$x^2 + y^2 + (z - a)^2 < b^2 + (c - d)^2$$

are inaccessible to the point P of the robot. It is known from elementary trigonometry that the maximum and minimum values of $-x\sin\theta + y\cos\theta$ are $\pm\sqrt{(x^2 + y^2)}$ and therefore, noting equation (7.14), we see that the region $x^2 + y^2 > b^2$ is also inaccessible to the robot.

The task of solving equations (7.13) to (7.15) for θ_1, θ_2, θ_3 is not undertaken here. The reader is referred to Problems 7.3 and 7.4 where guidance is given on the method of solution. For completeness the solution is stated below in full.

If $r = \sqrt{(x^2 + y^2)}$, $R = \sqrt{\{x^2 + y^2 + (z - a)^2 - b^2\}}$, $s = c^2 + d^2$ and α_1, α_2, β_1, β_2, δ are angles defined by

$$2cd\cos\delta = R^2 - s, \qquad 0 \le \delta \le \pi$$

$$r\sin\alpha_1 = y, \qquad r\cos\alpha_1 = x, \qquad r\sin\beta_1 = b, \qquad -\pi/2 \le \beta_1 \le \pi/2$$

$$R\sin\alpha_2 = d\sin\delta, \qquad R\cos\alpha_2 = -c - d\cos\delta,$$

$$R\sin\beta_2 = a - z, \qquad -\pi/2 \le \beta_2 \le \pi/2$$

then there are four possible solutions to the problem, given by:

$$\theta_1 = \alpha_1 - \beta_1, \qquad \theta_2 = N\alpha_2 + \beta_2 - \pi, \qquad \theta_3 = N\delta$$

$$\theta_1 = \alpha_1 + \beta_1 - \pi, \qquad \theta_2 = N\alpha_2 - \beta_2, \qquad \theta_3 = N\delta$$

where $N = \pm 1$. Multiples of 2π may be added to or subtracted from the expressions for θ_1, θ_2, θ_3 to obtain further solutions. It is usual to take those values of θ_1, θ_2, θ_3 that lie between $-\pi$ and π.

It is a straightforward task to find the solutions by means of a computer program. The following values are provided as a check for programs that readers may write:

x	y	z	θ_1 (deg)	θ_2 (deg)	θ_3 (deg)
5.5	3.5	4.0	15.53	−66.80	105.24
5.3	3.3	3.8	14.19	−68.16	109.61

The values of a, b, c, d are $a = 3.9$, $b = 1.9$, $c = 4.1$, $d = 5.9$, while the given solutions correspond to $\theta_1 = \alpha_1 - \beta_1$ and $N = 1$.

In conclusion we note that if the velocity of the point P is given then the quantities $\dot{\theta}_1$, $\dot{\theta}_2$, $\dot{\theta}_3$ can be found (see Problem 7.5).

Before considering the dynamics of a three-link robot we introduce Lagrange's equations and explain by means of an example how they are used.

7.5 LAGRANGE'S EQUATIONS

We have seen how the position of a dynamical system may be descibed by a suitable choice of variables (or coordinates). Examples are the (robot) angles θ_1, θ_2, θ_3 of Section 7.4 and Euler's angles θ, ϕ, ψ. A further example is provided by the variables θ_2, θ_3, θ_4 used to define the position of the RRRR mechanism (see Example 4.3). These (latter) variables, in contrast to θ_1, θ_2, θ_3 and θ, ϕ, ψ, are not geometrically independent since if θ_2 is given the values of θ_3, θ_4 can, as we have seen, be calculated.

In formulating Lagrange's equations we assume that the variables used to describe the position of a dynamical system are (geometrically) independent variables. The number of variables is taken to be n and they are denoted by q_r ($r = 1, 2, \ldots, n$).

The kinetic energy T of the system at any instant depends on the variables $q_1, q_2, \ldots, q_n$ and their derivatives $\dot{q}_1, \dot{q}_2, \ldots, \dot{q}_n$, and we may thus write

$$T = T(q_1, q_2, \ldots, q_n, \dot{q}_1, \dot{q}_2, \ldots, \dot{q}_n)$$

to show this dependence.

The expression for the rate of working $\mathrm{d}W/\mathrm{d}t$ of the forces acting on the dynamical system is composed of terms (see Section 6.10) such as $(\mathbf{F}_s + \mathbf{F}'_s) \cdot \mathbf{v}_s$ and $\boldsymbol{\omega} \cdot \mathbf{M}$. Now $\mathbf{v}_s$ and $\boldsymbol{\omega}$ depend linearly on the quantities $\dot{q}_1, \dot{q}_2, \ldots, \dot{q}_n$, and therefore $\mathrm{d}W/\mathrm{d}t$ will be given by an expression of the form

$$\frac{\mathrm{d}W}{\mathrm{d}t} = Q_1\dot{q}_1 + Q_2\dot{q}_2 + \ldots + Q_n\dot{q}_n$$

where the quantities Q_i ($i = 1, 2, \ldots, n$) are functions of the components of the forces and couples that are acting and the variables $q_1, q_2, \ldots, q_n$. The quantities Q_i, which are called the generalized components of force, play an important role in Lagrange's equations of motion. These equations, which are derived from Newton's laws of motion, state that

$$\frac{\mathrm{d}}{\mathrm{d}t}\left(\frac{\partial T}{\partial \dot{q}_r}\right) - \frac{\partial T}{\partial q_r} = Q_r \qquad (r = 1, 2, \ldots, n) \tag{7.16}$$

The reader should note that when the partial derivatives $\partial T/\partial \dot{q}_r$, $\partial T/\partial q_r$ are found the kinetic energy is treated as a function of the $2n$ independent variables $q_1, q_2, \ldots, q_n, \dot{q}_1, \dot{q}_2, \ldots, \dot{q}_n$. This particular point is elaborated in the discussion that now follows of the motion of a top.

For mathematical convenience we restate results obtained for the motion of a top just prior to and in Example 6.12, namely

$$\boldsymbol{\omega} = -\dot{\phi}\sin\theta\mathbf{s} + \dot{\theta}\mathbf{t} + (\dot{\phi}\cos\theta + \dot{\psi})\mathbf{c}$$

$$\mathbf{H}_O = -A\dot{\phi}\sin\theta\mathbf{s} + A\dot{\theta}\mathbf{t} + C(\dot{\phi}\cos\theta + \dot{\psi})\mathbf{c}$$

$$\mathbf{k} = -\sin\theta\mathbf{s} + \cos\theta\mathbf{c}$$

Now

$$2T = \boldsymbol{\omega} \cdot \mathbf{H}_O = A\dot{\phi}^2\sin^2\theta + A\dot{\theta}^2 + C(\dot{\phi}\cos\theta + \dot{\psi})^2 \tag{7.17}$$

The only force that contributes to $\mathrm{d}W/\mathrm{d}t$ is gravity (represented vectorially by $-Mg\mathbf{k}$) acting at G. But $\mathbf{v}_G = \boldsymbol{\omega} \times \underline{OG} = h\boldsymbol{\omega} \times \mathbf{c}$, and hence

$$\begin{aligned}\frac{\mathrm{d}W}{\mathrm{d}t} &= -Mg\mathbf{k} \cdot \mathbf{v}_G \\ &= -Mgh(-\sin\theta\mathbf{s} + \cos\theta\mathbf{c}) \cdot (-\dot{\phi}\sin\theta\mathbf{s} + \dot{\theta}\mathbf{t}) \times \mathbf{c} \\ &= Mgh\sin\theta\,\dot{\theta}\end{aligned}$$

If we put $q_1 = \phi$, $q_2 = \theta$, $q_3 = \psi$ then $Q_1 = Q_3 = 0$ and $Q_2 = Mgh\sin\theta$.

We have, using the expression (7.17) for T,

$$\frac{\partial T}{\partial \phi} = 0, \qquad \frac{\partial T}{\partial \dot{\phi}} = A\dot{\phi}\sin^2\theta + C(\dot{\phi}\cos\theta + \dot{\psi})\cos\theta$$

Thus, putting $r = 1$ in (7.16), we obtain

$$\frac{\mathrm{d}}{\mathrm{d}t}\{A\dot{\phi}\sin^2\theta + C(\dot{\phi}\cos\theta + \dot{\psi})\cos\theta\} = 0$$

Differentiating with respect to θ and $\dot{\theta}$ yields

$$\frac{\partial T}{\partial \theta} = A\dot{\phi}^2\sin\theta\cos\theta + C(\dot{\phi}\cos\theta + \dot{\psi})(-\dot{\phi}\sin\theta), \qquad \frac{\partial T}{\partial \dot{\theta}} = A\dot{\theta}$$

Putting $r = 2$ in (7.16) we have

$$A\ddot{\theta} - A\dot{\phi}^2\sin\theta\cos\theta + C(\dot{\phi}\cos\theta + \dot{\psi})\dot{\phi}\sin\theta = Mgh\sin\theta$$

Lastly

$$\frac{\partial T}{\partial \psi} = 0, \qquad \frac{\partial T}{\partial \dot{\psi}} = C(\dot{\phi}\cos\theta + \dot{\psi})$$

so that

$$\frac{\mathrm{d}}{\mathrm{d}t}(\dot{\phi}\cos\theta + \dot{\psi}) = 0$$

The three equations derived can be compared with those given in Problem 6.29.

The reader should note that Lagrange's equations may be used in a modified form (not given here) when the variables that describe the dynamical system under consideration are not (geometrically) independent.

7.6 DYNAMICS OF A THREE-LINK ROBOT

We have seen in Section 7.4 that the order in which the rotations about the three axes of the robot are performed is immaterial. Mathematically it is convenient to regard the rotations as being successively about axes 3, 2, 1 as a rotation about axis 3 does not affect the direction of axis 2. It follows from repeated use of (5.16) that if the direction of a unit vector fixed in link 3 is that of the unit vector $\mathbf{p}_3$ when $\theta_1 = \theta_2 = \theta_3 = 0$ then its direction after the three rotations is that of the unit vector $\mathbf{q}_3$, where

$$\mathbf{q}_3 = \mathbf{A}(\theta_1, \mathbf{n}_1)\mathbf{A}(\theta_2, \mathbf{n}_2)\mathbf{A}(\theta_3, \mathbf{n}_3)\mathbf{p}_3 \tag{7.18}$$

Similarly if $\mathbf{p}_2$, $\mathbf{q}_2$ define the directions of a unit vector fixed in link 2 before and after the rotations then

$$\mathbf{q}_2 = \mathbf{A}(\theta_1, \mathbf{n}_1)\mathbf{A}(\theta_2, \mathbf{n}_2)\mathbf{p}_2 \tag{7.19}$$

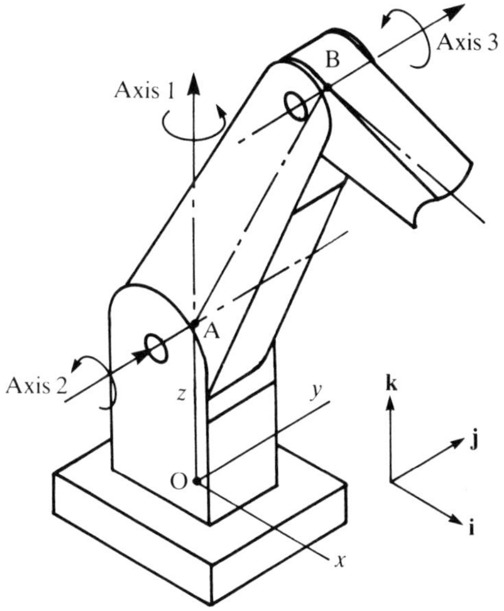

Fig. 7.6

The particular robot we consider is shown in Fig. 7.6. The arrangements of its links is slightly different from that of the robot in Figs 7.1 and 7.5. Axis 1 is still in the direction of Oz while axes 2 and 3 are parallel to each other and perpendicular to axis 1. However, axes 1 and 2 are now connected by a revolute joint at the point A which is on the axis Oz. Thus A is now a fixed

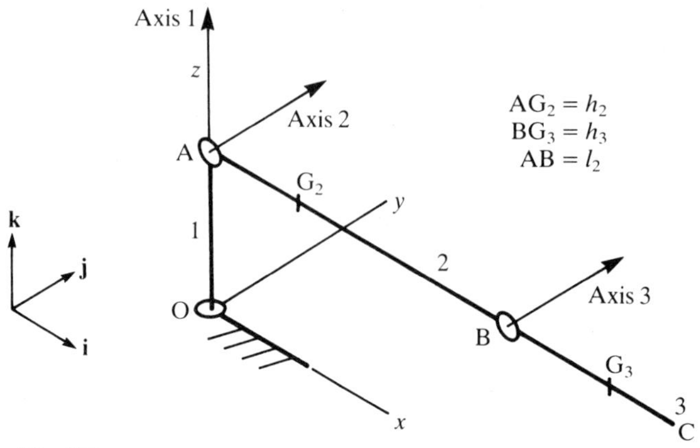

Fig. 7.7

point. The configuration (see Fig. 7.7) in which link 2, represented by AB, and link 3, represented by BC, are in line and perpendicular to Oz is taken to be the configuration $\theta_1 = \theta_2 = \theta_3 = 0$. The axis O$x$ is parallel to links 2 and 3 for this configuration, while Oy is in the (common) direction of axes 2 and 3.

As a preliminary to finding the kinetic energy of the robot we derive expressions for the components of the angular velocity $\boldsymbol{\omega}_2$ of link 2 with respect to axes fixed in link 2, and the components of the angular velocity $\boldsymbol{\omega}_3$ of link 3 with respect to axes fixed in link 3. Axes Aa_2, Ab_2, Ac_2 fixed in link 2, and axes Ba_3, Bb_3, Bc_3 fixed in link 3, are chosen so that for the configuration $\theta_1 = \theta_2 = \theta_3 = 0$ the directions of these axes are parallel to the unit vectors **i**, **j**, **k**. We take the associated unit vectors to be $\mathbf{a}_2$, $\mathbf{b}_2$, $\mathbf{c}_2$ and $\mathbf{a}_3$, $\mathbf{b}_3$, $\mathbf{c}_3$. By applying (7.18) and (7.19) we can obtain the components of these vectors with respect to **i**, **j**, **k** after the rotations.

Recalling the work of Section 7.4 we have $\mathbf{n}_1 = \mathbf{k}$, $\mathbf{n}_2 = \mathbf{n}_3 = \mathbf{j}$ and therefore, noting Example 5.2 once more,

$$\mathbf{A}(\theta_1, \mathbf{n}_1) = \mathbf{A}(\theta_1, \mathbf{k}) = \begin{bmatrix} \cos\theta_1 & -\sin\theta_1 & 0 \\ \sin\theta_1 & \cos\theta_1 & 0 \\ 0 & 0 & 1 \end{bmatrix}$$

$$\mathbf{A}(\theta_2, \mathbf{n}_2) = \mathbf{A}(\theta_2, \mathbf{j}) = \begin{bmatrix} \cos\theta_2 & 0 & \sin\theta_2 \\ 0 & 1 & 0 \\ -\sin\theta_2 & 0 & \cos\theta_2 \end{bmatrix}$$

giving

$$\mathbf{A}(\theta_1, \mathbf{k})\mathbf{A}(\theta_2, \mathbf{j}) = \begin{bmatrix} \cos\theta_1\cos\theta_2 & -\sin\theta_1 & \cos\theta_1\sin\theta_2 \\ \sin\theta_1\cos\theta_2 & \cos\theta_1 & \sin\theta_1\sin\theta_2 \\ -\sin\theta_2 & 0 & \cos\theta_2 \end{bmatrix} \tag{7.20}$$

Putting $\mathbf{p}_2$ in (7.19) successively equal to $[1, 0, 0]^T$, $[0, 1, 0]^T$, $[0, 0, 1]^T$ yields the following:

$$\begin{bmatrix} \mathbf{a}_2 \\ \mathbf{b}_2 \\ \mathbf{c}_2 \end{bmatrix} = \begin{bmatrix} \cos\theta_1 \cos\theta_2 & \sin\theta_1 \cos\theta_2 & -\sin\theta_2 \\ -\sin\theta_1 & \cos\theta_1 & 0 \\ \cos\theta_1 \sin\theta_2 & \sin\theta_1 \sin\theta_2 & \cos\theta_2 \end{bmatrix} \begin{bmatrix} \mathbf{i} \\ \mathbf{j} \\ \mathbf{k} \end{bmatrix} \tag{7.21}$$

The matrix of this transformation is the transpose of $\mathbf{A}(\theta_1, \mathbf{k})\mathbf{A}(\theta_2, \mathbf{j})$ and therefore the inverse relation, that is $\mathbf{i}, \mathbf{j}, \mathbf{k}$ expressed in terms of $\mathbf{a}_2, \mathbf{b}_2, \mathbf{c}_2$, can be written down directly.

Since $\mathbf{A}(\theta_3, \mathbf{n}_3) = \mathbf{A}(\theta_3, \mathbf{j})$ it follows, noting property P3 of the rotation matrix (see Chapter 5), that

$$\mathbf{A}(\theta_1, \mathbf{n}_1)\mathbf{A}(\theta_2, \mathbf{n}_2)\mathbf{A}(\theta_3, \mathbf{n}_3) = \mathbf{A}(\theta_1, \mathbf{k})\mathbf{A}(\theta_2 + \theta_3, \mathbf{j})$$

Hence to find the components of $\mathbf{a}_3, \mathbf{b}_3, \mathbf{c}_3$ we need only replace θ_2 by $\theta_2 + \theta_3$ in the transformation matrix of (7.21). Thus

$$\begin{bmatrix} \mathbf{a}_3 \\ \mathbf{b}_3 \\ \mathbf{c}_3 \end{bmatrix} = \begin{bmatrix} \cos\theta_1 \cos(\theta_2 + \theta_3) & \sin\theta_1 \cos(\theta_2 + \theta_3) & -\sin(\theta_2 + \theta_3) \\ -\sin\theta_1 & \cos\theta_1 & 0 \\ \cos\theta_1 \sin(\theta_2 + \theta_3) & \sin\theta_1 \sin(\theta_2 + \theta_3) & \cos(\theta_2 + \theta_3) \end{bmatrix} \begin{bmatrix} \mathbf{i} \\ \mathbf{j} \\ \mathbf{k} \end{bmatrix} \tag{7.22}$$

The reader should note that the choice of axes fixed in link 2 and link 3 is arbitrary, but it is advantageous to choose them so that they are parallel to $\mathbf{i}, \mathbf{j}, \mathbf{k}$ when $\theta_1 = \theta_2 = \theta_3 = 0$.

We deduce from (7.21) and (7.22) that

$$\mathbf{k} = -\sin\theta_2 \mathbf{a}_2 + \cos\theta_2 \mathbf{c}_2 = -\sin(\theta_2 + \theta_3)\mathbf{a}_3 + \cos(\theta_2 + \theta_3)\mathbf{c}_3 \tag{7.23}$$

The angular velocity $\boldsymbol{\omega}_1$ of link 1 is given by

$$\boldsymbol{\omega}_1 = \dot{\theta}_1 \mathbf{k}$$

Since axis 2 (in the direction of $\mathbf{b}_2$) is fixed relative to links 1 and 2 it follows, by virtue of the results contained in Section 6.9, that

$$\boldsymbol{\omega}_2 - \boldsymbol{\omega}_1 = \dot{\theta}_2 \mathbf{b}_2$$

Alternatively, noting (7.23),

$$\boldsymbol{\omega}_2 = -\dot{\theta}_1 \sin\theta_2 \mathbf{a}_2 + \dot{\theta}_2 \mathbf{b}_2 + \dot{\theta}_1 \cos\theta_2 \mathbf{c}_2$$

Likewise $\boldsymbol{\omega}_3 = \boldsymbol{\omega}_2 + \dot{\theta}_3 \mathbf{b}_3$ and therefore

$$\boldsymbol{\omega}_3 = -\dot{\theta}_1 \sin(\theta_2 + \theta_3)\mathbf{a}_3 + (\dot{\theta}_2 + \dot{\theta}_3)\mathbf{b}_3 + \dot{\theta}_1 \cos(\theta_2 + \theta_3)\mathbf{c}_3$$

In order to apply the formulae (6.36) and (6.37) for kinetic energy we need to know the inertia matrix at A (for link 2) with respect to the axes Aa_2, Ab_2, Ac_2, the velocity $\mathbf{v}_{G_3}$ of the centre of mass G_3 of link 3, and the inertia matrix at G_3 with respect to axes parallel to $\mathbf{a}_3, \mathbf{b}_3, \mathbf{c}_3$. We assume the following: Aa_2, Ab_2, Ac_2 are principal axes at A, the corresponding moments of inertia being A_2, B_2, C_2; and axes through G_3 in the directions of $\mathbf{a}_3$, $\mathbf{b}_3$, $\mathbf{c}_3$ are also principal axes, the associated moments of inertia being A_3, B_3, C_3. In addition we assume that the centre of mass G_2 is a point of the line AB distance h_2 from A, and G_3 is a point of BC distance h_3 from B. Denoting the kinetic energy of the links by T_1, T_2, T_3 we have, if C_1 is the moment of inertia of link 1 about Oz,

$$2T_1 = C_1\dot{\theta}_1^2$$
$$2T_2 = \dot{\theta}_1^2(A_2\sin^2\theta_2 + C_2\cos^2\theta_2) + B_2\dot{\theta}_2^2$$
$$2T_3 = m_3\mathbf{v}_{G_3}^2 + \dot{\theta}_1^2\{A_3\sin^2(\theta_2 + \theta_3) + C_3\cos^2(\theta_2 + \theta_3)\} + B_3(\dot{\theta}_2 + \dot{\theta}_3)^2$$

where m_3 is the mass of link 3.

The position vector r_{G_3} of G_3 with respect to O is given by

$$\mathbf{r}_{G_3} = \underline{OA} + l_2\mathbf{a}_2 + h_3\mathbf{a}_3$$

where l_2 is the distance between the points A and B. Hence

$$\begin{aligned}\mathbf{v}_{G_3} &= l_2\dot{\mathbf{a}}_2 + h_3\dot{\mathbf{a}}_3 \\ &= l_2\boldsymbol{\omega}_2 \times \mathbf{a}_2 + h_3\boldsymbol{\omega}_3 \times \mathbf{a}_3 \\ &= l_2(\dot{\theta}_1\cos\theta_2\mathbf{b}_2 - \dot{\theta}_2\mathbf{c}_2) + h_3\{\dot{\theta}_1\cos(\theta_2 + \theta_3)\mathbf{b}_3 - (\dot{\theta}_2 + \dot{\theta}_3)\mathbf{c}_3\}\end{aligned}$$

Noting, with the help of (7.21) and (7.22), that $\mathbf{c}_2 \cdot \mathbf{c}_3 = \cos\theta_3$ and $\mathbf{b}_2 \cdot \mathbf{c}_3 = \mathbf{c}_2 \cdot \mathbf{b}_3 = 0$, it is found that

$$\begin{aligned}\mathbf{v}_{G_3}^2 = {}& l_2^2(\dot{\theta}_1^2\cos^2\theta_2 + \dot{\theta}_2^2) + h_3^2\{\dot{\theta}_1^2\cos^2(\theta_2 + \theta_3) + (\dot{\theta}_2 + \dot{\theta}_3)^2\} \\ & + 2l_2h_3\{\dot{\theta}_1^2\cos\theta_2\cos(\theta_2 + \theta_3) + \dot{\theta}_2(\dot{\theta}_2 + \dot{\theta}_3)\cos\theta_3\}\end{aligned}$$

Thus

$$\begin{aligned}2T_3 = {}& \dot{\theta}_1^2\{A_3\sin^2(\theta_2 + \theta_3) + C_3\cos^2(\theta_2 + \theta_3)\} + B_3(\dot{\theta}_2 + \dot{\theta}_3)^2 \\ & + m_3\{l_2^2\cos^2\theta_2 + h_3^2\cos^2(\theta_2 + \theta_3) + 2l_2h_3\cos\theta_2\cos(\theta_2 + \theta_3)\}\dot{\theta}_1^2 \\ & + m_3(l_2^2 + h_3^2 + 2l_2h_3\cos\theta_3)\dot{\theta}_2^2 + m_3h_3^2\dot{\theta}_3^2 \\ & + 2m_3(h_3^2 + l_2h_3\cos\theta_3)\dot{\theta}_2\dot{\theta}_3\end{aligned}$$

It remains to find the generalized components of force. Couples of moments $L_1\mathbf{k}$, $L_2\mathbf{b}_2$, $L_3\mathbf{b}_3$ are applied respectively to links 1, 2, 3 while Oz is taken to be vertical. Recalling the expression $\boldsymbol{\omega} \cdot \mathbf{M}$ for the rate of working of a couple we see that the rate of working dW/dt of the couples and the forces of gravity on the robot is given by

$$\begin{aligned}\frac{dW}{dt} &= L_1\mathbf{k} \cdot (\dot{\theta}_1\mathbf{k}) + L_2\mathbf{b}_2 \cdot \boldsymbol{\omega}_2 + L_3\mathbf{b}_3 \cdot \boldsymbol{\omega}_3 - g\mathbf{k} \cdot (m_2\mathbf{v}_{G_2} + m_3\mathbf{v}_{G_3}) \\ &= L_1\dot{\theta}_1 + (L_2 + L_3)\dot{\theta}_2 + L_3\dot{\theta}_3 - m_2g\mathbf{k} \cdot \{h_2(\dot{\theta}_1\cos\theta_2\mathbf{b}_2 - \dot{\theta}_2\mathbf{c}_2)\} \\ &\quad - m_3g\mathbf{k} \cdot \{l_2(\dot{\theta}_1\cos\theta_2\mathbf{b}_2 - \dot{\theta}_2\mathbf{c}_2) + h_3[\dot{\theta}_1\cos(\theta_2 + \theta_3)\mathbf{b}_3 \\ &\quad - (\dot{\theta}_2 + \dot{\theta}_3)\mathbf{c}_3]\}\end{aligned}$$

Using (7.21) and (7.22) it is found that $\mathbf{k}\,.\,\mathbf{b}_2 = \mathbf{k}\,.\,\mathbf{b}_3 = 0$, $\mathbf{k}\,.\,\mathbf{c}_2 = \cos\theta_2$ and $\mathbf{k}\,.\,\mathbf{c}_3 = \cos(\theta_2 + \theta_3)$. Hence

$$\frac{dW}{dt} = L_1\dot{\theta}_1 + (L_2 + L_3)\dot{\theta}_2 + L_3\dot{\theta}_3 + g\cos\theta_2(m_2h_2 + m_3l_2)\dot{\theta}_2 + m_3gh_3\cos(\theta_2 + \theta_3)(\dot{\theta}_2 + \dot{\theta}_3)$$

giving for the generalized components of force

$$Q_1 = L_1$$
$$Q_2 = L_2 + L_3 + g\cos\theta_2(m_2h_2 + m_3l_2) + m_3gh_3\cos(\theta_2 + \theta_3)$$
$$Q_3 = L_3 + m_3gh_3\cos(\theta_2 + \theta_3)$$

It is now a straightforward matter to apply Lagrange's equations (see Problem 7.12):

$$\frac{d}{dt}\left(\frac{\partial T}{\partial \dot{\theta}_i}\right) - \frac{\partial T}{\partial \theta_i} = Q_i \qquad (i = 1, 2, 3)$$

We complete the work of this chapter by considering the kinematics of a device called the spherical wrist. This device, consisting of three links, is attached to the three-link robot analysed in Section 7.4. In order to facilitate the analysis we need a further theorem on rotations. We refer to the theorem as the decomposition theorem.

7.7 DECOMPOSITION THEOREM

This theorem states that:

If θ is a given angle and $\mathbf{l}$, $\mathbf{m}$, $\mathbf{n}$ are known unit vectors with $\mathbf{l}$, $\mathbf{m}$ perpendicular, then angles α, β, δ can be found such that

$$\mathbf{A}(\theta, \mathbf{n}) = \mathbf{A}(\alpha, \mathbf{l})\mathbf{A}(\beta, \mathbf{m})\mathbf{A}(\delta, \mathbf{l})$$

Alternatively:

A rotation of amount θ about a given axis ON can be decomposed into three rotations: the first about a given axis OL, the second about a given axis OM that is perpendicular to OL, and the third about the axis OL.

We give a proof of this theorem, in the course of which we show how the angles α, β, δ may be calculated.

☐ *Proofs* We choose axes Ox, Oy, Oz so that the unit vector $\mathbf{i}$ is in the direction of the vector $\mathbf{l}$, and $\mathbf{j}$ is in the direction of $\mathbf{m}$. Referred to the vectors $\mathbf{i}$, $\mathbf{j}$, $\mathbf{k}$ the elements of $\mathbf{A}(\theta, \mathbf{n})$ are given by a_{ij} $(i, j = 1, 2, 3)$. It is assumed that $a_{11} \neq \pm 1$ (see Problem 7.10).

The angles α, β, δ are defined by the following:

$$
\begin{aligned}
&\cos\beta = a_{11}, \qquad 0<\beta<\pi \\
&\sin\alpha\sin\beta = a_{21}, \qquad \cos\alpha\sin\beta = -a_{31}, \qquad -\pi<\alpha\le\pi \\
&\sin\beta\sin\delta = a_{12}, \qquad \sin\beta\cos\delta = a_{13}, \qquad -\pi<\delta\le\pi
\end{aligned} \tag{7.24}
$$

The angles α, β, δ exist by virtue of the property $\mathbf{AA}^{\mathrm{T}} = \mathbf{A}^{\mathrm{T}}\mathbf{A} = \mathbf{I}$ of the rotation matrix.

We now obtain expressions for a_{22}, a_{23}, a_{32}, a_{33} in terms of α, β, δ. It follows from the cofactor property of the rotation matrix that $a_{31} = a_{12}a_{23} - a_{13}a_{22}$ while $a_{11}a_{21} + a_{12}a_{22} + a_{13}a_{23} = 0$. These relations yield the following equations for a_{22} and a_{23}:

$$
\begin{aligned}
\sin\beta\sin\delta\, a_{23} - \sin\beta\cos\delta\, a_{22} &= -\cos\alpha\sin\beta \\
\sin\beta\cos\delta\, a_{23} + \sin\beta\sin\delta\, a_{22} &= -\sin\alpha\cos\beta\sin\beta
\end{aligned}
$$

from which we deduce, as $\sin\beta \neq 0$, that

$$
\begin{aligned}
a_{22} &= \cos\alpha\cos\delta - \sin\alpha\cos\beta\sin\delta \\
a_{23} &= -\cos\alpha\sin\delta - \sin\alpha\cos\beta\cos\delta
\end{aligned}
$$

The remaining quantities a_{32}, a_{33} are determined by the relations

$$
\begin{aligned}
a_{32} &= a_{13}a_{21} - a_{11}a_{23} = \sin\alpha\cos\delta + \cos\alpha\cos\beta\sin\delta \\
a_{33} &= a_{11}a_{22} - a_{12}a_{21} = \cos\alpha\cos\beta\cos\delta - \sin\alpha\sin\delta
\end{aligned}
$$

Elementary matrix multiplication now shows that

$$
\mathbf{A} = \begin{bmatrix} 1 & 0 & 0 \\ 0 & \cos\alpha & -\sin\alpha \\ 0 & \sin\alpha & \cos\alpha \end{bmatrix} \begin{bmatrix} \cos\beta & 0 & \sin\beta \\ 0 & 1 & 0 \\ -\sin\beta & 0 & \cos\beta \end{bmatrix} \begin{bmatrix} 1 & 0 & 0 \\ 0 & \cos\delta & -\sin\delta \\ 0 & \sin\delta & \cos\delta \end{bmatrix}
$$

or

$$
\mathbf{A}(\theta, \mathbf{n}) = \mathbf{A}(\alpha, \mathbf{i})\mathbf{A}(\beta, \mathbf{j})\mathbf{A}(\delta, \mathbf{i})
$$

It is evident from (7.24) that β could be replaced by $-\beta$ and α, δ by $\alpha \pm \pi$, $\delta \pm \pi$. The plus or minus sign is chosen so that the angles lie between $-\pi$ and π.

7.8 THE SPHERICAL WRIST

The robot shown in Fig. 7.8 is the three-link robot of Figs 7.1 and 7.5 with the addition of a device known as the spherical wrist. The six links of the robot are shown diagrammatically in Fig. 7.9. Link 4, represented by CD, is connected to link 3 at C by a revolute joint that permits relative rotation between the links about axis 4, which lies along the line of the collinear points B, C, D. Link 5 is joined to link 4 by a revolute joint at D, the axis of relative rotation (axis 5) being perpendicular to CD and DE and therefore

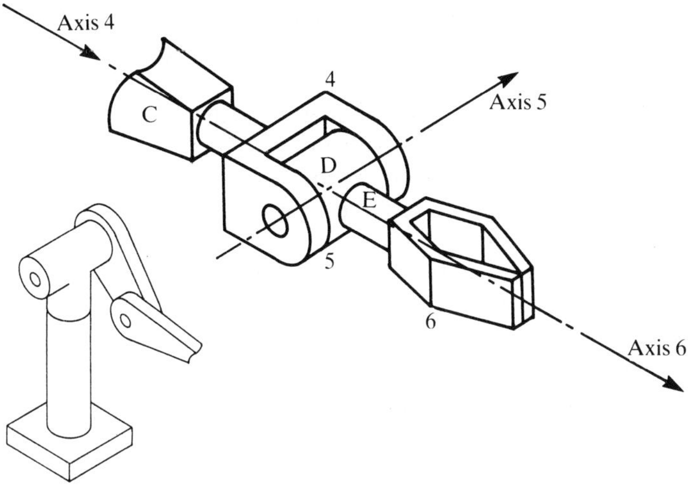
Axis 4
C
4
Axis 5
D
E
5
6
Axis 6

Fig. 7.8

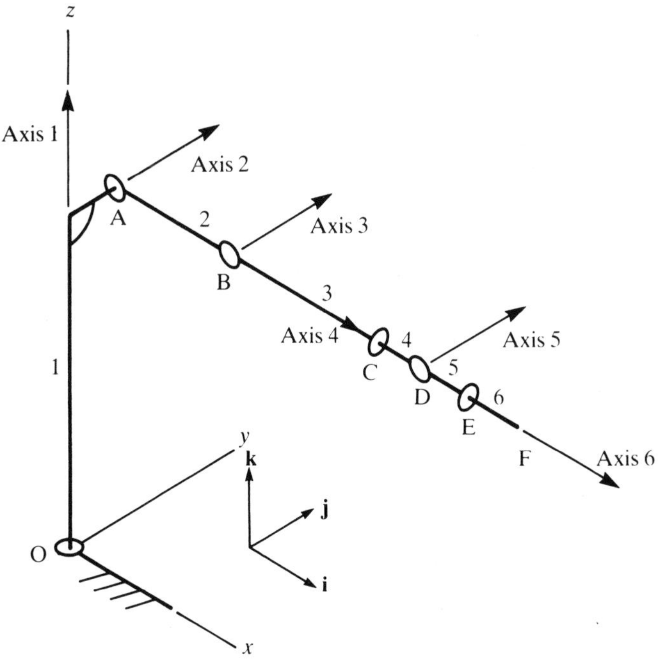
z
Axis 1
Axis 2
A
2
Axis 3
B
3
Axis 4
4
C
Axis 5
5
D
6
E
F
Axis 6
1
y
k
j
i
O
x

Fig. 7.9

parallel to axes 2 and 3. Finally link 6 is joined to link 5 by a revolute joint at E that permits relative rotation about axis 6, which coincides with the line of the collinear points D, E, F. Links 4, 5, 6 constitute the spherical wrist.

The notation of Section 7.4 is extended to deal with the additional links. The unit vectors in the directions of the axes 4, 5, 6, when the angles θ_i ($i = 1, 2, \ldots, 6$) of relative rotation are zero, are denoted by $\mathbf{n}_4$, $\mathbf{n}_5$, $\mathbf{n}_6$. Taking as in Section 7.4 $\mathbf{n}_1 = \mathbf{k}$ and $\mathbf{n}_2 = \mathbf{n}_3 = \mathbf{j}$, it follows that $\mathbf{n}_4 = \mathbf{n}_6 = \mathbf{i}$ and $\mathbf{n}_5 = \mathbf{j}$.

Further we let $\mathbf{a}_6$, $\mathbf{b}_6$, $\mathbf{c}_6$ be unit vectors fixed in link 6, their directions being those of $\mathbf{i}$, $\mathbf{j}$, $\mathbf{k}$ when $\theta_i = 0$. The purpose of the six-link robot is to orientate the unit vectors $\mathbf{a}_6$, $\mathbf{b}_6$, $\mathbf{c}_6$ in prescribed directions and at the same time to locate a point fixed in link 6 at a given position in space. There is no loss of generality in taking the point to be E because if the orientation of $\mathbf{a}_6$, $\mathbf{b}_6$, $\mathbf{c}_6$ together with the location in space of a point fixed in link 6 are known, the position of E is also (effectively) known.

The problem of finding the angles θ_i necessary to achieve the robot's objective can, as we now show, be split into two lesser problems, one of which has been solved in Section 7.4.

If $\mathbf{r}_\mathrm{D}$ and $\mathbf{r}_\mathrm{E}$ are the position vectors of the points D and E with respect to O it follows, since D, E, F are collinear, that

$$\mathbf{r}_\mathrm{D} = \mathbf{r}_\mathrm{E} - e\mathbf{a}_6$$

where e is the distance between the points D and E. Thus the position of D is known. Since the rotation about axis 4 does not affect the position of D we have, by virtue of (7.10),

$$\begin{aligned}\mathbf{r}_\mathrm{D} = \mathbf{A}(\theta_1, \mathbf{n}_1)\underline{\mathrm{OA}_0} &+ \mathbf{A}(\theta_1, \mathbf{n}_1)\mathbf{A}(\theta_2, \mathbf{n}_2)\underline{\mathrm{A}_0\mathrm{B}_0}\\ &+ \mathbf{A}(\theta_1, \mathbf{n}_1)\mathbf{A}(\theta_2, \mathbf{n}_2)\mathbf{A}(\theta_3, \mathbf{n}_3)\underline{\mathrm{B}_0\mathrm{D}_0}\end{aligned}$$

where D_0 is the position of D when $\theta_i = 0$ ($i = 1, 2, \ldots$). But $\mathbf{r}_\mathrm{D}$ is known, and therefore we can apply the analysis of Section 7.4, with $\underline{\mathrm{B}_0\mathrm{C}_0}$ replaced by $\underline{\mathrm{B}_0\mathrm{D}_0}$, to find θ_1, θ_2, θ_3.

Extending (7.18) we have

$$\mathbf{a}_6 = \mathbf{A}(\theta, \mathbf{n})\mathbf{A}(\theta_4, \mathbf{i})\mathbf{A}(\theta_5, \mathbf{j})\mathbf{A}(\theta_6, \mathbf{i})\mathbf{i}$$

where $\mathbf{a}_6$ and $\mathbf{i}$ are treated as column vectors and

$$\mathbf{A}(\theta, \mathbf{n}) = \mathbf{A}(\theta_1, \mathbf{n}_1)\mathbf{A}(\theta_2, \mathbf{n}_2)\mathbf{A}(\theta_3, \mathbf{n}_3)$$

Replacing $\mathbf{i}$ by $\mathbf{j}$ and $\mathbf{k}$ leads to the expressions for $\mathbf{b}_6$, $\mathbf{c}_6$. Now

$$\begin{bmatrix}\mathbf{a}_6\\ \mathbf{b}_6\\ \mathbf{c}_6\end{bmatrix} = \mathbf{L}\begin{bmatrix}\mathbf{i}\\ \mathbf{j}\\ \mathbf{k}\end{bmatrix}$$

where **L** is known. By virtue of results derived in Section 7.6 (see in particular (7.21) and (7.22)) **L** is the transpose of the matrix product

$$\mathbf{A}(\theta, \mathbf{n})\mathbf{A}(\theta_4, \mathbf{i})\mathbf{A}(\theta_5, \mathbf{j})\mathbf{A}(\theta_6, \mathbf{i})$$

and therefore

$$\mathbf{A}(\theta_4, \mathbf{i})\mathbf{A}(\theta_5, \mathbf{j})\mathbf{A}(\theta_6, \mathbf{i}) = \mathbf{A}(-\theta, \mathbf{n})\mathbf{L}^{\mathrm{T}}$$

where it is to be understood that the elements of the rotation matrices have been found with **i**, **j**, **k** as reference vectors. Once the elements of $\mathbf{A}(-\theta, \mathbf{n})\mathbf{L}^{\mathrm{T}}$ have been determined the angles θ_4, θ_5, θ_6 are found using the formulae (7.24) for α, β, δ given in the proof of the decomposition theorem.

The reader is referred to Problem 7.14 for an application of the results of this section.

PROBLEMS

7.1 Use the rotated axis theorem to show that a rotation about Ox of amount π followed by a rotation about Oz of amount $\pi/2$ is equivalent to a rotation about Oz of amount $\pi/2$ followed by a rotation about Oy of amount π.

Verify the result by finding the relevant rotation matrices.

7.2 In order to establish the rotated axis theorem it is sufficient to show that

$$\mathbf{A}(\phi, \mathbf{n}) = \mathbf{A}(\theta, \mathbf{l})\mathbf{A}(\phi, \mathbf{m})\mathbf{A}(-\theta, \mathbf{l})$$

Take **k** to be in the direction of **l** and take **j** to lie in the plane of **l** and **m**, so that **m** has components $[0, \cos\alpha, \sin\alpha]$ referred to **i**, **j**, **k**.

Show the following:

(a) $$\mathbf{A}(\phi, \mathbf{m}) = \cos\phi\mathbf{I} + (1 - \cos\phi)\begin{bmatrix} 0 \\ \cos\alpha \\ \sin\alpha \end{bmatrix}[0, \cos\alpha, \sin\alpha] + \sin\phi\begin{bmatrix} 0 & -\sin\alpha & \cos\alpha \\ \sin\alpha & 0 & 0 \\ -\cos\alpha & 0 & 0 \end{bmatrix}$$

(b) $\mathbf{n} = [-\sin\theta\cos\alpha, \cos\theta\cos\alpha, \sin\alpha]$.

Use (a) to obtain an expression for $\mathbf{A}(\theta, \mathbf{l})\mathbf{A}(\phi, \mathbf{m})\mathbf{A}(-\theta, \mathbf{l})$ and verify that it is equal to $\mathbf{A}(\phi, \mathbf{n})$, which should be expressed in a form similar to that for $\mathbf{A}(\phi, \mathbf{m})$.

7.3 Verify by direct substitution that

$$\theta = \alpha - \beta, \qquad \theta = \alpha + \beta - \pi$$

where

$$\sqrt{(p^2 + q^2)}\sin\alpha = q, \qquad \sqrt{(p^2 + q^2)}\cos\alpha = p,$$
$$\sqrt{(p^2 + q^2)}\sin\beta = r, \qquad -\pi/2 \le \beta \le \pi/2$$

are solutions of the equation

$$-p\sin\theta + q\cos\theta = r$$

Show, again by direct substitution, that

$$p\cos\theta + q\sin\theta = \begin{cases} \sqrt{(p^2 + q^2 - r^2)}, & \theta = \alpha - \beta \\ -\sqrt{(p^2 + q^2 - r^2)}, & \theta = \alpha + \beta - \pi \end{cases}$$

7.4 Show by squaring and adding corresponding sides of equations (7.13) to (7.15) that

$$x^2 + y^2 + (z - a)^2 = b^2 + c^2 + d^2 + 2cd\cos\theta_3$$

Solve equation (7.14) for θ_1 using the results contained in Problem 7.3.

Hence putting $\theta_3 = \delta$ show that equations (7.13), (7.15) are equivalent to

$$-d\sin\delta\sin\theta_2 + (c + d\cos\delta)\cos\theta_2 = \pm\sqrt{(x^2 + y^2 - b^2)}$$

and

$$(c + d\cos\delta)\sin\theta_2 + d\sin\delta\cos\theta_2 = a - z$$

and that they are consistent.

Obtain the complete solution to the equations (7.13) to (7.15).

7.5 Using equations (7.13) to (7.15) find expressions for x, y, z in terms of θ_1, θ_2, θ_3. Differentiate the expressions to show that

$$\dot{x} = -y\dot{\theta}_1 + (z - a)\cos\theta_1\,\dot{\theta}_2 - d\cos\theta_1\sin(\theta_2 + \theta_3)\,\dot{\theta}_3$$
$$\dot{y} = x\dot{\theta}_1 + (z - a)\sin\theta_1\,\dot{\theta}_2 - d\sin\theta_1\sin(\theta_2 + \theta_3)\,\dot{\theta}_3$$
$$\dot{z} = -(x\cos\theta_1 + y\sin\theta_1)\dot{\theta}_2 - d\cos(\theta_2 + \theta_3)\,\dot{\theta}_3$$

Deduce that

$$(x\cos\theta_1 + y\sin\theta_1)\dot{\theta}_1 = -\dot{x}\sin\theta_1 + \dot{y}\cos\theta_1$$

and hence that $\dot{\theta}_1$ is infinite or indeterminate when $x\cos\theta_1 + y\sin\theta_1 = 0$. Noting equation (7.14) and results contained in Problem 7.3, show this condition implies $x^2 + y^2 = b^2$.

Obtain an expression for $\dot{\theta}_3$ and find the positions of P for which it is infinite or indeterminate.

7.6 The position vector of a particle of mass m at time t is $r\mathbf{a}$, where $\mathbf{a} = \cos\theta\mathbf{i} + \sin\theta\mathbf{j}$. It is acted on by forces $R\mathbf{a}$, $S\mathbf{b}$, the vector $\mathbf{b}$ being defined by $\mathbf{b} = \mathbf{k} \times \mathbf{a}$. Find

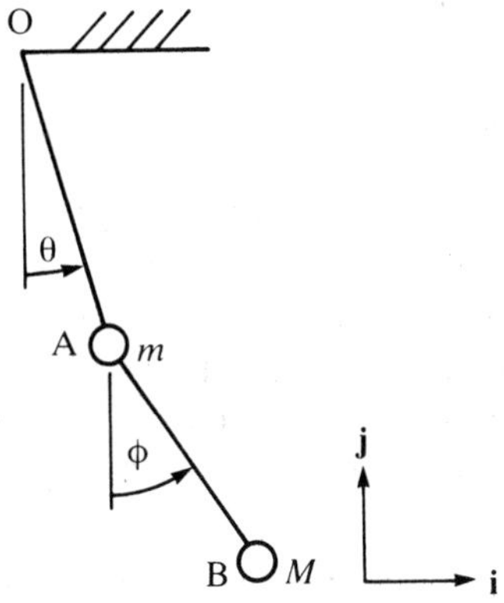

Fig. 7.10

(a) An expression for the kinetic energy of the particle
(b) The generalized components of force.
Obtain Lagrange's equations of motion for the particle.

7.7 Figure 7.10 shows a compound pendulum. The two masses m, M are attached to the strings OA, AB, where O is fixed, at the points A, B. Given that at time t the angles between OA, AB and the downward vertical are θ, ϕ, find
(a) The kinetic energy of the pendulum
(b) The generalized components of force.
Obtain Lagrange's equations of motion.

7.8 Find the generalized components of force for the problem solved in Example 6.16. Obtain Lagrange's equations of motion and compare these equations with the equations of motion derived by considering the motion of the centre of mass of the disc and the rate of change of its moment of momentum about C.
Note: if the disc is rolling without slipping, $\dot{x} - a\dot{\theta} = 0$, so that the coordinates x, θ are not independent and therefore Lagrange's equations cannot be applied in the form of (7.16).

7.9 Denoting the horizontal displacement of A in Problem 6.12 by $x\mathbf{i}$, obtain the kinetic energy of the rod in terms of x, θ and their derivatives. Determine the generalized components of force and hence derive Lagrange's equations. By putting $\ddot{x} = f$ obtain the equation for $\ddot{\theta}$ given in Problem 6.12.

7.10 In the proof of the decomposition theorem it was assumed that $a_{11} \neq \pm 1$.
Show that when $a_{11} = 1$ the matrix $\mathbf{A}(\theta, \mathbf{n})$ (referred to $\mathbf{i}, \mathbf{j}, \mathbf{k}$) is of the form

$$\begin{bmatrix} 1 & 0 & 0 \\ 0 & \cos\phi & -\sin\phi \\ 0 & \sin\phi & \cos\phi \end{bmatrix}$$

that is the rotation is about the axis OL. Take $\beta = 0$, $\alpha + \delta = \phi$.
What is the form of $\mathbf{A}(\theta, \mathbf{n})$ when $a_{11} = -1$?

7.11 By differentiating both sides of the matrix relation in (7.21) and eliminating **i**, **j**, **k** from the result, find the components of $\dot{\mathbf{a}}_2$, $\dot{\mathbf{b}}_2$, $\dot{\mathbf{c}}_2$ with respect to $\mathbf{a}_2$, $\mathbf{b}_2$, $\mathbf{c}_2$. Hence obtain the components of the angular velocity vector $\boldsymbol{\omega}_2$ with respect to $\mathbf{a}_2$, $\mathbf{b}_2$, $\mathbf{c}_2$. (Note the result contained in (5.31).)

7.12 Obtain Lagrange's equations for the three-link robot considered in Section 7.6.

7.13 Express the following matrices as the product of three matrices

(a) $\begin{bmatrix} 0 & -1 & 0 \\ 0 & 0 & 1 \\ -1 & 0 & 0 \end{bmatrix}$ (b) $\dfrac{1}{3}\begin{bmatrix} 2 & 2 & 1 \\ 1 & -2 & 2 \\ 2 & -1 & -2 \end{bmatrix}$.

7.14 Given the following:

$$a = 3.9, \quad b = 1.9, \quad c = 4.1, \quad d = 5.9, \quad e = 1.0$$

$$\underline{\mathrm{OE}} = 7\mathbf{i} + 3\mathbf{j} + 5\mathbf{k}, \qquad 3\mathbf{L} = \begin{bmatrix} 2 & 2 & -1 \\ 2 & -1 & 2 \\ 1 & -2 & -2 \end{bmatrix}$$

verify that $\theta_1 = 3.87°$, $\theta_2 = -73.92°$, $\theta_3 = 99.06°$, $\theta_4 = 89.99°$, $\theta_5 = 38.32°$, $\theta_6 = 28.78°$ are possible values of the angles θ_i $(i = 1, 2, \ldots, 6)$ for the robot in Section 7.8. Obtain other values.

Appendix A

POSITIONS OF THE CENTRES OF MASS OF SOME UNIFORM BODIES

Body	*Position of centre of mass*
Triangular plate*	Centroid of the triangle
Right circular cone, height h	At the point of the axis distant $3h/4$ from the vertex of the cone
Conical shell, height h	At the point of the axis distant $2h/3$ from of the vertex of the cone
Solid hemisphere, radius a	Point, distant $3a/8$ from centre, that lies on the radius perpendicular to the base of the hemisphere
Hemispherical shell, radius a	Point, distant $a/2$ from centre, that lies on the radius perpendicular to the base of the shell
Sector of circular plate, radius a angle subtended at the centre being 2θ	Point, $(2a\sin\theta)/3\theta$ from centre, that lies on the radius of symmetry
Arc of circle, radius a angle subtended at the centre being 2θ	Point, distant $a\sin\theta/\theta$ from centre, that lies on the radius of symmetry

* A plate is a body whose thickness is considered negligible

Appendix B

MOMENTS OF INERTIA (MI) OF UNIFORM BODIES OF MASS M

Body	*Axis*	*MI*
Rod, length l	perpendicular to rod through its midpoint	$Ml^2/12$
Rectangular plate*	in the plane of the plate bisecting sides of length l	$Ml^2/12$
Hoop, radius a	through the centre of the hoop perpendicular to its plane	Ma^2
Cylindrical shell, radius a	axis of symmetry of the shell	Ma^2
Hoop, radius a	a diameter of the hoop	$Ma^2/2$
Solid cylinder, radius a and length l	axis of symmetry of the cylinder	$Ma^2/2$
	through mass centre perp. to the axis of symmetry	$M(a^2 + l^2/3)/4$
Circular plate, radius a	a diameter	$Ma^2/4$
Spherical shell, radius a	a diameter	$2Ma^2/3$
Solid sphere, radius a	a diameter	$2Ma^2/5$
Rectangular block lengths of sides a, b, c	through the mass centre parallel to the sides of length a	$M(b^2 + c^2)/12$

* A plate is a body whose thickness is considered negligible

Body	*Axis*	*MI*
Right circular cone height h, base radius a	axis of cone	$3Ma^2/10$
	a diameter of the base of the cone	$M(3a^2 + 2h^2)/20$
Triangular plate, height h	base of triangle	$Mh^2/6$

Answers to problems

CHAPTER 1

1.1 (a) [9, 3, 5], [−5, −5, −3], [17, 6, 13], [−8, −1, 15]
(b) $\sqrt{6}$, 9, 14 (c) 50.6° (d) $(1/\sqrt{6})$[2, −1, 1]
(e) 35.26°, 114.1°, 65.91° (f) [−21, −4, 38], [−61, 128, −24].

1.2 (a) collinear; (b), (c) non-collinear.

1.3 (a), (b), (d) coplanar; (c) non-coplanar.

1.4 Verify $\mathbf{a}\,.\,\mathbf{b} = \mathbf{a}\,.\,\mathbf{c} = \mathbf{b}\,.\,\mathbf{c} = 0$, and determine whether or not $\mathbf{c} = \mathbf{a} \times \mathbf{b}$.
(a), (c) Left-handed, (b) right-handed.
(a) $3\mathbf{i} = [2, 2, 1]$, $3\mathbf{j} = [2, -1, -2]$, $3\mathbf{k} = [1, -2, 2]$
(b) $7\mathbf{i} = [2, -6, 3]$, $7\mathbf{j} = [3, -2, -6]$, $7\mathbf{k} = [6, 3, 2]$
(c) $11\mathbf{i} = [-9, -6, -2]$, $11\mathbf{j} = [-2, 6, -9]$, $11\mathbf{k} = [-6, 7, 6]$.
$\mathbf{A} = -\mathbf{a} - 7\mathbf{b} - 5\mathbf{c}$.

1.6 (a) $(\mathbf{A} \times \mathbf{B})\,.\,(\mathbf{C} \times \mathbf{D}) = -20$
(b) $2\mathbf{A} - \mathbf{B} + 2\mathbf{C} - \mathbf{D} = \mathbf{0}$
(c) $\mathbf{C} = 3\mathbf{A} - \mathbf{B} - \mathbf{A} \times \mathbf{B}$.

1.7 Find $\underline{AB} \times \underline{AC}$.

1.8 Find $\underline{EF}\,.\,\underline{EF}$ and $\underline{GH}\,.\,\underline{GH}$.

1.11 $2\underline{AD} = \underline{AB} + 2\underline{AC} + \underline{AB} \times \underline{AC}$.
Express $\underline{AE}$ in the same form as $\underline{AD}$ to show that D and E are on opposite sides of the plane ABC.
$\underline{OF} = [2, 8, 1]$.
Equation of ABC is $2x - 3y + 5z = 4$.

1.12 (a) Put $\mathbf{B} = \mathbf{j}$, $\mathbf{A} = \mathbf{C} = \mathbf{k}$ to show $\mu = 1$.
(b) Show $(\mathbf{A} \times \mathbf{B}) \times \mathbf{C} - \mathbf{A} \times (\mathbf{B} \times \mathbf{C}) = \mathbf{B} \times (\mathbf{C} \times \mathbf{A})$.

1.13 (a) $\mathbf{A}\,.\,\mathbf{B} \neq 0$.
(b) (i) $\mathbf{X} = [\alpha, \beta, \delta]$. Equations are $\beta + 2\delta = 1$, $\delta - \alpha = 1$,

$-2\alpha - \beta = 1$, $\delta = 1$. $\mathbf{X} = [0, -1, 1]$.
(ii) Take $\mathbf{D} = \mathbf{B}$, $\mathbf{E} = \mathbf{C}$ and show $\alpha = 1$.

1.14 Use the given result $\mathbf{X}^2 = \alpha^2\mathbf{A}^2 + \mathbf{C}^2$ to show $\alpha = \pm 1/2$. $\mathbf{X}_1 = [2, 1, -2]$, $\mathbf{X}_2 = [1, -2, 2]$.

1.15 Verify $\mathbf{A} . \mathbf{B} = \mathbf{C} . \mathbf{D} = \mathbf{A} . \mathbf{D} + \mathbf{B} . \mathbf{C} = 0$.
(a) $3\mathbf{X} = [2, -2, 1]$.
(b) $\mathbf{X} = [\alpha, \beta. \delta]$. The six equations for the components of $\mathbf{X}$ are (1) $10\beta + 5\delta = -5$, (2) $-7\delta - 10\alpha = -9$, (3) $-5\alpha + 7\beta = -8$, (4) $-4\beta - 5\delta = 1$, (5) $\delta + 4\alpha = 3$, (6) $5\alpha - \beta = 4$. Solve for α, using equations 5, 6 and 1.
Linear relations are:
7(equation 1) − 10(equation 3) = −5(equation 2)
(equation 4) + 5(equation 5) = 4(equation 6)
(equation 1) − (equation 3) + (equation 4) = (equation 6)

1.16 (b) $\mathbf{B}^2\mathbf{X} = \mathbf{B} \times \mathbf{A}$.

1.19 Substitute $\beta\mathbf{n}_1 + \mu\mathbf{n}_2$ into the equations of the planes and solve the resulting equations to find β and μ. Cartesian equations of the two planes are
$2x - y + 2z - 7 = 0$, $4x + 4y - 7z + 2 = 0$
$3\mathbf{n}_1 = [2, -1, 2]$, $4\mathbf{n}_2 = [4, 4, -7]$
To find β, μ substitute components of $\beta\mathbf{n}_1 + \mu\mathbf{n}_2$ into cartesian equations to obtain $27\beta - 10\mu - 63 = 0$, $-10\beta + 27\mu + 6 = 0$. $\beta = 1641/629$, $\mu = 468/529$.
Equation of line is
$629\mathbf{r} = [1302, -339, 730] + \alpha[-1, 22, 12]$
Otherwise: put $z = 0$ in cartesian equations to find common point $[13/6, -8/3, 0]$.

1.20 $\underline{OD} = 85[-0.7171, 0.2289, 0.6583]$,
$\underline{OE} = 85[-0.7510, -0.6603, 0.0064]$.
The values of α, β, μ for the problem are $\alpha = 0.3709$, $\beta = 0.5972$, $\mu = \pm 0.6195$.

1.22 $\underline{AB} \times \underline{AC} = [0, -6, 0]$, $\underline{AB} \times \underline{AD} = [-4, -10, -4]$.
If the lines AB, CD intersect then the points A, B, C, D would be coplanar and therefore the normals to the planes ABC, ABD would be parallel, which is not so.

1.23 $r = 1$, $p = -1$, $q = 1$, $\underline{OB} = [2, 6, -11]$, $\underline{OD} = [-4, -9, 2]$.

1.24 Put $\mathbf{r} = x\mathbf{i} + y\mathbf{j}$.
(a) Part of the parabola $y = 2x^2 - 1$ for which $|x| \leq 1$
(b) Spiral
(c) Closed curve symmetric about the x axis
(d) Ellipse $x^2/a^2 + y^2/b^2 = 1$.

1.25 $\mathbf{A}, \mathbf{B} = 0$, $\mathbf{A} \times \mathbf{B} = \mathbf{0}$.

1.27 $\mathbf{r} \cdot \ddot{\mathbf{r}} = 0$.

1.28 Show that the derivative of $\mathbf{a} \cdot \mathbf{b}$ is zero.
$\dot{\mathbf{c}} = \boldsymbol{\omega} \times \mathbf{c}$.

1.29 $\ddot{\mathbf{r}} = -e^{-t}(\mathbf{v} + g\mathbf{k})$ and hence $\ddot{\mathbf{r}}$ is in the direction of the vector $-(\mathbf{v} + g\mathbf{k})$.

1.30 $13a = 8$, $13b = 28$, $13c = -10$, $\dot{\mathbf{r}} = [6, 0, -6]$.
$\boldsymbol{\omega} = [1, 1, 1]$.

1.31 $(\mathrm{d}/\mathrm{d}t)(\mathbf{r}_i \cdot \mathbf{r}_j) = 0$. When $t = 0$, $\mathbf{r}_i = \mathbf{a}_i$, and therefore the angle between $\mathbf{r}_i$ and $\mathbf{r}_j$ is equal to the angle between $\mathbf{a}_i$ and $\mathbf{a}_j$.

1.32 $\mathbf{b} = [\cos\theta\cos\phi, \cos\theta\sin\phi, -\sin\theta]$, $\mathbf{c} = [-\sin\phi, \cos\phi, 0]$, $\mathbf{a} \cdot \mathbf{b} = 0$.
$\dot{\mathbf{b}} = -\dot{\theta}\mathbf{a} + \dot{\phi}\cos\theta\mathbf{c}$, $\dot{\mathbf{c}} = -\dot{\phi}(\sin\theta\mathbf{a} + \cos\theta\mathbf{b})$.
$\mathbf{i} = [\sin\theta\cos\phi, \cos\theta\cos\phi, -\sin\phi]$, $\mathbf{j} = [\sin\theta\sin\phi, \cos\theta\sin\phi, \cos\phi]$, $\mathbf{k} = [\cos\theta, -\sin\theta, 0]$, all referred to $\mathbf{a}$, $\mathbf{b}$, $\mathbf{c}$.
$\boldsymbol{\omega} = -\dot{\theta}\sin\phi\mathbf{i} + \dot{\theta}\cos\phi\mathbf{j} + \dot{\phi}\mathbf{k}$.

1.33 $\mu = -\beta\cos\alpha t$.

1.34 Show $(\boldsymbol{\omega}_1 - \boldsymbol{\omega}_2) \cdot (\mathbf{a} \times \mathbf{b}) = 0$.

1.35 $\boldsymbol{\omega} = \sec t\mathbf{j} + 2\mathbf{k}$.

1.36 All exist except **AB**, **BA**.

$$\mathbf{A} + \mathbf{B} = \begin{bmatrix} 2 & 3 & 6 & 7 \\ 2 & 5 & 7 & 8 \end{bmatrix}, \quad \mathbf{A}\mathbf{B}^{\mathrm{T}} = \begin{bmatrix} 6 & 8 \\ 8 & 10 \end{bmatrix}, \quad \mathbf{A}\mathbf{B}^{\mathrm{T}}\mathbf{C} = \begin{bmatrix} 14 & 2 \\ 18 & 2 \end{bmatrix}$$

$$\mathbf{A}^{\mathrm{T}}\mathbf{B} = \begin{bmatrix} 1 & 2 & 3 & 0 \\ 3 & 4 & 7 & 0 \\ 5 & 6 & 11 & 0 \\ 7 & 8 & 15 & 0 \end{bmatrix}, \quad \mathbf{C}^{\mathrm{T}}\mathbf{B}\mathbf{A}^{\mathrm{T}} = \begin{bmatrix} 14 & 18 \\ 2 & 2 \end{bmatrix}$$

1.37 (a) **A** and **B** must be conformable and square
(b) $\mathbf{AB} = \mathbf{BA}$ (c) $\mathbf{B}\mathbf{A}^n\mathbf{B}^{-1}$.

1.38 $\det\mathbf{A} = \det\mathbf{B} = -1$.

1.39 $\det\mathbf{A} = 1$, $\det\mathbf{B} = -1$, that is $\det\mathbf{A}$ or $\det\mathbf{B} \neq 0$.

$$(\mathbf{A}^2)^{-1} = \begin{bmatrix} 97 & -59 & -23 \\ -59 & 36 & 14 \\ -46 & 28 & 11 \end{bmatrix}, \quad (\mathbf{A}^{\mathrm{T}})^{-1} = \begin{bmatrix} -8 & 5 & 4 \\ 5 & -3 & -2 \\ 2 & -1 & -1 \end{bmatrix}$$

$$(\mathbf{AB})^{-1} = \begin{bmatrix} 127 & -78 & -31 \\ -205 & 126 & 50 \\ 148 & -91 & -36 \end{bmatrix}, \quad ((\mathbf{AB})^{\mathrm{T}})^{-1} = \begin{bmatrix} 127 & -205 & 148 \\ -78 & 126 & -91 \\ -31 & 50 & -36 \end{bmatrix}$$

1.40 (a) $\frac{1}{2}\begin{bmatrix} 1 & 1 \\ -1 & 1 \end{bmatrix}$ (b) $\frac{1}{2}\begin{bmatrix} -4 & 2 \\ 3 & -1 \end{bmatrix}$ (c) $\begin{bmatrix} 0 & -1 \\ 1 & 0 \end{bmatrix}$
(d) No inverse.

1.41 (a) No inverse.

(b) $\begin{bmatrix} -9 & 19 & -11 \\ 22 & -46 & 27 \\ -8 & 17 & -10 \end{bmatrix}$ (c) $\frac{1}{3}\begin{bmatrix} 7 & -2 & -1 \\ 1 & 1 & -1 \\ -5 & 1 & 2 \end{bmatrix}$

1.42 $\det(-\mathbf{I}) = (-1)^n$.

1.43 $\mathbf{D}^n = \begin{bmatrix} \alpha_1^n & 0 & 0 \\ 0 & \alpha_2^n & 0 \\ 0 & 0 & \alpha_3^n \end{bmatrix}$

Result true for n negative provided $\alpha_1\alpha_2\alpha_3 \neq 0$.

1.45 $x_1 = 0$, $x_2 = x_3$ (arbitrary).

1.47 (b) trace $\mathbf{ABC}$ = trace $((\mathbf{AB})\mathbf{C})$ = trace $(\mathbf{C}(\mathbf{AB}))$ = trace $\mathbf{CAB}$.

1.48 $\mathbf{A}^3 = \begin{bmatrix} 2 & -25 & -10 \\ -25 & 2 & -10 \\ 10 & 10 & -23 \end{bmatrix}$

1.49 $\mathbf{AA}^{\mathrm{T}} = \mathbf{I}$.

1.50 $\dot{\mathbf{D}}\mathbf{D}^{\mathrm{T}} = \dot{\theta}\begin{bmatrix} 0 & 0 & 0 \\ 0 & 0 & -1 \\ 0 & 1 & 0 \end{bmatrix}$, $\dot{\mathbf{E}}\mathbf{E}^{\mathrm{T}} = \dot{\beta}\begin{bmatrix} 0 & 0 & 1 \\ 0 & 0 & 0 \\ -1 & 0 & 0 \end{bmatrix}$, $\dot{\mathbf{F}}\mathbf{F}^{\mathrm{T}} = \dot{\alpha}\begin{bmatrix} 0 & -1 & 0 \\ 1 & 0 & 0 \\ 0 & 0 & 0 \end{bmatrix}$

Put $\mathbf{n}$ successively equal to [1, 0, 0], [0, 1, 0], [0, 0, 1].

CHAPTER 3

3.1 $\mathbf{r} = [-5, 0, -2] + \alpha[2, 1, 2]$.

3.2 No. If P, Q are of the same sign the relation $P\underline{CA} = Q\underline{BC}$ implies C is an internal point of AB, while if they are of opposite signs the relation shows C is an external point of AB.

3.3 $3\mathbf{r} = [4, 6, 5] + \alpha[1, 2, 2]$; $\surd(17)/9$.

3.4 (a) reduces to a couple; (c) reduces to a single force [0, 1, 1], line of action $\mathbf{r} = [8, -2, 2] + \alpha[0, 1, 1]$.

3.5 The point O_1 lies anywhere on the line $\mathbf{R}^2\mathbf{r} = \mathbf{R} \times \mathbf{M} + \beta\mathbf{R}$. $\mathbf{R}^2\alpha = \mathbf{M} \,.\, \mathbf{R}$.

3.6 $X\mathbf{r} = [0, -aY, 0] + \beta[X, Y, -Z]$.

3.9 $\underline{OQ_1} = [2, 1, 0]$, $3\underline{OQ_2} = [-1, 4, 8]$.

3.10 $\mathbf{F}_1 + \mathbf{F}_2 + \mathbf{F}_3 = \mathbf{0}$. Couple has moment $[4, -1, 1]$.

3.11 $0 \leq x < b$: $F = W(1 - b/l)$, $M = -Fx$

$b < x \leq l$: $F = -Wb/l$, $M = F(l - x)$

Shear force is discontinuous at the point $x = b$.

Bending moment has greatest magnitude at $x = b$.
Plot graph of $Wb(1 - b/l)$ against b.
$0 \le x \le b$: $M = -Wx(1 - b/l) - W_1x(1 - c/l)$
$b \le x \le c$: $M = -Wb(1 - x/l) - W_1x(1 - c/l)$
$c \le x \le l$: $M = -Wb(1 - x/l) - W_1c(1 - x/l)$

3.12 Force is $wl\mathbf{j}$ and couple is $(w/2)l^2\mathbf{k}$, where $\mathbf{i}$ is in the direction of $\underline{AB}$. $F = w(l - x)$, $M = (w/2)(l - x)^2$.

3.13 If AD = d, then bending moment is $-Wc(1 - d/l)$ if AC = $c < d$, and $-Wd(1 - c/l)$ if $c > d$.

3.14 $0 \le x \le c$: $$M = \frac{wx^2}{2} - \frac{wx}{2c}\{(2a + b)c - a(a + b)\}$$

$c \le x \le a + b$: $$M = \frac{w}{2}(a - x)(a + b - x)$$

3.15 Vertical reaction at E is $(F/4)(3\sqrt{3} - 1)$; reaction at A is $F/2$ horizontally and $(F/4)(5 - \sqrt{3})$ vertically up. AE is in tension $(F/4)(3 - \sqrt{3})$, while AF, AB are in compression each of amount $(F/2)(3\sqrt{3} - 4)$.

3.16 DE compression 4.5 kN, GH tension 3 kN, CH zero force, CJ tension 1.25 kN, DK tension 1.25 kN.

3.17 $20\sqrt{3}$ kN, $10\sqrt{3}$ kN, $20\sqrt{3}$ kN, 0.

CHAPTER 4

4.1 The distance between the points $[4, 9, -1]$ and $[6, 5, -1]$ is not the same as the distance between the points $[-6, 4, 10]$ and $[2, 2, 10]$.

4.2 $|\underline{A_iA_j}| = |\underline{B_iB_j}|$ for $i, j = 1, 2, 3, 4$, but
$\underline{A_1A_4} = \underline{A_1A_2} + 3\underline{A_1A_3} - \underline{A_1A_2} \times \underline{A_1A_3}$
$\underline{B_1B_4} = \underline{B_1B_2} + 3\underline{B_1B_3} + \underline{B_1B_2} \times \underline{B_1B_3}$

4.3 $\dot{\mathbf{v}}_{rel} = \ddot{\alpha}\mathbf{a} + \ddot{\beta}\mathbf{b} + \boldsymbol{\omega} \times (\dot{\alpha}\mathbf{a} + \dot{\beta}\mathbf{b})$.

4.5 $\dot{\theta}$ is the rate of change of angle between $\underline{BA}$ and Ox, that is between a direction fixed in the body and a direction fixed in space.

4.6
$$\mathbf{v}_P = -(c\omega\sin\theta - (v + a\omega)\sin(\theta + \phi))\mathbf{i} + (c\omega\cos\theta + (v + a\omega)\cos(\theta + \phi))\mathbf{j}$$
$$\mathbf{f}_P = \dot{v}(-\sin(\theta + \phi)\mathbf{i} + \cos(\theta + \phi)\mathbf{j}) - 2\omega v(\cos(\theta + \phi)\mathbf{i} + \sin(\theta + \phi)\mathbf{j}) + \mathbf{f}_Q$$
where
$$\mathbf{f}_Q = (a\cos(\theta + \phi) + c\cos\theta)(-\omega^2\mathbf{i} + \dot{\omega}\mathbf{j}) + (a\sin(\theta + \phi) + c\sin\theta)(-\dot{\omega}\mathbf{i} - \omega^2\mathbf{j})$$

4.8 $\dot{\theta}_3 = -5.92\,\text{rad s}^{-1}$, $\dot{\theta}_4 = 1.84\,\text{rad s}^{-1}$, $\ddot{\theta}_3 = 78.1\,\text{rad s}^{-2}$,
$\ddot{\theta}_4 = 395.7\,\text{rad s}^{-2}$.
$\mathbf{f}_G = -26.6\,\text{m s}^{-2}\mathbf{i} - 7.8\,\text{m s}^{-2}\mathbf{j}$.

4.9 $\theta_4 = 171.05°$, $\omega_4 = 2.26\,\text{rad s}^{-1}$,
$\mathbf{v}_{A_4} - \mathbf{v}_{A_3} = -0.62\,\text{m s}^{-1}\mathbf{s}_4$, $\mathbf{v}_{B_5} - \mathbf{v}_{B_4} = -0.036\,\text{m s}^{-1}\mathbf{s}_6$,
$\dot{\omega}_4 = 43.3\,\text{rad s}^{-2}$,
$\mathbf{f}_{A_4} - \mathbf{f}_{A_3} = 8.9\,\text{m s}^{-2}\mathbf{s}_4 - 2.81\,\text{m s}^{-2}\mathbf{t}_4$,
$\mathbf{f}_{B_5} - \mathbf{f}_{B_4} = -0.148\,\text{m s}^{-2}\mathbf{s}_4 - 0.163\,\text{m s}^{-2}\mathbf{t}_4$,
$\mathbf{f}_{B_6} = -4.27\,\text{m s}^{-2}\mathbf{s}_6$.

4.10 $\theta_4 = 45.29°$, $\theta_5 = 348.11°$, $\omega_4 = 47.3\,\text{rad s}^{-1}$, $\omega_5 = -9.9\,\text{rad s}^{-1}$,
$\mathbf{v}_{C_6} = -3.3\,\text{m s}^{-1}\mathbf{i}$, $\mathbf{v}_{A_4} - \mathbf{v}_{A_3} = -3.1\,\text{m s}^{-1}\mathbf{s}_4$, $\dot{\omega}_4 = -518.8\,\text{rad s}^{-2}$,
$\dot{\omega}_5 = 559\,\text{rad s}^{-2}$,
$\mathbf{f}_B = -179.2\,\text{m s}^{-2}\mathbf{s}_4 - 41.5\,\text{m s}^{-2}\mathbf{t}_4$,
$\mathbf{f}_{A_4} - \mathbf{f}_{A_3} = 148.6\,\text{m s}^{-2}\mathbf{s}_4 - 294.4\,\text{m s}^{-2}\mathbf{t}_4$, $\mathbf{f}_{C_6} = -91.26\,\text{m s}^{-2}\mathbf{i}$.

4.11 $\theta_4 = 14.64°$, $\theta_5 = 298.93°$, $\omega_4 = 30.84\,\text{rad s}^{-1}$, $\omega_5 = -4.45\,\text{rad s}^{-1}$,
$\mathbf{v}_{C_6} = -5.12\,\text{m s}^{-1}\mathbf{j}$, $\mathbf{v}_{A_4} - \mathbf{v}_{A_3} = 3.79\,\text{m s}^{-1}\mathbf{s}_4$, $\dot{\omega}_4 = -692\,\text{rad s}^{-2}$,
$\dot{\omega}_5 = -415\,\text{rad s}^{-2}$
$\mathbf{f}_{C_6} = 81.5\,\text{m s}^{-2}\mathbf{j}$, $\mathbf{f}_{A_4} - \mathbf{f}_{A_3} = 448\,\text{m s}^{-2}\mathbf{s}_4 + 234\,\text{m s}^{-2}\mathbf{t}_4$.

4.12 $\theta_4 = 95.70°$, $\omega_4 = 2.811\,\text{rad s}^{-1}$,
$\mathbf{v}_{B_6} = -1.190\,\text{m s}^{-1}\mathbf{i}$, $\mathbf{v}_{A_4} - \mathbf{v}_{A_3} = 0.252\,\text{m s}^{-1}\mathbf{s}_4$,
$\mathbf{v}_{B_5} - \mathbf{v}_{B_4} = 0.118\,\text{m s}^{-1}\mathbf{s}_4$, $\dot{\omega}_4 = -2.574\,\text{rad s}^{-2}$,
$\mathbf{f}_{B_5} - \mathbf{f}_{B_4} = 3.285\,\text{m s}^{-2}\mathbf{s}_4 + 0.663\,\text{m s}^{-2}\mathbf{t}_4$, $\mathbf{f}_{B_6} = 0.423\,\text{m s}^{-2}\mathbf{i}$.

4.14 $\mathbf{v}_C = 7.04\,\text{m s}^{-1}\mathbf{i} + 0.04\,\text{m s}^{-1}\mathbf{j}$, $\omega_6 = 50.7\,\text{rad s}^{-1}$.

4.17 $\sin(\theta_4 - \theta_3) \neq 0$ for any value of θ_2. Now $\sin(\theta_4 - \theta_3)$ varies continuously with θ_2 and therefore cannot change from a negative value to a positive value (or vice versa) as this requires it to take zero value.

4.19 PO $\sin\theta$.

4.22 Output shaft has angular speed of 703.4 rev min^{-1}.

4.23 Velocity ratios are −8.63, 3.67, 1.

4.24 $N_4 = 18$, $N_5 = 30$, $N_6 = 66$.

4.27 The values of θ when the follower dwells are equal to a and $a + (b/2)\pi$.

4.28 To find a coupler curve determine the components of $\underline{OC}$, noting $\underline{OC} = \underline{OA} + \underline{AC}$, referred to **i**, **j**. The vectors $\underline{OA}$, $\underline{AC}$ for a given θ_2 can be found using results given in the problem of position of the RRRR mechanism (Section 4.5).
To find y use the relation
$y\mathbf{j} - OQ\mathbf{i} = \underline{OD} = \underline{OA} + \underline{AC} - CD[\cos\theta_6, \sin\theta_6]$
where θ_6 is the angle between $\underline{DC}$ and **i**. Solve for θ_6 and then y.

CHAPTER 5

5.1 Put $\theta = \pm\pi$ in the relation (5.4).

5.2 If $\mathbf{r}$, $\alpha\mathbf{r}$ are the position vectors of P_1, P_2 before the rotation then $\mathbf{R}$, $\alpha\mathbf{R}$ are their position vectors after the rotation.

5.3 (a) $\mathbf{R}_1 = [-3, -2, -7]$, $\mathbf{R}_2 = [-1, 3, -4]$.
(b) $\mathbf{r}_1 \times \mathbf{r}_2 = [25, -19, -1]$, $\mathbf{R}_1 \times \mathbf{R}_2 = [29, -5, -11]$.
P goes to $2\mathbf{R}_1 + \mathbf{R}_2 - \mathbf{R}_1 \times \mathbf{R}_2$, i.e. $[-36, 4, -7]$, starts at $[-18, 29, -14]$.

5.4 Show that a particle initially at $[x, y, z]$ goes to $[y, x, -z]$ as the result of either pair of rotations.

5.5 Point of line at [1, 1, 3] goes to [1, 3, 1] and point at [4, 2, 5] goes to [3, 6, 0].
Equation of line after the rotation is $\mathbf{r} = [1, 3, 1] + \alpha[-2, -3, 1]$.
Point of line at $[-3, -3, 3]$ corresponds to $\alpha = 2$. Equation of line before rotation is $\mathbf{r} = [1, 1, 3] + \beta[-3, -1, -2]$, and hence putting $\beta = 2$ the coordinates of P are found to be $[-5, -1, -1]$.
Equation of rotated line when axis of rotation passes through [1, 5, 2] is $\mathbf{r} = [4, 3, 4] + \mu[-2, -3, 1]$.

5.6 (a) Rotation exists. Angle is π and axis is in the direction of [1, 1, 1].
(b), (c), (d) Rotations do not exist.

5.7 A goes from $[6, 2, -1]$ to $[2, -1, -6]$. B goes from $[1, 1, 9]$ to $[5, -7, 3]$. C goes from $[8, 1, -4]$ to $[0, 0, -9]$.
Equation of plane before rotation is $13x + 5y + 7z = 81$ and after it is $x - y - z = 9$.
Express $\underline{AP}$ in the form $\underline{AP} = \alpha\underline{AB} + \beta\underline{AC}$ and note that α, β are unaffected by the rotation. Position of P before the rotation is the point [4, 3, 2]. Note result in Problem 5.9(b) could be used.
Equation of plane after rotation about axis through $[-1, 2, 3]$ is $x - y - z = 1$.

5.9 (a) Express left-hand side and right-hand side in terms of $\mathbf{r}$, θ, $\mathbf{n}$, and simplify.
(b) Substitute for $\mathbf{R}$ in terms of $\mathbf{r}$, θ, $\mathbf{n}$ in right-hand side and reduce expression to $\mathbf{r}$.

5.10 (a) $\frac{1}{9}\begin{bmatrix} 4 & -4 & 7 \\ 8 & 1 & -4 \\ 1 & 8 & 4 \end{bmatrix}$ (b) $\frac{1}{3}\begin{bmatrix} 2 & 2 & 1 \\ -1 & 2 & -2 \\ -2 & 1 & 2 \end{bmatrix}$ (c) $\frac{1}{81}\begin{bmatrix} 17 & 56 & 56 \\ 56 & -49 & 32 \\ 56 & 32 & 49 \end{bmatrix}$

(d) $\begin{bmatrix} 0 & -1 & 0 \\ 0 & 0 & -1 \\ 1 & 0 & 0 \end{bmatrix}$ (e) $\frac{1}{3}\begin{bmatrix} 1 & 2 & 2 \\ 2 & -2 & 1 \\ 2 & 1 & -2 \end{bmatrix}$

5.11 (a) not a rotation matrix
(b) $\theta = 110.9°$, $[-3, 1, -3]$ (c) $\theta = 180°$, $[1, 2, 2]$
(d) $\theta = 180°$, $[1, -1, -3]$
(e), (f) not rotation matrices.

5.12 Condition is sufficient since $\mathbf{A}^2 = \mathbf{A}\mathbf{A}^{\mathrm{T}} = \mathbf{I}$ in addition to $\det\mathbf{A} = 1$.
trace $\mathbf{A} = 2\cos\theta + 1 = -1$, that is $\theta = \pi$.
Infinity of matrices since there is an infinite set of quantities l_1, l_2, l_3 with the property $l_1^2 + l_2^2 + l_3^2 = 1$.
Take $[l_1, l_2, l_3]$ successively equal to $[1, 0, 0]$, $[0, 1, 0]$, $[0, 0, 1]$ to obtain the matrices

$$\begin{bmatrix} 1 & 0 & 0 \\ 0 & -1 & 0 \\ 0 & 0 & -1 \end{bmatrix}, \begin{bmatrix} -1 & 0 & 0 \\ 0 & 1 & 0 \\ 0 & 0 & -1 \end{bmatrix}, \begin{bmatrix} -1 & 0 & 0 \\ 0 & -1 & 0 \\ 0 & 0 & 1 \end{bmatrix}$$

5.13 (a) If $\theta = k\pi$ then $\mathbf{A}(\theta, \mathbf{n}) = \cos k\pi\mathbf{I} + (1 - \cos k\pi)\mathbf{B}$, which is a symmetric matrix since $\mathbf{I}$, $\mathbf{B}$ are symmetric matrices.

(b) If $\mathbf{A}(\theta, \mathbf{n})$ is symmetric then $2\sin\theta \begin{bmatrix} 0 & -l_3 & l_2 \\ l_3 & 0 & -l_1 \\ -l_2 & l_1 & 0 \end{bmatrix} = 0.$

5.14 $\mathbf{Ar} = \mathbf{r} = \mathbf{Ir}$. Hence $(\mathbf{A} - \mathbf{I})\mathbf{r} = \mathbf{0}$.
(a) $[1, -1, 1]$ (b) $[-1, 1, 1]$ (c) $[1, 2, -2]$.

5.15 (a) $\theta = \pi/2$, $[2, -2, 1]$ (b) $\theta = \pi$, $[2, 2, 1]$
(c) $\theta = 2\arctan\sqrt{17}$, $[3, 2, 2]$ (d) $\theta = \pi/2$, $[-2, 1, -2]$.

$$\frac{1}{9}\begin{bmatrix} 4 & -7 & -4 \\ -1 & 4 & -8 \\ 8 & 4 & 1 \end{bmatrix}, \frac{1}{9}\begin{bmatrix} -1 & 8 & 4 \\ 8 & -1 & 4 \\ 4 & 4 & -7 \end{bmatrix}, \frac{1}{9}\begin{bmatrix} 1 & 4 & 8 \\ 8 & -4 & 1 \\ 4 & 7 & -4 \end{bmatrix},$$

$$\frac{1}{9}\begin{bmatrix} 4 & 4 & 7 \\ -8 & 1 & 4 \\ 1 & -8 & 4 \end{bmatrix}$$

5.16 Replace $\mathbf{N}$ in (c) by $(\mathbf{r}_1 \times \mathbf{r}_2)/|\mathbf{r}_1 \times \mathbf{r}_2|$, noting that $\mathbf{R}_1 + \mathbf{r}_1 = \mathbf{R}_2 + \mathbf{r}_2 = \mathbf{0}$.
Vector is $[1, -1, 1]$.

5.17 $\alpha = 2$, $\beta = -1/2$, $\theta = 2\arctan\sqrt{2}$, $\sqrt{2}\mathbf{n} = [-1, 0, -1]$.

5.18 $\mathbf{A}(\beta, \mathbf{j})\mathbf{A}(\alpha, \mathbf{i}) = \begin{bmatrix} \cos\beta & \sin\alpha\sin\beta & \sin\beta\cos\alpha \\ 0 & \cos\alpha & -\sin\alpha \\ -\sin\beta & \sin\alpha\cos\beta & \cos\alpha\cos\beta \end{bmatrix}$

5.20 Point of plane situated at $[2, 3, -2]$ goes to the point $[8, 2, 3]$ as a result of the first rotation and then to the point $[10, 7, 6]$ as a result of the second rotation. Normal to the plane $[1, 1, 1]$ becomes $[-1, 1, -1]$ after the two rotations. Equation of the plane is $x - y + z = 9$;

translation is [8, 4, 8]; rotation, given by $\cos\theta = -5/6$, $\surd(11)\mathbf{n} = [-1, -3, 1]$, is about an axis through [10, 7, 6].

5.21 Diagonal in the direction of the vector [1, 1, 1] lies along the y axis after the two rotations. Find the matrix that represents the combined rotations.

5.22 Altering the order of two rotations does not change the angle of rotation of the single rotation equivalent to the combined rotations.

5.28 (a) Since $a_{21}^2 + a_{22}^2 + a_{23}^2 = 1$ it follows that $a_{21} = a_{23} = 0$. Use the cofactor property to show $a_{12} = a_{32} = 0$, $a_{11} = a_{33}$, $a_{13} = -a_{31}$. Define ϕ by $a_{11} = \cos\phi$, $a_{13} = \sin\phi$.
(b) Proceed as in (a).

5.29 (a) Displacement equivalent to a translation [10, −2, 4] followed by a rotation of amount π about an axis, in the direction of [1, −1, −3], that passes through the point [5, 3, 7].
(b) Translation [1, 0, −3] followed by a rotation of amount $\pi/2$ about an axis in the direction of [6, 3, 2] that passes through [3, 1, 2].

5.30 $\boldsymbol{\omega} = [2, -2, 1]$. Put $\boldsymbol{\omega} = \alpha\mathbf{v}_1 \times \mathbf{v}_2$, where α is to be found.

5.31 $\boldsymbol{\omega} = [1, 1, 1]$. Put $\boldsymbol{\omega} = \alpha(\mathbf{v}_3 - \mathbf{v}_1) \times (\mathbf{v}_2 - \mathbf{v}_1)$.

5.32 Magnitude is $\surd 249$.

5.33 (a) $\dot{r}\mathbf{a} + r\omega\mathbf{b}$ (b) 0.
$\mathbf{a} = \cos\omega t\mathbf{i} + \sin\omega t\mathbf{j}$, $\mathbf{b} = -\sin\omega t\mathbf{i} + \cos\omega t\mathbf{j}$

5.34 In (a) and (b) $\dot{\mathbf{A}} = \ddot{\mathbf{A}} = \mathbf{0}$.

5.35 $(\underline{OQ_i} - \underline{OQ_0}) \cdot \boldsymbol{\omega} = 0$ for all i.
Least speed is length of perpendicular from the origin O to the plane.

CHAPTER 6

6.1 70 N, 63.6 N.
Mass of 1.48 kg at angular position of 222°.

6.3 $124.2\,\text{rad}\,\text{s}^{-1}$, 23 N.
Masses are 66.6 g and 38.6 g at angular positions 255° and 296°.

6.4 Take $\theta_1 = 0$, then $\theta_2 = 143°$, $\theta_3 = 233°$ or $\theta_2 = 217°$, $\theta_3 = 127°$. (The possible positions of masses m_2, m_3 are mirror images in the line $\theta_1 = 0$.)
Magnitudes equal to 74 N.
Balancing masses each of 9.4 g at angles 167°, 347° when $\theta_2 = 217°$, $\theta_3 = 127°$. Second solution when $\theta_2 = 143°$, $\theta_3 = 233°$ is mirror image of first solution in line $\theta_1 = 0$.

6.5 Masses removed are 2.47 g at 143.8° and 3.49 g at 354.2°. Magnitudes are 41.6 N at X, 10.1 N at Y.

6.6 $\mathbf{F}_{21} = 41.1\,\text{N}\,\mathbf{i} + 16.7\,\text{N}\,\mathbf{j}$, $\mathbf{F}_{41} = 3.6\,\text{N}\,\mathbf{i} - 4.6\,\text{N}\,\mathbf{j}$. Torque is 0.24 N m.

6.7 $\mathbf{F}_{12} = -4557\,\text{N}\,\mathbf{i} - 12\,981\,\text{N}\,\mathbf{j}$, $\mathbf{F}_{14} = 1272\,\text{N}\,\mathbf{i} - 3408\,\text{N}\mathbf{j}$ (recall $\mathbf{F}_{ij}$ is the force exerted by link i on link j at a joint). Torque is 365 Nm.

6.8 $\omega_2 = 50\,\text{rad}\,\text{s}^{-1}$, $\dot{\omega}_2 = 0$, $\dot{\omega}_4 = 1302\,\text{rad}\,\text{s}^{-2}$, $\dot{\omega}_3 = 585.9\,\text{rad}\,\text{s}^{-2}$.
$\mathbf{F}_{12} = -1181\,\text{N}\,\mathbf{i} - 739\,\text{N}\,\mathbf{j}$.
Torque is 11.7 N m.

6.9 $\omega_2 = 105\,\text{rad}\,\text{s}^{-1}$, $\omega_4 = 39.2\,\text{rad}\,\text{s}^{-1}$, $\dot{\omega}_2$ is negligible, $\dot{\omega}_4 = -1033\,\text{rad}\,\text{s}^{-2}$.
$\mathbf{F}_{12} = 304\,\text{N}\,\mathbf{i} - 3312\,\text{N}\,\mathbf{j}$, $\mathbf{F}_{14} = -2047\,\text{N}\,\mathbf{i} - 1165\,\text{N}\,\mathbf{j}$.
Torque is -207 N m.
Note force exerted by link 3 on link 4 is perpendicular to PA.

6.10 $\mathbf{F}_{12} = -135\,\text{N}\,\mathbf{i} + 23\,\text{N}\,\mathbf{j}$, $\mathbf{F}_{14} = -\mathbf{F}_{12}$, $\mathbf{F}_{23} = \mathbf{F}_{34} = \mathbf{F}_{12}$.
Torque is 13.4 N m.

6.12 Equations are:

$$F_1 = M(f - l\cos\theta\,\ddot{\theta} + l\sin\theta\,\dot{\theta}^2)$$
$$F_2 + Mg = -Ml(\sin\theta\,\ddot{\theta} + \cos\theta\,\dot{\theta}^2)$$
$$3(F_1\cos\theta + F_2\sin\theta) = Ml\ddot{\theta}$$

6.13 Mass $2m$.
Primary forces including that due to mass $2m$ and secondary forces would be equivalent to couples.

6.14 Magnitude is
$\sqrt{\{(\alpha - 1.2)^2 + 0.04 + (\alpha - 1.9)\cos\omega t(0.8\sin\omega t - 2.8\cos\omega t)\}}$.

6.15 Primary forces reduce to a couple whose moment is
$m\omega^2 rd\{(3\sin\omega t + \cos\omega t)\mathbf{i} + (-3\cos\omega t + \sin\omega t)\mathbf{j}\}$.
Secondary forces are in equilibrium.

6.16 (a) $\omega_3 = -151.35\,\text{rad}\,\text{s}^{-1}$, $\dot{\omega}_3 = 46\,849\,\text{rad}\,\text{s}^{-2}$, $-10\,690\,\text{m}\,\text{s}^{-2}\mathbf{i}$.
(b) $\mathbf{F}_{21} = 9615\,\text{N}\,\mathbf{i} + 2228\,\text{N}\,\mathbf{j}$.
Approximate acceleration is $-10\,617\,\text{m}\,\text{s}^{-2}\mathbf{i}$.
Replace connecting rod by mass 0.505 kg at A and 0.112 kg at B. These two masses have the same mass and same centre of mass as the rod. The reciprocating mass at B is 0.472 kg. Approximate expression for $\mathbf{F}_{21}$ is $9581\,\text{N}\,\mathbf{i} + 1878\,\text{N}\,\mathbf{j}$. Tension in light rod connecting the masses placed at A, B is 5069 N.

6.17 Sum of moments of inertia is $2\Sigma\, m_s \mathbf{r}_s^2$, which is independent of the choice of axes.

6.19 Use the transformation law (6.31) noting $\mathbf{L}\mathbf{L}^{\text{T}} = \mathbf{I}$.

6.20 Use the conditions that particles must have the same mass and same centre of mass as the rod and that their (common) moment of inertia about an axis perpendicular to the rod through its midpoint is $Ml^2/12$.

6.21 $Ma^2/4$.

6.22 Inertia matrix is $\frac{Ma^2}{3}\begin{bmatrix} 1 & 0 & 0 \\ 0 & 7 & 0 \\ 0 & 0 & 8 \end{bmatrix}$.

6.23 α, $\pi/2 + \alpha$, where $(a^2 - b^2)\tan 2\alpha = ab$ and $0 < a < \pi/2$.

6.24 (b) $2Ma^2$, $11Ma^2$, $11Ma^2$.

(c) Inertia matrix is $\frac{Ma^2}{9}\begin{bmatrix} 28 & 9 & 9 \\ 9 & 28 & -9 \\ 9 & -9 & 28 \end{bmatrix}$.

Principal moments of inertia are $(37/9)Ma^2$, $(37/9)Ma^2$, $(10/9)Ma^2$.

6.25 Angular velocity of the gimbal is $\dot{\alpha}\mathbf{a} + p(\cos\alpha\mathbf{b} - \sin\alpha\mathbf{c})$ and that of the rotor is $\dot{\alpha}\mathbf{a} + p\cos\alpha\mathbf{b} + (n - p\sin\alpha)\mathbf{c}$.
Reactions of bearings equivalent to a couple since centre of mass of rotor and gimbal are at rest. Couple is
$\dot{\alpha}((C - A - B)p\sin\alpha - C'n)\mathbf{b} + ((B - A - C)p\dot{\alpha}\cos\alpha + C'\dot{n})\mathbf{c}$.

6.26 Couple required is $A\Omega\dot{\theta}\sin 2\theta\mathbf{k}$, where $\mathbf{k}$ is parallel to the shaft. Steady state, defined by θ = constant, given by the roots of $\sin 2\theta = \theta$. Solutions are $\theta = 0$, 0.948 rad.

6.27 Forces at P, Q constitute a couple. Force at P is $C\omega\cos\alpha((0.35\omega\sin\alpha + n)/(2r))\mathbf{a}$.
Differentiate relation $\mathbf{k} \cdot \mathbf{c} = \sin\alpha$.

6.30 Take $\boldsymbol{\omega} = \omega\mathbf{k}$. Dynamic reactions of bearings are equivalent to a couple of moment $\pm(M\omega^2/24)[l^2 - 3r^2]\sin 2\theta\mathbf{c}$, where $\mathbf{c}$ is a unit vector perpendicular to the axis of the cylinder and axis of rotation (sign to be suitably chosen).

6.31 Maximum acceleration when $2100\cos\theta + 1800\cos 2\theta = 0$. Solution is $\theta = 1.078$ rad, giving $\ddot{\theta} = 7.6\,\text{rad s}^{-2}$.
Torque M due to load is equal to $1725/\pi$. Maximum and minimum rotational speeds occur when $2100\sin\theta + 900\sin 2\theta = M$. Check given values by substitution. Percentage fluctuation is 0.065%.

6.32 Torque exerted by electric motor is $A/3$. Total work done is $A(\theta_2 - \theta_1)/6 + A(4\sqrt{2})/9 = 1.053A$.

6.33 $dT/dt = A\omega_1\dot{\omega}_1 + B\omega_2\dot{\omega}_2 + C\omega_3\dot{\omega}_3$. Substitute for $\dot{\omega}_1$, $\dot{\omega}_2$, $\dot{\omega}_3$ using Euler's equations.

6.34 Differentiate the given expression with respect to time and use the equations from Problem 6.29 to show derivative is zero.

6.36 65.4 N s.
Reaction of hinges is an impulsive couple of magnitude $32.7h$ N s m (h is the height of the door in metres).

CHAPTER 7

7.1 Put $\theta = \pi/2$, $\mathbf{l} = \mathbf{k}$, $\phi = \pi$, $\mathbf{m} = \mathbf{i}$ in (7.1) so that $\mathbf{n} = \mathbf{j}$.

$$\mathbf{A}(\pi/2, \mathbf{k}) = \begin{bmatrix} 0 & -1 & 0 \\ 1 & 0 & 0 \\ 0 & 0 & 1 \end{bmatrix}, \quad \mathbf{A}(\pi, \mathbf{i}) = \begin{bmatrix} 1 & 0 & 0 \\ 0 & -1 & 0 \\ 0 & 0 & -1 \end{bmatrix},$$

$$\mathbf{A}(\pi, \mathbf{j}) = \begin{bmatrix} -1 & 0 & 0 \\ 0 & 1 & 0 \\ 0 & 0 & -1 \end{bmatrix}$$

Products have value $\begin{bmatrix} 0 & 1 & 0 \\ 1 & 0 & 0 \\ 0 & 0 & -1 \end{bmatrix}$.

7.2 Note

$$\mathbf{A}(\theta, \mathbf{l})\begin{bmatrix} 0 & -\sin\alpha & \cos\alpha \\ \sin\alpha & 0 & 0 \\ -\cos\alpha & 0 & 0 \end{bmatrix}\mathbf{A}(-\theta, \mathbf{l})$$

$$= \begin{bmatrix} 0 & -\sin\alpha & \cos\theta\cos\alpha \\ \sin\alpha & 0 & \sin\theta\cos\alpha \\ -\cos\theta\cos\alpha & -\sin\theta\cos\alpha & 0 \end{bmatrix}$$

7.4 Solve equation $(c + d\cos\delta)\sin\theta_2 + d\sin\delta\cos\theta_2 = a - z$ by putting $p = -(c + d\cos\delta)$, $q = d\sin\delta$, $r = a - z$.
Deduce $-d\sin\delta\sin\theta_2 + (c + d\cos\delta)\cos\theta_2 = -\surd(x^2 + y^2 - b^2)$ when $\theta_2 = \alpha_2 - \beta_2$, and therefore $\theta_1 = \alpha_1 + \beta_1 - \pi$ must be paired with $\theta_2 = N\alpha_2 - \beta_2$.

7.5 $\dot{\theta}_3$ is infinite or indeterminate when $\sin\theta_3 = 0$ as $-cd\sin\theta_3\, \dot{\theta}_3 = x\dot{x} + y\dot{y} + (z - a)\dot{z}$.
When $\sin\theta_2 = 0$, $\cos\theta_3 = \pm 1$ so that $x^2 + y^2 + (z - a)^2 = b^2 + (c \pm d)^2$, and therefore P is on one of the surfaces that divide space into regions accessible and inaccessible to the robot.

7.6 (a) $2T = m(\dot{r}^2 + r^2\dot{\theta}^2)$.
(b) R and rS; $m(\ddot{r} - r\dot{\theta}^2) = R$ and $m(2\dot{r}\dot{\theta} + r\ddot{\theta}) = S$.

7.7 (a) $2T = ml^2\dot{\theta}^2 + M\{L^2\dot{\phi}^2 + l^2\dot{\theta}^2 + 2lL\cos(\phi - \theta)\dot{\theta}\dot{\phi}\}$
where $l = |\underline{\mathrm{OA}}|$, $\mathrm{L} = |\underline{\mathrm{AB}}|$.
(b) $-(M + m)gl\sin\theta$, $-MgL\sin\phi$;
$(m + M)l\ddot{\theta} + ML\cos(\phi - \omega)\ddot{\theta} - ML\sin(\phi - \omega)\dot{\phi}^2 = -(M + m)$
$g\sin\theta$ $L\ddot{\phi} + l\cos(\phi - \theta)\ddot{\theta} + l\sin(\phi - \theta)\dot{\theta}^2 = -g\sin\phi$

7.8 $Mg(\sin\alpha - \mu\cos\alpha)$, $\mu Mga\cos\alpha$.
Lagrange's equations are identical with the equations of motion.

7.9 $2T = M\{\dot{x}^2 + (4/3)l^2\dot{\theta}^2 - 2l\cos\theta\dot{x}\dot{\theta}\}$.
Generalized components of force are F_1, $-Mgl\sin\theta$.

7.10 Note, since $a_{11} = 1$, that $a_{12} = a_{13} = 0$. Use cofactor property to show $a_{21} = a_{31} = 0$. Define ϕ by $\sin\phi = a_{32}$, $\cos\phi = a_{33}$.
When $a_{11} = -1$ the matrix $\mathbf{A}(\theta, \mathbf{n})$ is of the form

$$\begin{bmatrix} -1 & 0 & 0 \\ 0 & -\cos\phi & \sin\phi \\ 0 & \sin\phi & \cos\phi \end{bmatrix}$$

which is equal to the matrix product

$$\begin{bmatrix} 1 & 0 & 0 \\ 0 & -1 & 0 \\ 0 & 0 & -1 \end{bmatrix}\begin{bmatrix} -1 & 0 & 0 \\ 0 & 1 & 0 \\ 0 & 0 & -1 \end{bmatrix}\begin{bmatrix} 1 & 0 & 0 \\ 0 & \cos\phi & -\sin\phi \\ 0 & \sin\phi & \cos\phi \end{bmatrix}$$

$$= \mathbf{A}(\pi, \mathbf{i})\mathbf{A}(\pi, \mathbf{j})\mathbf{A}(\phi, \mathbf{i})$$

7.11 $$\begin{bmatrix} \dot{\mathbf{a}}_2 \\ \dot{\mathbf{b}}_2 \\ \dot{\mathbf{c}}_2 \end{bmatrix} = \begin{bmatrix} 0 & \dot{\theta}_1\cos\theta_2 & -\dot{\theta}_2 \\ -\dot{\theta}_1\cos\theta_2 & 0 & -\dot{\theta}_1\sin\theta_2 \\ \dot{\theta}_2 & \dot{\theta}_1\sin\theta_2 & 0 \end{bmatrix}\begin{bmatrix} \mathbf{a}_2 \\ \mathbf{b}_2 \\ \mathbf{c}_2 \end{bmatrix}$$

Note $\begin{bmatrix} \mathbf{a}_2 \\ \mathbf{b}_2 \\ \mathbf{c}_2 \end{bmatrix} = \mathbf{A}_2^{\mathrm{T}}\mathbf{A}_1^{\mathrm{T}}\begin{bmatrix} \mathbf{i} \\ \mathbf{j} \\ \mathbf{k} \end{bmatrix}$.

7.12 Lagrange's equations are

$$\frac{\mathrm{d}}{\mathrm{d}t}\{\dot{\theta}_1[C_1 + A_2\sin^2\theta_2 + (C_2 + m_3l_2^2)\cos^2\theta_2 + A_3\sin^2(\theta_2 + \theta_3)$$
$$+ (C_3 + m_3h_3^2)\cos^2(\theta_2 + \theta_3) + 2m_3l_2h_3\cos\theta_2\cos(\theta_2 + \theta_3)]\} = Q_1$$

$$\{B_2 + B_3 + m_3(l_2^2 + h_3^2 + 2l_2h_3\cos\theta_3)\}\ddot{\theta}_2 + \{B_3 + m_3(h_3^2 + l_2h_3\cos\theta_3\}\ddot{\theta}_3$$
$$+ \frac{1}{2}\dot{\theta}_1^2\{(C_2 - A_2 + m_3l_2^2)\sin 2\theta_2 + (C_3 - A_3 + m_3h_3^2)\sin 2(\theta_2 + \theta_3)$$
$$+ 2m_3l_2h_3\sin(2\theta_2 + \theta_3)\} - m_3l_2h_3\sin\theta_3(2\dot{\theta}_2 + \dot{\theta}_3)\dot{\theta}_3 = Q_2$$

$$\{B_3 + m_3h_3^2 + m_3l_2h_3\cos\theta_3\}\ddot{\theta}_2 + (B_3 + m_3h_3^2)\ddot{\theta}_3 + m_3l_2h_3\sin\theta_3\dot{\theta}_2^2$$
$$+ \dot{\theta}_1^2\sin(\theta_2 + \theta_3)\{\cos(\theta_2 + \theta_3)(C_3 - A_3 + m_3h_3^2)$$
$$+ m_3l_2h_3\cos\theta_2\} = Q_3$$

7.13 (a) $$\begin{bmatrix} 1 & 0 & 0 \\ 0 & -1 & 0 \\ 0 & 0 & -1 \end{bmatrix}\begin{bmatrix} 0 & 0 & -1 \\ 0 & 1 & 0 \\ 1 & 0 & 0 \end{bmatrix}\begin{bmatrix} 1 & 0 & 0 \\ 0 & 0 & -1 \\ 0 & 1 & 0 \end{bmatrix}$$

Corresponds to $\beta = -\pi/2$, $\alpha = \pi$, $\delta = \pi/2$.

(b) $\frac{1}{15}\begin{bmatrix} \sqrt{5} & 0 & 0 \\ 0 & -2 & -1 \\ 0 & 1 & -2 \end{bmatrix}\begin{bmatrix} 2 & 0 & \sqrt{5} \\ 0 & 3 & 0 \\ -\sqrt{5} & 0 & 2 \end{bmatrix}\begin{bmatrix} \sqrt{5} & 0 & 0 \\ 0 & 1 & -2 \\ 0 & 2 & 1 \end{bmatrix}$

Corresponds to $\cos\beta = 2/3$, $\sin\beta = (\sqrt{5})/3$, $\sin\alpha = 1/\sqrt{5}$, $\cos\alpha = -2/\sqrt{5}$, $\sin\delta = 2/\sqrt{5}$, $\cos\delta = 1/\sqrt{5}$ (see (7.24)). There are alternatives in (a), (b).

7.14 Other solutions (not all):

θ_1	θ_2	θ_3	θ_4	θ_5	θ_6
3.87	−73.92	99.06	−90.06	−38.32	−107.22
−143.43	131.03	99.06	−171.54	69.97	129.26
3.87	48.97	−99.06	39.26	78.47	109.49

Index